Methods in Molecular Biology

Series Editor
John M. Walker
School of Life and Medical Sciences
University of Hertfordshire
Hatfield, Hertfordshire, AL10 9AB, UK

For further volumes:
http://www.springer.com/series/7651

Ebolaviruses

Methods and Protocols

Edited by

Thomas Hoenen • Allison Groseth

Friedrich-Loeffler-Institut, Greifswald - Insel Riems, Germany

Humana Press

Editors
Thomas Hoenen
Friedrich-Loeffler-Institut
Greifswald - Insel Riems
Germany

Allison Groseth
Friedrich-Loeffler-Institut
Greifswald - Insel Riems
Germany

ISSN 1064-3745 ISSN 1940-6029 (electronic)
Methods in Molecular Biology
ISBN 978-1-4939-7115-2 ISBN 978-1-4939-7116-9 (eBook)
DOI 10.1007/978-1-4939-7116-9

Library of Congress Control Number: 2017940477

Cover illustration: Electron micrograph of an ebolavirus-infected cell, provided by Prof. Takeshi Noda (Kyoto University, Japan).

Printed on acid-free paper

This Humana Press imprint is published by Springer Nature
The registered company is Springer Science+Business Media LLC
The registered company address is: 233 Spring Street, New York, NY 10013, U.S.A.

Dedication

The editors wish to dedicate this book to Michael P. Kiley (1942–2004), whose own groundbreaking work during the early years taught us much of what we know today about the molecular biology of ebolaviruses. We hope he would have been proud of all the discoveries for which his own work paved the way.

Preface

While filoviruses as a group were first discovered 50 years ago, and ebolaviruses have been known since 1976, many of these viruses' secrets remained hidden until the use of molecular biological techniques became widespread at the end of the 1980s. Since then, advances, first in sequencing and cloning and then in life cycle modeling and reverse genetics, animal modeling, and most recently in imaging techniques, have provided successive layers of insight into the biology and pathogenesis of these agents. Further, as increasing numbers of outbreaks have been identified, culminating with the recent ebolavirus epidemic of unprecedented proportions in West Africa that threatened to spread internationally to an extent that was previously unimaginable, there has been a heightened interest in the development and advancement of techniques for diagnostics, high-throughput antiviral screening, and vaccine development. It has also made clear the need for a better understanding of the ecology of these agents and the development of tools for understanding the immune response in animal species other than laboratory standard animals.

This book seeks to provide a sampling of key methods that have supported these advancements in the field of ebolavirus molecular biology. In retrospect, it seems clear that, as with many other virus families, much of what we know of ebolavirus biology is the direct result of the widespread implementation of molecular biological methods, with the advancements in our knowledge mirroring closely the development and availability of new techniques. Similarly it is clear that the individual protocols within this volume can be appreciated as part of the continuing spectrum of assays that fuel filovirus research, and we hope that access to these detailed protocols, outlining many of the important techniques currently being used by researchers, will also help to propel new research forward into the next decade. Of course such an endeavor is not possible without the support of a large group of outstanding experts who are willing to openly share their developments with one another, and we are fortunate that this is the case in the filovirus research community. Indeed, this collaborative spirit was clearly demonstrated recently during the international response to the ebolavirus epidemic in West Africa, which saw filovirus research groups from across the globe coming together to apply their skills and experiences, including openly sharing their results and protocols, to directly help each other and the affected people in West Africa.

We are very happy that all generations of filovirus researchers have come together in the making of this book, including one of the researchers who co-discovered filoviruses more than 50 years ago, many senior researchers who have been instrumental in moving the filovirus field forward over the last 20–30 years, as well as young investigators who are just now bringing new ideas and techniques into the field. We would like to thank all the contributors to this book, and particularly Prof. Takeshi Noda (Kyoto University, Japan) for providing the electron micrograph of an ebolavirus-infected cell for the cover image. Finally, we hope that together we have produced a book that will not only serve as a guide to the next generations of filovirus researchers but also help bring experts from other areas into the filovirus research arena.

Greifswald - Insel Riems, Germany — *Thomas Hoenen*
Greifswald - Insel Riems, Germany — *Allison Groseth*

Contents

Contributors

BRIAN R. AMMAN • *Virus Host Ecology, Viral Special Pathogens Branch, Division of High Consequence Pathogens and Pathology, National Center for Emerging and Zoonotic Infectious Diseases, Centers for Disease Control and Prevention, Atlanta, GA, USA*

NATARAJAN AYITHAN • *Department of Microbiology, University of Iowa, Iowa City, IA, USA*

LOGAN BANADYGA • *Laboratory of Virology, Division of Intramural Research, National Institute of Allergy and Infectious Diseases, National Institutes of Health, Hamilton, MT, USA*

CHRISTOPHER F. BASLER • *Center for Microbial Pathogenesis, Institute for Biomedical Sciences, Georgia State University, Atlanta, GA, USA*

STEPHAN BECKER • *Institute of Virology, Philipps University Marburg, Marburg, Germany*

NADINE BIEDENKOPF • *Institute of Virology, Philipps University Marburg, Marburg, Germany*

KRISTINA BRAUBURGER • *Department of Microbiology and National Emerging Infectious Diseases Laboratories, School of Medicine, Boston University, Boston, MA, USA; Department of Biology, Lund University, Lund, Sweden*

RACHEL B. BROUILLETTE • *Department of Microbiology, University of Iowa, Iowa City, IA, USA*

ERIK P. CARTER • *Microbiology and National Emerging Infectious Diseases Laboratories, School of Medicine, Boston University, Boston, MA, USA*

JEREMY E. COFFIN • *Viral Vector Core Facility, University of Iowa, Iowa City, IA, USA*

JOHN H. CONNOR • *Microbiology and National Emerging Infectious Diseases Laboratories, Boston University School of Medicine, Boston, MA, USA; Biomedical Engineering Department, Boston University, Boston, MA, USA*

TESSA CRESSEY • *Department of Microbiology and National Emerging Infectious Diseases Laboratories, School of Medicine, Boston University, Boston, MA, USA*

ROBERT W. CROSS • *Department of Microbiology and Immunology, University of Texas Medical Branch, Galveston, TX, USA; Galveston National Laboratory, University of Texas Medical Branch, Galveston, TX, USA*

GEORGE G. DAABOUL • *Electrical and Computer Engineering Department, Boston University, Boston, MA, USA*

OLGA DOLNIK • *Institute of Virology, Philipps University Marburg, Marburg, Germany*

HIDEKI EBIHARA • *Laboratory of Virology, Division of Intramural Research, National Institute of Allergy and Infectious Diseases, National Institutes of Health, Hamilton, MT, USA; Department of Molecular Medicine, Mayo Clinic, Rochester, MN, USA*

DARRYL FALZARANO • *Vaccine and Infectious Disease Organization – International Vaccine Centre (VIDO-InterVac), University of Saskatchewan, Saskatoon, SK, Canada*

Friederike Feldmann • *Rocky Mountain Veterinary Branch, Division of Intramural Research, National Institute of Allergy and Infectious Diseases, National Institutes of Health, Rocky Mountain Laboratories, Hamilton, MT, USA*

Heinz Feldmann • *Laboratory of Virology, Division of Intramural Research, National Institute of Allergy and Infectious Diseases, National Institutes of Health, Rocky Mountain Laboratories, Hamilton, MT, USA*

Claire Marie Filone • *National Biological Threat Characterization Center, National Biodefense Analysis & Countermeasures Center, Science & Technology Directorate, Department of Homeland Security, Frederick, MD, USA*

David S. Freedman • *NanoView Diagnostics Inc., Boston, MA, USA*

Wakako Furuyama • *Division of Global Epidemiology, Research Center for Zoonosis Control, Hokkaido University, Sapporo, Japan*

Thomas W. Geisbert • *Galveston National Laboratory, University of Texas Medical Branch, Galveston, TX, USA; Department of Microbiology and Immunology, University of Texas Medical Branch, Galveston, TX, USA*

Bryan D. Griffin • *National Microbiology Laboratory, Public Health Agency of Canada, Winnipeg, MB, Canada; Department of Microbiology, University of Manitoba, Winnipeg, MB, Canada*

Allen Grolla • *Special Pathogens Program, National Microbiology Laboratory, Public Health Agency of Canada, Winnipeg, MB, Canada*

Allison Groseth • *Friedrich-Loeffler-Institut, Greifswald-Insel Riems, Germany*

Elaine Haddock • *Division of Intramural Research, National Institute of Allergy and Infectious Diseases, National Institutes of Health, Hamilton, MT, USA*

Thomas Hoenen • *Friedrich-Loeffler-Institut, Greifswald - Insel Riems, Germany*

Kathleen H. Holt • *Viral Vector Core Facility, University of Iowa, Iowa City, IA, USA*

Hans Dieter Klenk • *Institut für Virologie, Philipps–Universität Marburg, Marburg, Germany*

Gary Kobinger • *Department of Infectious Diseases, Université Laval, Québec, QC, Canada; Department of Microbiology, University of Manitoba, Winnipeg, MB, Canada; Department of Laboratory Medicine, University of Pennsylvania, Philadelphia, PA, USA*

Andrea Kroeker • *National Microbiology Laboratory, Public Health Agency of Canada, Winnipeg, MB, Canada; Department of Microbiology, University of Manitoba, Winnipeg, MB, Canada*

Thomas G. Ksiazek • *Department of Pathology, University of Texas Medical Branch, Galveston, TX, USA; Galveston National Laboratory, University of Texas Medical Branch, Galveston, TX, USA*

Priya Luthra • *Center for Microbial Pathogenesis, Institute for Biomedical Sciences, Georgia State University, Atlanta, GA, USA*

Andrea Marzi • *Laboratory of Virology, Division of Intramural Research, National Institute of Allergy and Infectious Diseases, National Institutes of Health, Hamilton, MT, USA*

Mathieu Mateo • *INSERM U1111—CNRS UMR5308, Centre International de Recherche en Infectiologie (CIRI), Université Claude Bernard Lyon 1, Ecole*

Normale Supérieure de Lyon, Lyon, France; UBIVE, Institut Pasteur, CIRI, Lyon, France

WENDY MAURY • *Department of Microbiology, University of Iowa, Iowa City, IA, USA*

DAVID MILLER • *National Biological Threat Characterization Center, National Biodefense Analysis & Countermeasures Center, Science & Technology Directorate, Department of Homeland Security, Frederick, MD, USA*

CHAD E. MIRE • *Galveston National Laboratory, University of Texas Medical Branch, Galveston, TX, USA; Department of Microbiology and Immunology, University of Texas Medical Branch, Galveston, TX, USA*

HIROKO MIYAMOTO • *Division of Global Epidemiology, Research Center for Zoonosis Control, Hokkaido University, Sapporo, Japan*

ELKE MÜHLBERGER • *Department of Microbiology and National Emerging Infectious Diseases Laboratories, School of Medicine, Boston University, Boston, MA, USA*

EMILY V. NELSON • *Department of Microbiology and National Emerging Infectious Diseases Laboratories, School of Medicine, Boston University, Boston, MA, USA*

TAKESHI NODA • *Institute for Virus Research, Kyoto University, Kyoto, Japan*

JUDITH OLEJNIK • *Department of Microbiology and National Emerging Infectious Diseases Laboratories, School of Medicine, Boston University, Boston, MA, USA*

XIANGGUO QIU • *National Microbiology Laboratory, Public Health Agency of Canada, Winnipeg, MB, Canada; Department of Microbiology, University of Manitoba, Winnipeg, MB, Canada*

OLIVIER REYNARD • *INSERM U1111—CNRS UMR5308, Centre International de Recherche en Infectiologie (CIRI), Université Claude Bernard Lyon 1, Ecole Normale Supérieure de Lyon, Lyon, France*

STEVEN M. SCHERR • *Mechanical Engineering Department, Boston University, Boston, MA, USA*

KRISTINA M. SCHMIDT • *Department of Microbiology and National Emerging Infectious Diseases Laboratories, School of Medicine, Boston University, Boston, MA, USA; Institute for Novel and Emerging Infectious Diseases, Federal Research Institute for Animal Health, Friedrich-Loeffler-Institut, Greifswald – Insel Riems, Germany*

GORDIAN SCHUDT • *Institute of Virology, Philipps University Marburg, Marburg, Germany*

AMY J. SCHUH • *Virus Host Ecology, Viral Special Pathogens Branch, Division of High Consequence Pathogens and Pathology, National Center for Emerging and Zoonotic Infectious Diseases, Centers for Disease Control and Prevention, Atlanta, GA, USA*

DANA P. SCOTT • *Rocky Mountain Veterinary Branch, Division of Intramural Research, National Institute of Allergy and Infectious Diseases, National Institutes of Health, Rocky Mountain Laboratories, Hamilton, MT, USA*

ELIF Ç. SEYMOUR • *Biomedical Engineering Department, Boston University, Boston, MA, USA*

PATRICK L. SINN • *Department of Pediatrics, University of Iowa, Iowa City, IA, USA; Viral Vector Core Facility, University of Iowa, Iowa City, IA, USA*

WERNER SLENCZKA • *Institut für Virologie, Philipps–Universität Marburg, Marburg, Germany*

AYATO TAKADA • *Division of Global Epidemiology, Research Center for Zoonosis Control, Hokkaido University, Sapporo, Japan*

JONATHAN S. TOWNER • *Virus Host Ecology, Viral Special Pathogens Branch, Division of High Consequence Pathogens and Pathology, National Center for Emerging and Zoonotic Infectious Diseases, Centers for Disease Control and Prevention, Atlanta, GA, USA*
M. SELIM ÜNLÜ • *Biomedical Engineering Department, Boston University, Boston, MA, USA; Mechanical Engineering Department, Boston University, Boston, MA, USA; Electrical and Computer Engineering Department, Boston University, Boston, MA, USA; Photonics Center, Boston University, Boston, MA, USA*
VERONICA VINE • *Laboratory of Virology, Division of Intramural Research, National Institute of Allergy and Infectious Diseases, National Institutes of Health, Rocky Mountain Laboratories, Hamilton, MT, USA*
VICTORIA WAHL-JENSEN • *National Biological Threat Characterization Center, National Biodefense Analysis & Countermeasures Center, Science & Technology Directorate, Department of Homeland Security, Frederick, MD, USA*
REIKO YOSHIDA • *Division of Global Epidemiology, Research Center for Zoonosis Control, Hokkaido University, Sapporo, Japan*
MARKO ZIVCEC • *Laboratory of Virology, Rocky Mountain Laboratories, Division of Intramural Research, National Institute Allergy and Infectious Disease, National Institutes of Health, Hamilton, MT, USA; Viral Special Pathogens Branch, Division of High Consequence Pathogens and Pathology, National Center for Emerging and Zoonotic Infectious Diseases, Centers for Disease Control and Prevention, Atlanta, GA, USA*

Part I

Introduction

Chapter 1

Marburg- and Ebolaviruses: A Look Back and Lessons for the Future

Hans Dieter Klenk and Werner Slenczka

Abstract

Since the discovery of Marburg virus 50 years ago, filoviruses have reemerged in the human population more than 40 times. Already the first episode was as dramatic as most of the subsequent ones, but none of them was as devastating as the West-African Ebola virus outbreak in 2013–2015. Although progress toward a better understanding of the viruses is impressive, there is clearly a need to improve and strengthen the measures to detect and control these deadly infections.

Key words Marburgvirus outbreaks, Ebolavirus outbreaks, Ecology of filoviruses, Discovery of filoviruses

1 The Discovery of Marburg Virus in 1967

Almost half a century has passed since the discovery of filoviruses. In August 1967, laboratory workers at the Behringwerke in Marburg who had processed organs of African green monkeys came down with what appeared to be an infectious disease. With myalgia, malaise, headache, and fever, the symptoms were not particularly alarming at the beginning, but became more severe at the end of the first week. Nausea, vomiting, and diarrhea suggested now to health care practitioners that the diagnosis might be dysentery or typhoid fever requiring treatment at the university hospital. At admission, conjunctivitis, exanthema and enanthema were observed, but shigella or salmonella were not found. During the second week temperature fell to 38 °C, and signs of liver destruction, leukopenia, thrombocytopenia, petechia, and more severe forms of hemorrhagic diathesis developed in about a quarter of the patients. They started to bleed from all body orifices and needle punctures which proved to be a *signum mali ominis.* In the course of the epidemic, 20 persons all of which had contact with monkey tissues were hospitalized in Marburg, and two members of the

Thomas Hoenen and Allison Groseth (eds.), *Ebolaviruses: Methods and Protocols*, Methods in Molecular Biology, vol. 1628, DOI 10.1007/978-1-4939-7116-9_1, © Springer Science+Business Media LLC 2017

medical staff became infected while attending these patients. At the same time, six patients with similar symptoms were hospitalized in Frankfurt, some of which also had done work on monkeys at the Paul-Ehrlich-Institut. Furthermore, in Belgrade, Yugoslavia, a veterinarian performing an autopsy on dead monkeys became infected along with his wife who nursed him during the first few days of the illness. Altogether, there were 32 cases with seven fatalities at the three sites. Among the survivors, relapses with hepatitis, orchitis, and uveitis, with virus persisting in semen and in the anterior eye chamber, were typical during the convalescent phase. Mental confusion and paresthesia were indicative of cerebral involvement. In one case, sexual transmission was observed 120 days after the primary disease [1–3].

Immediately after the outbreak, it was clear to the physicians at the hospitals in Marburg and Frankfurt that they were observing a new infectious disease, and the microbiologists at the Institute of Hygiene in Marburg set out on a search for the infectious agent, soon to be joined by scientists from other places within and outside Germany. Diagnostic analysis covering most of the agents known at that time to cause hemorrhagic fever failed to reveal the etiology of the outbreak. Still in August 1967, serological testing for leptospirosis was positive with some cases, but attempts to isolate leptospirae from the blood of guinea pigs inoculated with patient samples were not successful. When the high pathogenicity of the agent became apparent, the diagnostic search was halted for a short period, because it seemed too dangerous to continue this work under the poor safety conditions of the laboratory in Marburg at that time. By the middle of September, it was clear, however, that the agent was not highly contagious. Only a few cases of secondary infection and no new cases had occurred during the previous 2 weeks. Therefore, experiments with guinea pigs were resumed that showed now that an agent could be passed with increasing pathogenicity in these animals. They came down with high fever, hepatitis, and hemorrhage, similar to the disease in humans. However, attempts to identify the agent by microscopic analysis failed. Meanwhile, reconvalescent human and guinea pig sera had been obtained. With these sera, virus-specific inclusion bodies were detected in liver and spleen cells of infected animals by immunofluorescence analysis. Blood samples from these animals were sent to the Bernhard-Nocht-Institut in Hamburg where long filamentous virus particles never observed before were detected in the electron microscope on November 20, 1967. Thus, Marburg virus (MARV) was identified as the infectious agent 3 months after the outbreak had begun. The studies leading to the discovery were published in 1967 and 1968 [4–6] and soon confirmed by others [7, 8].

2 Marburg Virus Outbreaks Between 1975 and 2014

MARV reemerged several times since 1967. Except for three cases that resulted from laboratory accidents in Russia [9, 10], all of these infections occurred directly in or could be traced back to tropical Africa (Table 1, Fig. 1). Whereas often only a few cases were involved [10–14], more severe outbreaks have also been observed. Thus, several hundred people were involved with case-fatality rates of up to 90%, when the virus reemerged in the Democratic Republic of the Congo (DRC) in 1998 [15], in Angola in 2004 [16], and in Uganda in 2012 [10]. In total, MARV outbreaks have resulted in 474 cases of hemorrhagic fever and 378 deaths [10] (Table 1).

3 The Emergence of Ebolavirus in 1976

In 1976, almost a decade after MARV had emerged, another filovirus was discovered, when two hemorrhagic fever outbreaks occurred simultaneously in Zaire, now the DRC, and in southern Sudan. Originally it was thought that both outbreaks were caused by the same virus, which was called Ebola virus after a river in north-western Zaire. But it was recognized much later that viruses belonging to two different species of the *Ebolavirus* genus were involved: Ebola virus (EBOV) was responsible for the outbreak in Zaire and Sudan virus (SUDV) for the outbreak in Sudan [17].

In June and July 1976, the first cases were reported from Nzara in western Equatorial Province of Sudan, now South Sudan, a small town close to the rain forest zone. The outbreak was strongly associated with index cases in a single cotton factory and spread to close relatives (67 cases). The epidemic spread then to the neighboring areas of Maridi, Tembura, and Juba. High levels of transmission occurred in the hospital of Maridi, a teaching center for student nurses (213 cases). The outbreak lasted until November, during which time approximately 15 generations of person-to-person transmissions occurred. Transmission required close contact with an acute case and was, in Maridi, usually associated with nursing patients. In total, there were 284 probable and confirmed cases in the Sudan outbreak with a case-fatality rate of 53% [18–21].

The second outbreak started by the end of August in equatorial rain forest areas of north-western Zaire. In total, there were 318 probable and confirmed cases and 280 deaths. The presumed index case came to Yambuku Mission Hospital for treatment of acute malaria where he received an injection of chloroquine. Like in Maridi, the hospital played a crucial role in the spread of the disease, and for more than 25% of the infected persons the only risk factor elucidated were injections at Yambuku Mission Hospital. After the

Table 1

Filovirus outbreaks

Year	Virus	Location	Cases (CFR)	Epidemiology
1967	MARV	Germany, Yugoslavia, *ex* Uganda	32 (22%)	Exposure to imported laboratory monkeys
1975	MARV	South Africa *ex* Zimbabwe	3 (33%)	
1976	SUDV	South Sudan (Nzara, Maridi, Tembura)	284 (53%)	
1976	EBOV	DRC (Yambuku)	318 (88%)	
1977	EBOV	DRC (Tandala)	1 (100%)	
1979	SUDV	South Sudan (Nzara, Yambio)	34 (65%)	
1980	MARV	Kenya (Nzaia, Nairobi)	2 (50%)	Exposure to bat-infested cave
1987	MARV	Kenya (Kisumu)	1 (100%)	Exposure to bat-infested cave
1989	RESTV	USA (Reston VA) *ex* Philippines	4 (–)	Exposure to imported laboratory monkeys
1992	RESTV	Italy (Siena) *ex* Philippines	–	Imported laboratory monkeys, no human exposure
1994	EBOV	Gabon (Ogooue-Invindo)	52 (60%)	Exposure to bush meat
1994	TAFV	Ivory Coast (Tai forest)	1 (–)	Necropsy on chimpanzee
1995	TAFV	Liberia	1 (–)	
1995	EBOV	DRC (Kikwit)	315 (81%)	
1996	EBOV	Gabon (Mayibout)	37 (57%)	Exposure to bush meat
1996	EBOV	Gabon (Booue)	62 (74%)	Exposure to bush meat
1996	RESTV	USA (Alice TX) *ex* Philippines	–	Imported laboratory monkeys, no human infection
1998–2000	MARV	DRC (Durba, Watsa)	154 (83%)	Exposure to bat-infested cave
2000–2001	SUDV	Uganda (Gulu)	425 (53%)	
2001–2002	EBOV	Gabon (Ogooue-Invindo) RC (Cuvette)	124 (78%)	Exposure to bush meat
2002–2003	EBOV	Gabon (Ogooue-Invindo) RC (Cuvette)	143 (90%)	Exposure to bush meat
2003	EBOV	RC (Mbomo, Mbandza)	35 (83%)	Exposure to bush meat
2004	SUDV	South Sudan (Yambio)	17 (42%)	
2004–2005	MARV	Angola (Uige)	252 (90%)	

(continued)

Table 1 (continued)

Year	Virus	Location	Cases (CFR)	Epidemiology
2005	EBOV	RC (Etoumbi, Mbomo)	12 (89)	
2007	EBOV	DRC (Kasai Occidental)	32 (44%)	
2007	MARV	Uganda (Kamwenge)	4 (25%)	Exposure to bat-infested mine
2007	BDBV	Uganda (Bundibugyu)	147 (25%)	
2008	RESTV	Philippines	6 (–)	Exposure to pigs
2008	MARV	Netherlands, USA *ex* Uganda	2 (50%)	Exposure to bat-infested cave
2008–2009	EBOV	DRC (Kasai Occidental)	32 (44%)	
2011	SUDV	Uganda (Nakasimata)	1 (100%)	
2012	MARV	Uganda (Kabale, Ibanda)	20 (45%)	
2012	SUDV	Uganda (Kibaale)	24 (71%)	
2012	SUDV	Uganda (Luwero)	6 (50%)	
2012	BDBV	DRC (Isiro)	62 (55%)	
2014	MARV	Uganda (Mpigi)	1 (100%)	
2014	EBOV	DRC (Equateur Province)	66 (70%)	Exposure to bush meat
2013–2015	EBOV	Guinea, Liberia, Sierra Leone	28,599 (40%)	Single introduction from exposure to bats?

Data from [10, 17] and WHO Epidemic and Pandemic Alert Response
Cases resulting from laboratory accidents are not listed
CFR case-fatality rate

outbreak in the hospital there was subsequent dissemination in surrounding villages to people caring for sick relatives or having other forms of close contact. The epidemic which lasted until the end of October 1976 spread relatively slowly in the infested area, and all villages involved were within 60 km of Yambuku [19, 22, 23]. Except for a few smaller outbreaks in the 1970s (17), EBOV and SUDV did not reemerge for almost 20 years.

4 The Reston Virus Outbreak in 1989

In 1989, a virus of a third ebolavirus species, Reston virus (RESTV), emerged as the causative agent of an outbreak among cynomolgus monkeys in Reston VA, USA. The animals had been imported from the Philippines into the United States [24]. A similar epizootic was observed in 1992, when cynomolgus monkeys were imported into Italy from the same supplier who had shipped the

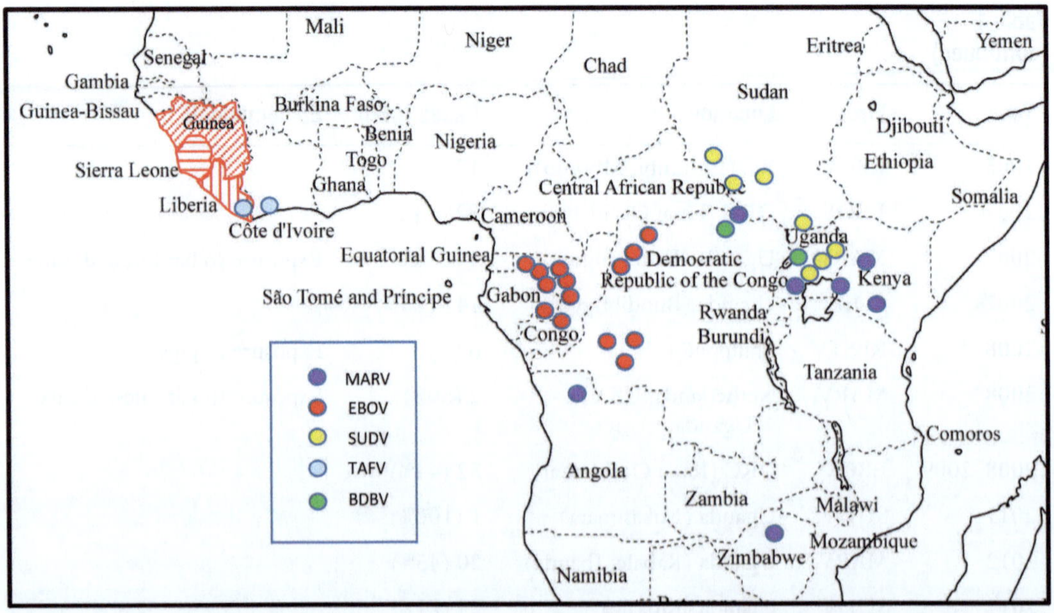

Fig. 1 Locations of filovirus outbreaks in Africa. The sites of MARV, EBOV, SUDV, TAFV, and BDBV outbreaks in Africa and the sites where outbreaks exported to South Africa, Europe and North America originated are shown. The exact locations are indicated in Table 1

1989 monkeys into the US. In contrast to EBOV and SUDV, RESTV seems not to be pathogenic for humans, since several animal caretakers who were infected during the outbreak in Reston did not show disease symptoms [25]. Similarly, during the 2008 RESTV outbreak in the Philippines individuals with occupational exposure to pigs were also found to be seropositive for RESTV but without a history of significant associated illness [26].

5 Ebolaviruses Emerge and Reemerge in Africa, 1994–2014

In 1994, a virus belonging to a fourth ebolavirus species, Taï Forest virus (TAFV), was isolated from a scientist who performed a necropsy in Taï forest, Ivory Coast, on a wild chimpanzee whose troop had undergone increased mortality, presumably due to infection with this virus [27]. A year later, there may have been another human TAFV case in Liberia [28].

In January 1995, EBOV reemerged causing a large outbreak in the DRC. Until August, the official end of the epidemic, 315 cases had occurred of which 244 died. The center of the epidemic was Kikwit and the surrounding areas in the Bandundu region of the south-western DRC. It is believed that the outbreak started from a charcoal worker who had infected several members of his household already in December 1994. The first patient at Kikwit

General Hospital was a male laboratory worker on whom two laparotomies had been performed after a false diagnosis of typhoid fever with intestinal perforation. Three days after the second laparotomy the patient died showing massive intra-abdominal hemorrhages. Four days after the first laparotomy, members of the medical staff started to come down with fever, headache, muscle aches, and hemorrhages. About three quarters of the first 70 patients were health care workers [29, 30]. With only 1.6% sequence variation in the glycoprotein gene the virus was very similar to EBOV of the 1976 outbreak [31].

At the time of the Kikwit outbreak and in the following decade multiple EBOV outbreaks were observed in the northern border regions of Gabon and the Republic of Congo (RC) [10, 32]. Most human infections have been associated there with the hunting and handling of animal carcasses (mainly great apes), and EBOV has decimated the gorilla and chimpanzee populations in these areas. SUDV reemerged in 2000–2001 in the Gulu district in northern Uganda. With 425 cases and 53% mortality, this was the largest filovirus outbreak documented until then. In 2007–2008, there has been a major EBOV epidemic in Kasai Occidental province, DRC, and several smaller SUDV outbreaks have been observed between 2004 and 2012 in South Sudan and Uganda (10). In 2007, a virus belonging to a new ebolavirus species was identified, named Bundibugyo virus (BDBV) after the region in Uganda where it emerged. BDBV appeared again in 2012 in the eastern part of the DRC. Altogether BDBV was responsible for 211 recorded cases and 71 deaths [33–36].

Except for the RESTV episodes and the recent EBOV epidemic in West Africa, there had been 23 ebolavirus outbreaks between 1976 and 2014 with 2469 recorded cases and 1641 deaths, in total (10, Table 1). The available evidence indicates that EBOV and SUDV outbreaks occur in distinct geographical areas. All SUDV outbreaks observed so far cluster in northern Uganda and in the most southern part of South Sudan. In contrast, EBOV emerges in the rain forest areas of Gabon, RC, and western DRC, but also in West Africa as we know now (Fig. 1). Thus, there appear to be ecological niches that might depend on different animal reservoirs.

6 The Search for the Filovirus Reservoir

Already at the onset of the MARV outbreak in 1967 its zoonotic background was obvious. All patients with primary infections in Marburg as well as in Frankfurt and Belgrade had direct contact with blood, organs, and cell cultures from African green monkeys that were needed for the production of poliomyelitis vaccines. The animals had been imported from Uganda and appeared to be healthy, but they were killed soon after their arrival in Marburg and Frankfurt. However, a subsequent study in which monkeys were experimentally infected with MARV revealed 100% lethality [37]. When EBOV and SUDV emerged in 1976 there was no hint for a zoonotic infection, but chimpanzees, gorillas, and other nonhuman primates were clearly involved in later outbreaks. However, since EBOV was also highly pathogenic for these animals, the concept was soon abandoned that they are the natural filovirus reservoir.

For many years, the search for the reservoir was not successful, until it focused on bats. First it was recognized that the patients infected with MARV in the 1980s had visited caves at Mt. Elgon in Kenya that housed large colonies of these animals [11, 12]. Subsequent studies indicated that the experimental infection of insectivorous and fruit bats allowed EBOV replication without causing disease [38]. Further evidence was obtained when EBOV-specific antibodies were detected in animals caught in Gabon and the RC in the field [39]. At the same time MARV-specific antibodies and RNA have also been found in bats [40], and the isolation of MARV from *Rousettus aegyptiacus* bats caught in a mine in Uganda that had been the site of a small human outbreak provided definite proof for the bat reservoir of this virus [41]. Finally, there is evidence that bats are the reservoir for RESTV in the Philippines [42] and for a novel filovirus, Lloviu virus, recently discovered in Spain [43].

7 Lessons from the Ebola Virus Outbreak in West Africa, 2013–2015

As pointed out above and summarized in Table 1, filoviruses emerged between 1967 and 2014 periodically with increasing frequency. Because of their high case-fatality rate most of these outbreaks were dramatic. However, they never involved more than a few hundred people and were usually confined to relatively small, remote areas and a limited time period. With the recent EBOV outbreak in West Africa, the situation has changed completely. Since its presumptive onset due to a zoonotic transmission from a bat to a young boy late in 2013 [44], the outbreak has grown steadily in terms of infected people and geographic spread. Besides Guinea, large parts and densely populated urban areas of neighboring Liberia and Sierra Leone became infested. When the outbreak finally came to an end almost 2 years after its start, more than 28,000 disease cases and 11,000 deaths, including many victims among health care workers and physicians, had been reported. The damage to the economy of the affected countries and the burden on international aid organizations were enormous. The concept that the virus may persist in semen and other body niches is strengthened by the outbreak. With the large numbers of survivors it can therefore not be excluded that sporadic cases will still emerge in the affected areas. Thus, continuous vigilance is necessary.

The outbreak was not only unprecedented in its size, but it was also unexpected because, apart from the two TAFV cases in 1994 and 1995, there had been little evidence for the presence of Ebola viruses in West Africa. It has to be mentioned, however, that in a seroepidemiological study carried out in 1984, 1.8% of 556 sera from the Eastern Province of Sierra Leone proved to be Ebola virus-positive. Interestingly, the samples had been collected at the Eastern Clinic Mobai, located at the border to Guinea just across from Gueckedou where the 2013–15 outbreak originated [45]. Thus, the impact of seroepidemiological studies should not be underestimated.

Although it is impossible to predict, when and where such outbreaks will occur in the future and whether they will be caused by ebolaviruses or marburgviruses, we definitely have to be better prepared for them. In the last decades there has been remarkable progress in the research on filoviruses. As described in other chapters of this volume, structure and replication of the viruses have been elucidated, mechanisms underlying pathogenesis and host defense are understood, and techniques for rapid diagnosis under field conditions have been developed. Some of the accomplishments are listed in Table 2. They include the development of promising vaccines and therapeutics that contributed, however, only marginally to the control of this outbreak, since they were still at an experimental stage during most of the epidemic period. To develop these regimens to a stage where they can be used effectively in a forthcoming emergency is mandatory.

Table 2

Progress in molecular filovirus research

1992–1993	Complete genome sequences of MARV and EBOV	[46–48]
1998–2008	Structure of viral proteins elucidated	[49–51]
1998	Minigenome system developed	[52]
2000	Adenovirus-based EBOV vaccine developed	[53]
2001	Infectious recombinant EBOV developed	[54]
2004	VSV-based EBOV and MARV vaccines developed	[55]
2011	Niemann–Pick C1 protein essential for EBOV entry	[56]
2012	Monoclonal antibodies for EBOV therapy	[57]
2014	Virus-like particles capable of multiple infectious cycles	[58]

Without doubt, invasion of EBOV into a socioeconomic environment characterized by densely populated megacities, high population mobility and poor medical infrastructure greatly contributed to the unprecedented dynamics of the West African outbreak. On the other hand, filoviruses, like many RNA viruses, have a high genetic flexibility that might facilitate mutations affecting host range, pathogenicity, transmissibility, or tenacity and, thus, promote human adaptation. Such mutations, which may also contribute to the increased spread of the virus, have not been observed in past outbreaks, but it will be interesting to find out if they have occurred now.

References

1. Martini GA, Knauf HG, Schmidt HA et al (1968) Über eine bisher unbekannte, von Affen überschleppte Infektionskrankheit. Dtsch Med Wochenschr 93:559–571
2. Stille W, Bölle E, Holm E et al (1968) Über eine durch *Cercopithecus aetiops* übertragene Infektionskrankheit. Dtsch Med Wochenschr 93:572–582
3. Slenczka W, Klenk HD (2007) Forty years of Marburg virus. J Infect Dis 196(Suppl 2): S131–S135
4. Siegert R, Shu HI, Slenczka W et al (1967) Zur Aetiologie einer unbekannten, von Affen ausgegangenen menschlichen Infektionskrankheit. Dtsch Med Wochenschr 92:2443–2470
5. Siegert R, Shu HL, Slenczka W et al (1968) The etiology of an unknown human infection transmitted by monkeys. Ger Med Mon 13:1–2
6. Slenczka W, Shu HL, Piepenburg G et al (1968) Antigen-Nachweis des "Marburg-Virus" in den Organen infizierter Meerschweinchen durch Immunfluoreszenz. Dtsch Med Wochenschr 93:612–616
7. Kunz C, Hofmann H, Kovac W et al (1968) Biologische und morphologische Charakteristika des in Marburg aufgetretenen Hämorrhagischen Fiebers. Wien Klin Wochenschr 80:161–162
8. Kissling RE, Robinson RQ, Murphy FA et al (1968) Agent of disease contracted from green monkeys. Science 160:888–890
9. Nikiforov VV, Turovskii YI, Kalinin PP et al (1994) The laboratory case of Marburg hemorrhagic fever (in Russian). J Mikrobiol Epidemiol Immunol 3:104–106
10. Banadyga L, Ebihara H (2015) Epidemiology and pathogenesis of filovirus infections. In: Pattnaik AK, Whitt MA (eds) Biology and pathogenesis of rhabdo- and filoviruses. World Scientific Publ., Singapore, pp 453–486
11. Smith DH, Johnson BK, Isaacson M et al (1982) Marburg virus disease in Kenya. Lancet 1:816–820
12. Johnson ED, Johnson BK, Silverstein D et al (1996) Characterization of a new Marburg virus isolated from a 1987 fatal case in Kenya. Arch Virol Suppl 11:101–104
13. Timen A, Koopmans MPG, Vossen ACTM et al (2009) Reponse to imported Marburg hemorrhagic fever to the Netherlands (2009). Emerg Infect Dis 15:1171–1175
14. Centers for Disease Control and Prevention (2009) Imported case of Marburg hemorrhagic fever – Colorado. MMWR 58: 1377–1381
15. Bausch DG, Borchert M, Grein T et al (2003) Risk factors for Marburg hemorrhagic fever, Democratic Republic of the Congo. Emerg Infect Dis 9:1531–1537
16. Towner JS, Khristova MI, Sealy TK et al (2006) Marburg virus genomics and association with a large hemorrhagic fever outbreak in Angola. J Virol 80:6497–6516
17. Feldmann H, Klenk HD (1996) Marburg and Ebola viruses. Adv Virus Res 47:1–52
18. Mohd el Tahir B (1978) The hemorrhagic fever outbreak in Maridi, western Equatoria, southern Sudan. In: Pattyn SR (ed) Ebola virus hemorrhagic fever. Elsevier/North-Holland, Amsterdam, pp 125–137
19. Bowen ETW, Lloyd G, Harris WD et al (1977) Viral hemorrhagic fever in southern Sudan and northern Zaire. Lancet 1:571–573
20. Francis DP, Smith DH, Highton RB et al (1978) Ebola fever in the Sudan, 1976: epidemiological aspects of the disease. In: Pattyn SR (ed) Ebola virus hemorrhagic fever. Elsevier/North-Holland, Amsterdam, pp 129–135

21. Smith DH, Francis D, Simpson DIH et al (1978) The Nzara outbreak of hemorrhagic fever. In: Pattyn SR (ed) Ebola virus hemorrhagic fever. Elsevier/North-Holland, Amsterdam, pp 137–141
22. Breman JG, Piot P, Johnson KM et al (1978) The epidemiology of Ebola hemorrhagic fever in Zaire 1976. In: Pattyn SR (ed) Ebola virus hemorrhagic fever. Elsevier/North-Holland, Amsterdam, pp 103–124
23. Johnson KM, Lang JV, Webb PA et al (1977) Isolation and partial characterization of a new virus causing hemorrhagic fever in Zaire. Lancet 1:569–571
24. Jahrling PB, Geisbert TW, Galgard DW et al (1990) Preliminary report: isolation of Ebola virus from monkeys imported to USA. Lancet 335:502–505
25. Peters CJ, Johnson FD, Jahrling PB et al (1993) Filoviruses. In: Morse SP (ed) Emerging viruses. Oxford University Press, Oxford, pp 159–175
26. Miranda ME, Miranda NL (2011) Reston ebolavirus in humans and animals in the Philippines: a review. J Infect Dis 204(Suppl 3):S757–S760
27. Le Guenno B, Formenty P, Wyers M et al (1995) Isolation and partial characterization of a new strain of Ebola virus. Lancet 345: 1271–1274
28. Le Guenno B, Formenty P, Boesch C (1999) Ebola virus outbreaks in the Ivory Coast and Liberia, 1994–1995. Curr Top Microbiol Immunol 235:77–84
29. World Health Organization (1995) Ebola hemorrhagic fever. Wkly Epidemiol Rec 70: 149–152
30. World Health Organization (1995) Ebola hemorrhagic fever. Wkly Epidemiol Rec 70: 241–242
31. Sanchez A, Trappier S, Mahy BWJ et al (1996) The virion glycoproteins of Ebola viruses are encoded in two reading frames and are expressed through transcriptional editing. Proc Natl Acad Sci U S A 93:3603–3607
32. Georges AJ, Leroy FB, Renaut AA et al (1999) Ebola virus hemorrhagic fever outbreaks in Gabon, 1994–1997. J Infect Dis 179(Suppl I):S65–S75
33. Towner JS, Sealy TK, Khristova ML et al (2008) Newly discovered ebola virus associated with hemorrhagic fever outbreak in Uganda. PLoS Pathog 4:e1000212
34. Wamala JF, Lukwago L, Malimbo M et al (2010) Ebola virus hemorrhagic fever associated with novel virus strain, Uganda, 2007–2008. Emerg Infect Dis 16:1087–1092
35. McNeil A, Farnon EC, Wamala JF et al (2010) Proportion of deaths and clinical features of Bundibugyo Ebola virus infection, Uganda. Emerg Infect Dis 16:1969–1972
36. Albarino CG, Shoemaker ML, Khristova ML et al (2013) Genomic analysis of filoviruses associated with four viral hemorrhagic fever outbreaks in Uganda and the Democratic Republic of the Congo in 2012. Virology 442:97–100
37. Haas R, Maass G (1971) Experimental infection of monkeys with the Marburg virus. In: Martini GA, Siegert R (eds) Marburg virus disease. Springer Verlag, Heidelberg, pp 136–143
38. Swanepoel R, Lemon PA, Burt F et al (1996) Experimental inoculation of plants and animals with Ebola virus. Emerg Infect Dis 2:321–325
39. Pourrut X, Delicat A, Rollin PE et al (2007) Spatial and temporal pattern of Zaire ebolavirus antibody prevalence in the possible reservoir bat species. J Infect Dis 196(Suppl 2):S176–S183
40. Towner JS, Pourrut X, Albarino CG et al (2007) Marburg virus infection detected in a common African bat. PLoS One 2:e764
41. Towner JS, Amman BR, Sealy TK et al (2009) Isolation of genetically diverse Marburg viruses from Egyptian fruit bats. PLoS Pathog 5:e1000536
42. Taniguchi S, Watanabe S, Masangkay JS et al (2011) Reston Ebolavirus antibodies in bats, the Philippines. Emerg Infect Dis 17: 1559–1560
43. Negredo A, Palacios G, Vazquez-Moron S et al (2011) Discovery of an ebolavirus-like virus in Europe. PLoS Pathog 7:e1002304
44. Baize S, Pennetier D, Pharm D et al (2014) Emergence of Zaire Ebola virus in Guinea. N Engl J Med 371:1418–1425
45. Slenczka W, Rietschel M, Hoffmann C et al (1984) Seroepidemiologische Untersuchungen über das Vorkommen von Antikörpern gegen Marburg und Ebola Virus in Afrika. Mitt Österr Ges Tropenmedizin Parasitol 6:53–60
46. Feldmann H, Mühlberger E, Randolf A et al (1992) Marburg virus, a filovirus: messenger RNAs, gene order, and regulatory elements of the replication cycle. Virus Res 24:1–9
47. Bukreyev AA, Volchkov VE, Blinov VM et al (1993) The complete nucleotide sequence of the Popp (1967) strain of Marburg virus: a comparison with the Musoke (1980) strain. Arch Virol 140:1589–1600
48. Sanchez A, Kiley MP, Holloway BP et al (1993) Sequence analysis of the Ebola virus genome: organization, genetic elements, and compari-

son with the genome of the Marburg virus. Virus Res 29:215–240

49. Weissenhorn W, Calder A, Lee KH et al (1998) The central structural feature of the membrane fusion protein subunit from the Ebola virus glycoprotein is a long triple stranded coiled coil. Proc Natl Acad Sci U S A 95:6032–6036
50. Dessen A, Volchkov V, Dolnik O et al (2000) Crystal structure of the matrix protein VP40 from Ebola virus. EMBO J 19:4228–4236
51. Lee JE, Fusco ML, Hessell AJ et al (2008) Structure of the Ebola virus glycoprotein bound to an antibody from a human survivor. Nature 454:177–182
52. Mühlberger E, Weik M, Volchkov V et al (1999) Comparison of transcription and replication strategies of Marburg and Ebola virus by using artificial replication systems. J Virol 73:2333–2342
53. Sullivan NJ, Sanchez A, Rollin PE et al (2000) Development of a preventive vaccine for Ebola virus infection in primates. Nature 408: 605–609
54. Volchkov VE, Volchkova VA, Mühlberger E et al (2001) Recovery of infectious Ebola virus from cDNA: transcriptional RNA editing of the GP gene controls viral cytotoxicity. Science 291:1965–1969
55. Jones SM, Feldmann H, Ströher U et al (2005) Live attenuated recombinant vaccine protects non-human primates against either Ebola virus or Marburg virus. Nat Med 11:786–790
56. Qiu X, Audet J, Wong G et al (2012) Successful treatment of ebola virus-infected cynomolgus macaques with monoclonal antibodies. Sci Transl Med 4:138ra181
57. Carette JE, Raaben M, Wong AL et al (2011) Ebola virus entry requires the cholesterol transporter Niemann-Pick C1. Nature 477: 340–343
58. Watt A, Moukambi F, Banadyga L, Groseth A, Callison J, Herwig A, Ebihara H, Feldmann H, Hoenen T (2014) A novel life cycle modeling system for Ebola virus shows a genome length-dependent role of VP24 in virus infectivity. J Virol 88:10511–10524

Chapter 2

Forty Years of Ebolavirus Molecular Biology: Understanding a Novel Disease Agent Through the Development and Application of New Technologies

Allison Groseth and Thomas Hoenen

Abstract

Molecular biology is a broad discipline that seeks to understand biological phenomena at a molecular level, and achieves this through the study of DNA, RNA, proteins, and/or other macromolecules (e.g., those involved in the modification of these substrates). Consequently, it relies on the availability of a wide variety of methods that deal with the collection, preservation, inactivation, separation, manipulation, imaging, and analysis of these molecules. As such the state of the art in the field of ebolavirus molecular biology research (and that of all other viruses) is largely intertwined with, if not driven by, advancements in the technical methodologies available for these kinds of studies. Here we review of the current state of our knowledge regarding ebolavirus biology and emphasize the associated methods that made these discoveries possible.

Key words Ebola virus, Molecular biology, Virus life cycle, Entry, Budding, Replication, Antiviral mechanisms, Host cell interaction, Reverse genetics

1 Introduction

Molecular biology seeks to understand biological phenomena at a molecular level, and achieves this through the study of DNA, RNA, proteins, and/or other macromolecules (e.g., those involved in the modification of these substrates). Indeed, viruses themselves can be viewed as molecular tools, as they themselves represent little more than (relatively) simple assemblages of the above. As such they provide an opportunity to study interactions between these kinds of molecules in a system with a dramatically reduced level of complexity compared to other models. Similarly, as obligate parasites, their intimate interaction with the host cell also provides an opportunity to investigate these host cell processes themselves, in addition to the virus' interactions with them.

With ebolaviruses only having been discovered in 1976, this field has benefitted almost throughout its history from the

Thomas Hoenen and Allison Groseth (eds.), *Ebolaviruses: Methods and Protocols*, Methods in Molecular Biology, vol. 1628,
DOI 10.1007/978-1-4939-7116-9_2,

availability of molecular methods, and as such can really have been said to have grown up in the era of molecular biology. The result is a continuously shifting focus toward new research areas that is intertwined with, and one could even suggest driven by, advancements in the available technical methodologies.

In this chapter we seek to provide a review of the current state of our knowledge regarding ebolavirus biology, while highlighting some of the hallmark advancements that have been made over the years, and the associated methods that made these discoveries possible, so as to provide a frame of reference within which the individual methods that are presented later in this volume can be more fully appreciated.

2 Phase I: Identifying a New Enemy (1978–1985)

As a novel emerging infectious disease, many of the earliest studies of ebolaviruses unsurprisingly focused on the pathogenic potential of these agents, and on establishing the epidemiological information needed for disease control. These early studies already included the application of the newly developed enzyme-linked immunosorbent assay (ELISA) assay technology to determine antibody prevalence in the affect region prior to the first outbreaks [1], and also outside the outbreak region [2]. However, within a few years research began to focus on the most basic of questions: that of identifying what these viruses were and what they were made of. Fortunately, many of the basic tools for just such investigations had been recently developed. The earliest basic science report on ebolavirus biology was based on transmission electron microscopy imaging, developed already in the 1930s, which allowed the authors to show that ebolaviruses have a physical structure that consists of filamentous particles nearly identical to those of marburgviruses [3] (Fig. 1a). Further, polyacrylamide gel electrophoresis could be used to identify several major virion structural proteins, and also to define the sizes for these proteins based on comparison to Vesicular Stomatitis Virus (VSV) [4]. These first studies also determined that the ebolavirus glycoprotein $GP_{1,2}$ is heavily glycosylated and unusually large [4]. It was not until 5 years later that the same group, again relying on the fundamental method of gel electrophoresis, ultimately succeeded in identifying all seven of the structural proteins known today, namely the RNA-dependent RNA polymerase (L), glycoprotein ($GP_{1,2}$), nucleoprotein (NP), virion protein (VP) 40, VP35, VP30, and VP24, and assigning them their current naming based on their sizes and/or their presumed functions [5]. Further, this study produced the first estimates of protein abundance in particles, values that remain reasonably consistent (i.e., all estimates within fourfold) with the most recent estimates based on electron

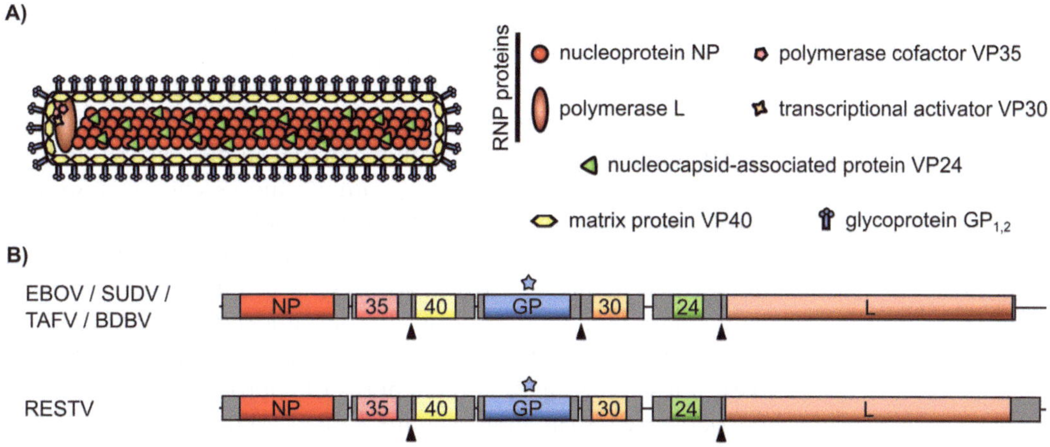

Fig. 1 Ebola virus structure. (**a**) Virion structure. Ebola virus particles contain a central nucleocapsid, which is made up of the viral RNA genome and the ribonucleoprotein complex (RNP) proteins, i.e., the nucleoprotein NP, the polymerase L, the polymerase cofactor VP35, and the transcriptional activator VP30. Also associated with the nucleocapsid is the protein VP24. Nucleocapsids are surrounded by the so-called matrix space, in which VP40 is located, which drives budding. Embedded in the host-cell-derived envelope is the glycoprotein $GP_{1,2}$, which facilitates particle entry. (**b**) Genome structure. The genome organizations of Ebola virus (EBOV), Sudan virus (SUDV), Taï Forest Virus (TAFV), Bundibugyo Virus (BDBV), and Reston Virus (RESTV) are shown. Genes are indicated with their respective names (with VP35, VP40, VP30, and VP24 abbreviated as 35, 40, 30, and 24, respectively). Untranslated regions are shown as *grey boxes*. Gene overlaps are indicated by *triangles*, and the editing site in the GP gene is highlighted by a *star*

tomography [6] (if one allows for the poor staining and thus predictable underestimation of $GP_{1,2}$ levels by Coomassie staining due to its extensive glycosylation).

At the same time analyses of the nucleic acid of Ebola virus were being conducted and showed that it existed as a nonsegmented unit (as observed in gel electrophoresis), was composed of single-stranded RNA (based on a combination of NaOH and RNase susceptibility) and was in a negative-sense orientation (based on the lack of infectiousness of the purified RNA) [7]. The RNA was also shown to be longer than VSV, although the calculated molecular weight of ~4.0×10^6 to 4.2×10^6 Da would suggest a somewhat shorter (~12 kb) genome than the 19 kb that we now know to be the case. Still, taken together these two studies, alongside evidence that these viruses were serologically distinct from rhabdoviruses [8], which they structurally most closely resembled, strongly suggested that marburgviruses and ebolaviruses are a separate taxon distinct from any other viruses known at that time [9]. Finally, agarose gel electrophoresis, combined with in vitro translation experiments, allowed six of the seven ebolavirus mRNAs (the exception being L, which is in low abundance) to be identified, and was used to determine that ebolavirus transcripts are in fact monocistronic, and that NP is the first gene in the gene order, as with other nonsegmented negative sense RNA viruses [10].

While much of this work now appears rather trivial, in an era reliant on classical biochemical approaches, and before the widespread implementation of (reverse transcription (RT)-) PCR, cloning/expression techniques, and sequencing, these represented major findings. Indeed, these fundamental studies relying on little more than electron microscopy, immunofluorescence analysis, and basic biochemical analysis and gel electrophoresis can be said to have been largely responsible for the classification of ebolaviruses as a sister genus to marburgviruses, which together form the *Filoviridae* family, as well as for our understanding that these unique pathogens are negative-sense single-stranded RNA viruses with unusually long genome lengths of approximately 19 kb that encode seven structural proteins (NP, VP35, VP40, GP, VP30, VP24, and L).

3 Phase II: Focusing on Sequences (1986–1995)

With a workable view of the genetic and protein makeup of the virus particles already in place by the mid/late-80s, the next phase of research into ebolavirus molecular biology shifted its focus to the functions of individual proteins and genetic elements in driving various essential steps of the virus lifecycle. This area remained a subject of intense interest for the next 10–15 years, and saw the development of Sanger sequencing making a huge contribution to developing a more refined view of the virus genome. Similarly, advancements in tools related to antibody detection, expression systems, and reporter systems became available to fuel these kinds of studies.

The first genetic elements identified in the virus were conserved 3′ terminal sequences, which based on their conservation between ebolaviruses and marburgviruses and the function of similarly positioned elements in other virus families were proposed to be responsible for polymerase binding and packaging [11]. Indeed, we now know that these elements represent highly conserved terminal hairpin-forming bipartite promoter regions [12].

Subsequent work involving cloning and sequencing of NP mRNAs or terminal fragments of the viral RNA using polyadenylation and RT-PCR-based approaches, similar to those sometimes now used for rapid amplification of cDNA ends (RACE) sequencing of genome ends, then allowed further identification of the transcriptional "start" and "stop" signals responsible for regulating the initiation and termination of transcription [13]. Again, while these sequences have been slightly refined over the years to give a consensus representative of all gene products and ebolavirus species, they have remained essentially unchanged since these early days of filovirus molecular biology research. Further, the authors could determine the sequence and amino acid composition of the entire NP protein, giving us our first look up close at an ebolavirus protein [13].

Soon to follow was first a partial [14] and then a complete sequence of GP [15]. Here we also saw the first hints of the influence that data repositories such as EMBL, SwissPROT, and GenBank, and the bioinformatics tools that they host, would eventually have in the field, and indeed on molecular biology in general. The availability of these tools allowed the first in silico studies of ebolaviruses to be conducted, and lead to the suggestion that the ebolavirus $GP_{1,2}$ possesses a retrovirus-like immunosuppressive motif [14, 16], a finding that when followed up years later yielded evidence that this motif may be functional and influence CD4 and CD8 T cell biology [17]. The availability of these new bioinformatics tools also allowed for the first time genetic and protein level comparisons between ebolaviruses and marburgviruses [18, 19]. The interest in applying bioinformatics also extended to studies of evolutionary rate and pressures, and to assessments looking to understand the origin and evolution of filoviruses [20].

However, in the era of sequencing and genomics there was still room for classical biochemistry-based techniques, including protein sequencing, which was used to establish the protein sequences of NP, VP35, and VP40, and established their order as the first three transcripts encoded by the genome, something that was at that time problematic given that theoretical predictions suggested that VP35 should actually encode the larger of the two proteins, although this is not experimentally observed. Further it was established that VP30 was in fact a distinct protein unique to filoviruses, and not a cleavage product of NP, as had sometimes been suggested [21]. Then, at long last, came the first full-length sequence of an Ebola virus genome, which definitively confirmed the genome organization, although it was already suspected based on an earlier complete sequence for the closely related marburgvirus [22], and established the sequences of the previously missing intergenic regions [23] (Fig. 1b). It also clearly paved the way for the sequencing of complete genomes for other ebolavirus species in the years to follow [24–27], and thus much of our appreciation of the genetic/phylogenetic structure of the family as we know it today—with a clear separation of the family of *Filoviridae* not only into two distinct genera, *Ebolavirus* and *Marburgvirus*, but further subdivision of the ebolaviruses into five species: *Zaire ebolavirus*, *Sudan ebolavirus*, *Reston ebolavirus*, *Taï Forest ebolavirus*, and the recently discovered *Bundibugyo ebolavirus*.

This time span saw also several notable public health events in the timeline of ebolavirus history, including the importation of a new ebolavirus species, *Reston ebolavirus*, into the USA and Italy, as well as the discovery of the another new pathogenic ebolavirus species, *Taï Forest ebolavirus*, in Ivory Coast, and what was until recently the largest and most deadly recorded outbreak of ebolavirus, in Kikwit in 1995. Thus, while ebolavirus research during this time period had clearly been heavily focused on sequencing, these events encouraged renewed efforts to adapt molecular methods, including those based on electron microscopy and various antibody reactivity-based methods, to develop new and broader specificity diagnostic methods for these viruses (e.g., [28–36]).

4 Phase III: It's All About Function (1995–Present)

The next years were marked by a clear transition away from sequencing and toward the first experiments looking into more functional aspects of ebolavirus biology, and the molecular and cell biological basis for these functions. In retrospect one can see that this transition was clearly triggered by the accumulation of a critical mass of genetic information, and the forthcoming development of a variety of molecular tools/approaches needed to conduct these kinds of experiments. However, it was also clearly fueled by some fascinating discoveries that resulted, at least in part, from examinations of the newly available genetic information itself.

4.1 The Many Forms of GP

Possibly one of the most fascinating of such discoveries was also the first, and concerned the ability of the GP gene to undergo transcriptional editing at a stretch of 7 U residues to produce the full length $GP_{1,2}$ mRNA transcript, while in the majority of cases a soluble version of the glycoprotein was produced (sGP) [37, 38] (Fig. 2). Indeed, recently it has been confirmed that other editing products already suggested by this study can in fact also be generated, and lead to the production of an additional small soluble GP (ssGP) [39]. Further studies showed the additional production of a stable C-terminal cleavage product of sGP, known as delta-peptide [40], as well as shed forms of both GP_1 (produced as a result of disulfide bond instability) [41] and the full length $GP_{1,2}$ (produced by tumor necrosis factor α-converting enzyme (TACE) cleavage to yield $GP_{1,2\Delta TM}$) [42]. Interest in the molecular basis for the development of this unusual diversity of GP products, and also the potential relevance of these different forms of GP, helped motivate a transition in ebolavirus research as a whole away from genomics and into studies of protein function in subsequent years. However, even now these questions remain at best partially answered, and thus the biology of these soluble glycoproteins, and their relevance for virus infection, is still being studied today.

The first proposed function of sGP synthesis was simply to serve as a means of regulating $GP_{1,2}$ expression, and thus controlling excessive $GP_{1,2}$-mediated cytotoxicity [43], as a result of the overlapping open reading frame organization between these two proteins, which strongly favors sGP synthesis. This is supported by some studies that suggest that $GP_{1,2}$-mediated cytotoxicity is not a

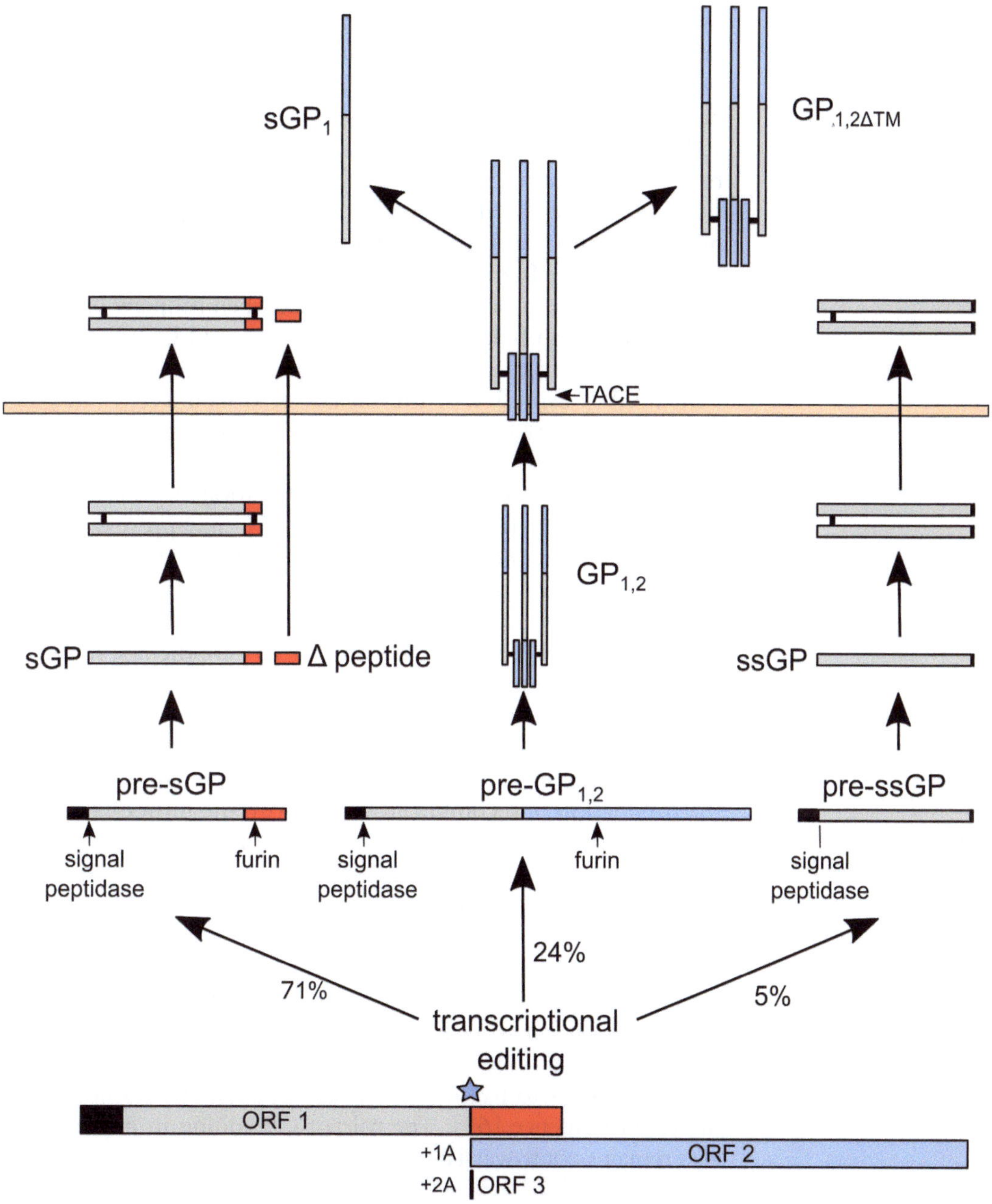

Fig. 2 Glycoprotein processing. The Ebola virus glycoprotein gene contains three open reading frames (ORFs), which can be accessed through transcriptional editing. In cases where no transcriptional editing occurs, only ORF 1 is transcribed, which leads to the expression of pre-sGP, which is processed by signal-peptidase and furin to yield the secreted glycoprotein sGP and Δ peptide. If the second ORF is accessed by insertion of a single nucleotide during transcription of the editing site (highlighted by a *star*), pre-$GP_{1,2}$ is produced, which is processed by the same proteases into the mature $GP_{1,2}$, the only Ebola virus glycoprotein that is membrane-associated. However, upon cleavage by TACE a secreted version of this protein ($GP_{1,2\Delta TM}$) can also be produced. Further, soluble GP_1 can also be released from $GP_{1,2}$. Finally, if two nucleotides are inserted in the editing site during transcription, resulting in the third ORF being transcribed after the editing site, pre-ssGP is produced. This protein is processed by signal peptidase only, resulting in the small secreted glycoprotein ssGP

major factor when expression is moderate [44]. More recently, immunological effects such as the inhibition of neutrophil function [45, 46] and antigenic subversion of antibody responses toward nonneutralizing epitopes [47] have also been described. Further, sGP has been suggested to possess an anti-inflammatory function that may help mitigate negative effects of $GP_{1,2}$ on vascular integrity [48]. Similarly, delta-peptide has been proposed to have a modulatory role, in this case by inhibiting the entry of virus into target cells [49]. In the case of $GP_{1,2\Delta TM}$, evidence for a function has only recently been obtained and suggest when properly presented as a trimer, it can bind to and activate uninfected DCs and macrophages [50], and that its shedding directly modulates the levels of EBOV $GP_{1,2}$ expressed at the surface of virus-infected cells, and thus its availability for incorporation into virus particles [51].

There was also tremendous interest starting at this time in the biology of the surface glycoprotein itself. Various aspects including its fusion activity [52], structure [53–55], and processing [56] were studied by a number of different groups, which all contributed to our current view of $GP_{1,2}$ as a heterotrimeric, class I fusion protein which is processed from a single polypeptide by furin cleavage. Intriguingly, already at this time there were the first hints that $GP_{1,2}$ cleavage by furin is not necessary for function and infectivity, in contrast to the situation with many other viruses [57]. However, these findings would not be directly confirmed for many more years until the development of full-length clone systems allowed the production of recombinant viruses with the furin cleavage site knocked out [58, 59]. The role of $GP_{1,2}$ in pathogenesis is more controversial, with early studies indicating that its expression alone causes vascular leakage from vessel explants, and thus that it is the main viral determinant of Ebola pathogenicity [60]. However, more recently studies using recombinant viruses in which the $GP_{1,2}$ from Ebola virus (EBOV, species *Zaire ebolavirus*) and the apathogenic Reston virus (RESTV, species *Reston ebolavirus*) were exchanged have shown that while EBOV $GP_{1,2}$ enhances virulence and contributes to pathological changes, alone it does not confer increased virulence [61]. The role of GP editing in pathogenesis also remains controversial [43, 62].

4.2 Uncovering an Unusual Mechanism of Entry

The mechanism by which Ebola virus enters target cells was for many years a source of tremendous confusion within the field, with numerous binding proteins, including Folate Receptor Alpha [63], a wide variety of C-type lectins (including dendritic cell-specific intercellular adhesion molecule-3-grabbing non-integrin (DC-SIGN) [64, 65], liver/lymph node-specific ICAM-3 grabbing non-integrin (L-SIGN) [64], other human macrophage calcium-dependent lectins [66]), DC-SIGN receptor [65], and most recently, Tim-1 [67], being identified. However, in many cases binding to these surface receptors, while beneficial, is nonessential and/or their relevance is restricted to certain cell types. How such a situation is possible may have been clarified by recent evidence identifying Niemann-Pick C1 (NPC-1), which acts at the levels of the endosome, at a stage after both surface attachment and uptake, as an apparently critical receptor for ebolavirus entry [68, 69].

The mechanism of uptake of ebolavirus into host cells also was a controversial topic, with various studies suggesting the involvement of clathrin-mediated endocytosis [70, 71], caveolin-mediated endocytosis [72], or macropinocytosis [73]. However, a major limitation of these studies was that they were performed using virus pseudotypes (VSV, HIV, etc. carrying the ebolavirus $GP_{1,2}$ glycoprotein), which differ significantly in their morphology compared to ebolavirus particles, a factor that could potentially affect uptake by some or all of these pathways. More recently, researchers have begun performing experiments with infectious ebolavirus, or structurally similar ebolavirus-like particles, and these appear to largely demonstrate virus uptake via macropinocytosis [74–76]. However, some studies continue to demonstrate that other pathways could also still be playing a role, and that in fact multiple different mechanisms may actually be used in concert [77–79].

Equally controversial has been the role of protease cleavage of the viral glycoprotein. It appears clear that endosomal cleavage of GP_1 into a low molecular weight form is necessary to obtain a fusion competent conformation. So far, the only proteases identified as participating in this process are the endosomal cysteine proteases Cathepsin L (CatL) and Cathepsin B (CatB) [80, 81]. Interestingly, there appear to be differences in processing in different cells types [82], as well as differences in Cathepsin sensitivity among filovirus species [83–85]. Further, a recent in vivo study has also demonstrated that despite the demonstrated importance of Cathepsin cleavage in a number of cell culture systems, during ebolavirus infection in the mouse model neither CatB nor CatL appear to be necessary, and CatB/L double knockout mouse embryonic fibroblast cells remain equally susceptible to virus infection [85], suggesting a role for other proteases under at least some circumstances. Indeed, recent studies with a thermolysin-trimmed GP_1 [86], which structurally resembles that produced by Cathepsin cleavage, appear to support that, at least in principle, a similar effect could be achieved by digestion with other cellular proteases.

Overall, while several issues remain to be fully clarified, what these findings have led to is a current model of virus entry (Fig. 3) in which no specific critical cell surface receptor is required, but rather a number of attachment factors serve to concentrate virus particles at the cell surface to enhance their uptake, which then occurs primarily via macropinocytosis. Following endosome acidification, trimming of GP_1 by Cathepsins, or possibly also other proteases, exposes the receptor binding domain, and following

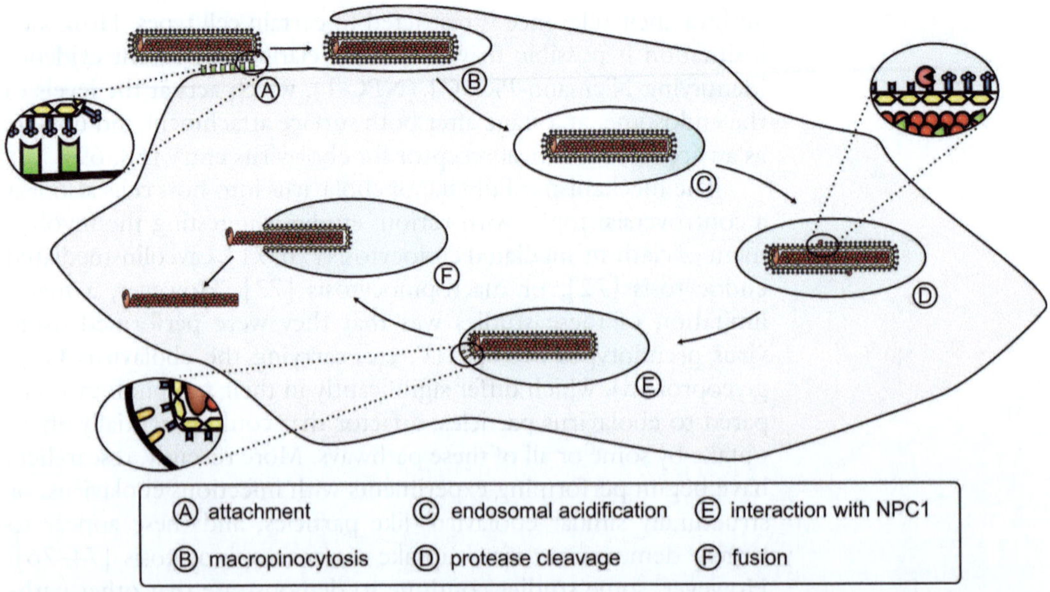

Fig. 3 Model of Ebola virus entry. At the cell surface Ebola viruses attach to C-type lectins through $GP_{1,2}$, or to phosphatidylserine receptors (e.g., Tim-1) through interactions with the viral envelope (**a**). Uptake into endosomes occurs through macropinocytosis (**b**). After endosomal acidification (**c**) host proteases cleave $GP_{1,2}$ to yield a 19 kDa form (**d**), which can then interact with the cellular receptor NPC-1 (**e**), ultimately resulting in membrane fusion (**f**) and release of nucleocapsids into the cytoplasm

binding to the NPC-1 receptor, fusion occurs to release the viral nucleocapsid into the cytoplasm.

4.3 The Mechanics of Genome Replication and Transcription

The late 90s also saw the first studies of transcription and replication mechanics for filoviruses being conducted (Fig. 4). These studies relied on the development of minigenome systems, which use reporter-expressing viral genome analogues to model viral transcription and replication, and produced findings that define our understanding of the basic viral components necessary for these processes (i.e., NP, VP35, VP30, and L) [87]. While the roles of NP (nucleoprotein), VP35 (polymerase cofactor), and L (polymerase) were easy to rationalize based on better studied analogous virus systems, the need for VP30, which lacks clear homologues in other systems, remained a mystery for several years. Eventually, however, VP30 was shown to serve as a cofactor for EBOV transcription, which is needed to overcome a hairpin structure in the 3′ viral noncoding region [88]. The unusual existence of this additional RNP component has led to much interest in the role of VP30 over recent years, with several further studies focusing on the role of its phosphorylation state in the regulation of RNA synthesis. In particular it has been demonstrated that phosphorylation of VP30 regulates its interactions with both VP35 and NP, and must be dynamic in order to allow it to regulate both

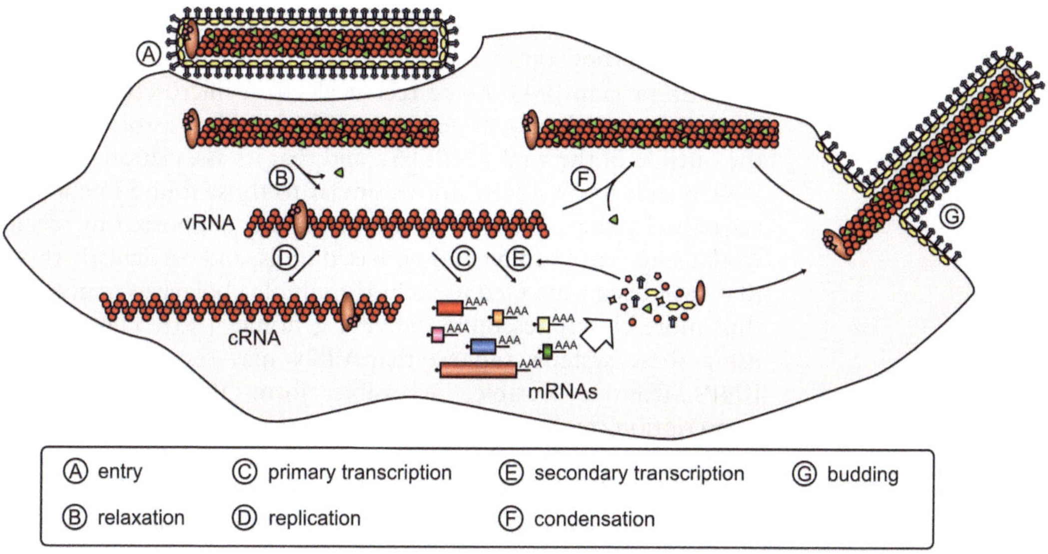

Fig. 4 Model of Ebola virus genome replication and transcription. After entry (**a**) into the cell, nucleocapsids relax (**b**) due to dissociation of VP24. This allows the viral polymerase complex that entered the cell within the virus particle to perform primary transcription (**c**), which produces the viral proteins necessary for genome replication (**d**) and further, secondary transcription (**e**) and protein production. Produced VP24 then leads to a condensation of newly formed ribonucleoprotein complexes into packaging-competent nucleocapsids (**f**), which are transported to the surface, where budding takes place in a process driven by VP40

transcription and replication [89, 90]. Recently, it has also been demonstrated that VP30 is capable of directly binding to the 3'-end of the ebolavirus genome RNA, and that this binding also stabilizes VP35/L RNA binding [90]. As a result, it has been suggested that phosphorylation allows VP30 to modulate the composition of the RNP complex, in order to form either a transcriptase or replicase complex [89], and thereby regulates these two activities of the viral polymerase. Of practical significance, this peculiar dependence of the ebolavirus transcriptional mechanism on VP30 has in recent years also facilitated the development of "biologically contained" Ebola viruses that lack this viral protein in their genome, and whose replication is thus restricted to VP30-expressing cells, allowing them to be studied at lower levels of containment [91]. Most recently, VP30 has also been shown to have a novel function as a trans-acting factor for RNA editing of the GP gene [92]. This process seems to be both sequence and RNA structure dependent and provides some of the first mechanistic information about how this important process in the EBOV lifecycle is being regulated.

Recent discoveries in the field have also clarified the role of the other viral protein unique to filoviruses, VP24, and shown that it too plays an important role in the regulation of viral RNA synthesis. Early biochemical studies suggested that VP24 might function as a minor matrix protein, due to its apparent localization in the matrix space in salt dissociation experiments [93], as well as a number of properties considered potentially consistent with function as a matrix protein [94]. More recent electron microscopy analyses have, however, demonstrated that VP24 is in fact associated with the outside of the viral RNPs [6], and that its association with the RNP results in rigid RNP forms similar to those found in ebolavirus particles [95]. This observation is further supported by recent work using various minigenome-based tools, and particularly those that have been expanded to include multiple ebolavirus genes, and thus more closely resemble the viral genomes [96]. The findings using these systems suggest that VP24 may serve to condense RNPs from a flexible, accessible form that can undergo transcription/replication to a rigid packaging-competent form, and that this condensation also serves to lock the polymerase at the genome terminus so that it remains competent to perform primary transcription upon virus infection [96]. Thus this also appears to explain previous observations that VP24 serves as an inhibitor of viral RNA synthesis when overexpressed from plasmids [97, 98], and that it is necessary for production of nucleocapsids capable of undergoing primary transcription [99].

Finally, studies have also started to address the question of where exactly in the host cell ebolavirus genome replication and transcription take place. Both IFA-based and reporter virus studies have recently been used to demonstrate inclusion bodies as the sites of virus RNA synthesis, thereby clarifying a long-standing question about whether these structures are biologically relevant, or simply represent functionally dead masses of accumulated protein [100, 101].

4.4 Understanding Particle Morphogenesis and Budding

The biosafety requirements for work on live EBOV have resulted in the development of a large collection of reverse genetics-based tools to study and model aspects of the viral lifecycle (e.g., various monocistronic and multicistronic minigenome and transcription and replication-competent virus-like particle (trVLP) assays) [87, 96, 99, 102, 103]. While these systems were first developed as simple genome analogues to look at viral RNA synthesis, the more complex of these systems are now finding applications for studying other steps in the virus lifecycle, including morphogenesis and budding. Further, as the number of biosafety level 4 facilities worldwide and the availability of high-tech imaging platforms (including live cell imaging) increases, detailed immunofluorescence labeling and reporter viruses studies are now also being used to study the movement of virus proteins and RNAs during the virus lifecycle and are providing important insights. For example, a recent study used VP30-GFP labeled viruses to look at the transport of nucleocapsids within the cytoplasm from sites of replication to the sites of budding at the cell surface and the dependence of this process on the actin cytoskeleton [104]. Interestingly, recent single-particle tracking studies have shown that actin also directs the movement of the matrix protein (VP40) to the plasma membrane as well [105], where it must eventually meet up with these packaging-ready nucleocapsids. Clearly these kinds of studies demonstrate how powerful these advanced imaging techniques are likely to become in the coming years for looking at these poorly defined interfaces between the different stages in the virus lifecycle, in particular the process of morphogenesis. However, at present our understanding of this process for ebolaviruses still remains limited.

Unlike RNP transport and morphogenesis, the budding process of ebolaviruses has actually been quite intensely studied. The major contributor to budding is the matrix protein VP40, and studies of its function in virus budding began in earnest in 2000. Already these early studies demonstrated that recombinantly expressed VP40 was capable of independently binding to membranes [106], and that it induced the release of lipid enveloped particles [107, 108]. Further, unlike expression of $GP_{1,2}$, which leads to the production of pleomorphic particles, VP40 expression drives the production of particles having an authentic filamentous structure [109]—all features expected of a bona fide viral matrix protein.

At the same time overlapping proline-rich late domains of both the PT/SAP and PPxY types common to many other viral matrix proteins were identified in VP40, although in ebolaviruses they display an unusual overlapping arrangement (i.e., 7-PTAPPEY-13). As with similar domains in other viruses these were then soon shown to directly support virus budding by mediating interaction with WW domain containing proteins, including the yeast homolog of Nedd4, mammalian Nedd4, Tsg101, and most recently ITCH [110–114], and thereby to facilitate interaction with the Vps4 pathway upon which ebolavirus budding is also dependent [111]. However, while these interactions have been shown to support budding, production of a recombinant ebolaviruses lacking these late-domain motifs has demonstrated that such a virus is still viable, and thus that alternate mechanisms of virus release must also exist [115]. Potentially consistent with this finding, a third YPx_nL/I type late domain (18-YPARSNSTI-26) was recently identified in ebolavirus VP40 and appears to mediate an additional interaction with the ESCRT-III protein Alix [114, 116], which might also contribute to budding, although further studies are still needed to confirm these findings.

Crystal structures were also obtained early during the study of VP40 biology and have revealed a unique duplicated two-domain structure connected by a flexible hinge that controls the transition between monomeric and multimeric states [117, 118]. Others then further demonstrated that hexamerization of VP40 into ring-like structures might be triggered by membrane binding [106,

119], an observation that is supported by biochemical evidence that oligomerization is indeed a prerequisite for the budding activity of VP40 [120].

Another significant area that until recently had remained poorly studied concerns how VP40 is actually able to drive membrane deformation during the budding process. However, a number of recent studies using biochemical and biophysical approaches have demonstrated an intrinsic ability of VP40 to penetrate into lipid bilayers and induce membrane curvature consistent with virus particle formation [105, 121] in a manner that also appears to be dependent on the lipid composition of the membranes themselves [122]. As such these studies provide the first insight into the mechanics of vesiculation as a key step in the budding process.

4.5 Interactions with the Innate Immune Response

In addition to being a time of great interest in virus biology, the late 1990s also saw the first studies begin to investigate the effects of virus infection on the immune system, and specifically interactions with the interferon (IFN) system [123, 124]. It was only a few years later that it was determined that VP35 is responsible for actively inhibiting the IFN production pathway [125], and that this occurs through interaction with IFN regulatory factor 3 (IRF3) [126]. This process was then further shown to be dependent both on trimerization of VP35 through a coil-coiled domain located in the N-terminal portion of the protein [127], and specific basic residues located in the C-terminus of the protein [128], which also harbors a dsRNA-binding domain [129]. Later studies showed that the ability of VP35 to inhibit IFN production is due to direct binding of VP35 to IKKε and TANK-binding kinase 1 (TBK1) in a manner that blocks subsequent IRF3 and IRF7 interaction, and thus their subsequent activating phosphorylation [130]. Recent co-crystal structures of the VP35 RNA-binding domain together with dsRNA have provided interesting insights into the mechanism of dsRNA binding, which for ebolavirus appears to occur efficiently both along the backbone of the dsRNA and at the terminal-free ends, and may block detection of dsRNA templates by RNA helicases, such as retinoic acid inducible gene I (RIG-I) [131].

The relevance of IFN antagonism for virus infection and pathogenesis was quickly recognized, and was highlighted by the finding that unlike immunocompetent mice, IFN receptor α knockout mice are highly susceptible to ebolavirus infection [132]. In addition, the production of recombinant viruses in which the IFN inhibitory activity of VP35 is abolished showed significant attenuation both in vitro [133, 134] and in vivo [135]. Microarray analyses have also suggested that the extent of IFN suppression may correlate with the virulence of different filovirus species [136], a finding that was recently supported by the implementation of highly standardized IFN antagonism assays [137].

More recently, a second point of interference with the IFN system has been described for ebolaviruses, this time with respect to inhibition of IFN signaling and the resulting production of interferon stimulated genes (ISGs) [138]. This work further showed that inhibition was mediated by direct binding of VP24 to karyopherin-α1, and that this blocks signal transducer and activator of transcription 1 (STAT1) nuclear accumulation, which is essential for subsequent activation of ISG transcription [138]. Further studies eventually showed that VP24 is in fact capable of binding to all members of the nucleoprotein interactor 1 (NPI-1) subfamily (i.e., karyopherins-α1, -α5, and -α6) [139]. Since the VP24-binding site appears to lie within the STAT1-binding site on karyopherin-α, it has been suggested that VP24 may inhibit STAT1 translocation by competing for the same binding site on karyopherin-α [139, 140]. At the same time, biochemical analysis, in conjunction with recently obtained X-ray crystallographic structures and deuterium exchange mass spectrometry analyses, indicate that VP24 can also directly bind to STAT1 itself [141].

Interestingly, an additional point of action for VP24 has also been recently reported, with data suggesting that VP24 can block IFN-stimulated phosphorylation of p38-α in some, but not all, cell lines [142]. While this pathway is known to be involved in the IFN response to other viruses [143, 144], the relevance and further details of this mechanism still need to be established. In any event, this multifaceted approach of targeting both IFN production and signaling is by no means uncommon and again suggests the critical importance of controlling this aspect of the innate immune response for virus survival.

In addition to blocking the production of IFN itself and its subsequent signaling to produce interferon stimulated genes (ISGs) there is some evidence that ebolaviruses are able to specifically counteract the activities of individual ISGs as well. The best studied of these effects is the inhibition of Tetherin (BST-2) by ebolavirus $GP_{1,2}$ [145], which seems to occur via an unusual mechanism that does not involve blocking Tetherin's cell surface expression [146, 147]. Recently this mechanism has been shown to be dependent on the $GP_{1,2}$ transmembrane domain [148], as well as an intact receptor-binding domain and correct N-glycan processing [149], and has been proposed to involve $GP_{1,2}$s ability to block interaction between VP40 and Tetherin [150].

In addition to interfering with the IFN pathway, ebolavirus infection has been shown to interfere with a variety of other pathways and functions related to innate immune defense. In particular, VP35 has also been described to interfere with and even actively suppress activation of the protein kinase R (PKR) pathway in order to avoid translational shutoff [151, 152]. Further, VP35 has been shown to act as a suppressor of RNA silencing, a function that is dependent upon its dsRNA-binding activity [153]. Finally, ebolavirus infection has been recently shown to lead to sequestration of stress granule proteins into inclusion bodies, which might help the virus avoid yet another antiviral response, as neither canonical stress granule formation, nor the associated translational arrest, are observed in ebolavirus-infected cells [154]. In addition to all these mechanisms, which inhibit immunity on the cellular level, there are also a number of mechanisms by which ebolaviruses interfere with the immune system on an organismal level. These aspects are discussed in Chapter 3, which addresses the molecular and clinical pathogenesis of Ebola virus.

5 Final Remarks

Taking a step back it is clear that ebolavirus research has seen a number of "trends" over the years with sudden, although sometimes unfortunately short-lived, interest in particular areas. Equally evident is that these shifts in focus (at least as evidenced by publication output) can be driven by specific situations, such as outbreaks, but can also be strongly influenced by the development and accessibility of new technologies. Certainly one of the biggest trends in the last decade or so has been the entry into the field of a wide variety of experts with diverse specialties, something which is in no small part facilitated by advancements in molecular biology approaches that allow studies on biosafety level 4 pathogens like ebolavirus to be conducted outside a high containment environment. As such the number of publications has exploded and the variety of research being conducted has diversified accordingly. It can only be hoped that this trend will continue into the future as new tools and techniques become available to support continued investigations into the many remaining open questions regarding filovirus biology. Finally, it should be noted that it is still the case that the vast majority of ebolavirus studies focus solely on the *Zaire ebolavirus* species, and while it is generally believed that most of the basic biological functions are conserved among all ebolavirus species, this is also clearly not always the case. Thus, there is still much work to do in identifying molecular differences that can help us explain the differences in pathogenicity and geographical distribution observed between different ebolaviruses.

Acknowledgments

The authors wish to thank Heinz Feldmann for his insightful comments on the manuscript.

References

1. Van der Groen G, Pattyn SR (1979) Measurement of antibodies to Ebola virus in human sera from N.W.-Zaire. Ann Soc Belg Med Trop 59(1):87–92
2. Saluzzo JF, Gonzalez JP, Herve JP, Georges AJ, Johnson KM (1980) Preliminary note on the presence of antibodies to Ebola virus in the human population in the eastern part of the Central African Republic. Bull Soc Pathol Exot Filiales 73(3):238–241
3. Ellis DS, Simpson IH, Francis DP, Knobloch J, Bowen ET, Lolik P, Deng IM (1978) Ultrastructure of Ebola virus particles in human liver. J Clin Pathol 31(3):201–208
4. Kiley MP, Regnery RL, Johnson KM (1980) Ebola virus: identification of virion structural proteins. J Gen Virol 49(2):333–341
5. Elliott LH, Kiley MP, McCormick JB (1985) Descriptive analysis of Ebola virus proteins. Virology 147(1):169–176
6. Beniac DR, Melito PL, Devarennes SL, Hiebert SL, Rabb MJ, Lamboo LL, Jones SM, Booth TF (2012) The organisation of Ebola virus reveals a capacity for extensive, modular polyploidy. PLoS One 7(1):e29608. doi:10.1371/journal.pone.0029608
7. Regnery RL, Johnson KM, Kiley MP (1980) Virion nucleic acid of Ebola virus. J Virol 36(2):465–469
8. Van der Groen G, Elliot LH (1982) Lack of cross reactivity of rhabdovirus antibodies with Marburg and Ebola antigens in the indirect immunofluorescent antibody test. Ann Soc Belg Med Trop 62(1):67–68
9. Kiley MP, Bowen ET, Eddy GA, Isaacson M, Johnson KM, McCormick JB, Murphy FA, Pattyn SR, Peters D, Prozesky OW, Regnery RL, Simpson DI, Slenczka W, Sureau P, van der Groen G, Webb PA, Wulff H (1982) Filoviridae: a taxonomic home for Marburg and Ebola viruses? Intervirology 18(1–2): 24–32
10. Sanchez A, Kiley MP (1987) Identification and analysis of Ebola virus messenger RNA. Virology 157(2):414–420
11. Kiley MP, Wilusz J, McCormick JB, Keene JD (1986) Conservation of the 3′ terminal nucleotide sequences of Ebola and Marburg virus. Virology 149(2):251–254
12. Weik M, Enterlein S, Schlenz K, Muhlberger E (2005) The Ebola virus genomic replication promoter is bipartite and follows the rule of six. J Virol 79(16):10660–10671. doi:10.1128/JVI.79.16.10660-10671.2005
13. Sanchez A, Kiley MP, Holloway BP, McCormick JB, Auperin DD (1989) The nucleoprotein gene of Ebola virus: cloning, sequencing, and in vitro expression. Virology 170(1):81–91
14. Volchkov VE, Blinov VM, Netesov SV (1992) The envelope glycoprotein of Ebola virus contains an immunosuppressive-like domain similar to oncogenic retroviruses. FEBS Lett 305(3):181–184
15. Will C, Muhlberger E, Linder D, Slenczka W, Klenk HD, Feldmann H (1993) Marburg virus gene 4 encodes the virion membrane protein, a type I transmembrane glycoprotein. J Virol 67(3):1203–1210
16. Becker Y (1995) Retrovirus and filovirus "immunosuppressive motif" and the evolution of virus pathogenicity in HIV-1, HIV-2, and Ebola viruses. Virus Genes 11(2–3): 191–195
17. Yaddanapudi K, Palacios G, Towner JS, Chen I, Sariol CA, Nichol ST, Lipkin WI (2006) Implication of a retrovirus-like glycoprotein peptide in the immunopathogenesis of Ebola and Marburg viruses. FASEB J 20(14):2519–2530. doi:10.1096/fj.06-6151com
18. Bukreyev A, Volchkov VE, Blinov VM, Netesov SV (1993) The GP-protein of Marburg virus contains the region similar to the 'immunosuppressive domain' of oncogenic retrovirus P15E proteins. FEBS Lett 323(1–2):183–187
19. Bukreyev AA, Volchkov VE, Blinov VM, Netesov SV (1993) The VP35 and VP40 proteins of filoviruses. Homology between Marburg and Ebola viruses. FEBS Lett 322(1):41–46
20. Suzuki Y, Gojobori T (1997) The origin and evolution of Ebola and Marburg viruses. Mol Biol Evol 14(8):800–806
21. Elliott LH, Sanchez A, Holloway BP, Kiley MP, McCormick JB (1993) Ebola protein analyses for the determination of genetic organization. Arch Virol 133(3–4):423–436
22. Feldmann H, Muhlberger E, Randolf A, Will C, Kiley MP, Sanchez A, Klenk HD (1992) Marburg virus, a filovirus: messenger RNAs, gene order, and regulatory elements of the replication cycle. Virus Res 24(1):1–19
23. Sanchez A, Kiley MP, Holloway BP, Auperin DD (1993) Sequence analysis of the Ebola virus genome: organization, genetic elements, and comparison with the genome of Marburg virus. Virus Res 29(3):215–240
24. Ikegami T, Calaor AB, Miranda ME, Niikura M, Saijo M, Kurane I, Yoshikawa Y, Morikawa S (2001) Genome structure of Ebola virus subtype Reston: differences among Ebola subtypes. Brief report. Arch Virol 146(10): 2021–2027

25. Groseth A, Stroher U, Theriault S, Feldmann H (2002) Molecular characterization of an isolate from the 1989/90 epizootic of Ebola virus Reston among macaques imported into the United States. Virus Res 87(2):155–163
26. Sanchez A, Rollin PE (2005) Complete genome sequence of an Ebola virus (Sudan species) responsible for a 2000 outbreak of human disease in Uganda. Virus Res 113(1):16–25. doi:10.1016/j.virusres.2005.03.028
27. Towner JS, Sealy TK, Khristova ML, Albarino CG, Conlan S, Reeder SA, Quan PL, Lipkin WI, Downing R, Tappero JW, Okware S, Lutwama J, Bakamutumaho B, Kayiwa J, Comer JA, Rollin PE, Ksiazek TG, Nichol ST (2008) Newly discovered ebola virus associated with hemorrhagic fever outbreak in Uganda. PLoS Pathog 4(11):e1000212. doi:10.1371/journal.ppat.1000212
28. Geisbert TW, Jahrling PB (1995) Differentiation of filoviruses by electron microscopy. Virus Res 39(2–3):129–150
29. Becker S, Feldmann H, Will C, Slenczka W (1992) Evidence for occurrence of filovirus antibodies in humans and imported monkeys: do subclinical filovirus infections occur worldwide? Med Microbiol Immunol (Berl) 181(1):43–55
30. Geisbert TW, Jahrling PB (1990) Use of immunoelectron microscopy to show Ebola virus during the 1989 United States epizootic. J Clin Pathol 43(10):813–816
31. Geisbert TW, Rhoderick JB, Jahrling PB (1991) Rapid identification of Ebola virus and related filoviruses in fluid specimens using indirect immunoelectron microscopy. J Clin Pathol 44(6):521–522
32. Rollin PE, Ksiazek TG, Jahrling PB, Haines M, Peters CJ (1990) Detection of Ebola-like viruses by immunofluorescence. Lancet 336(8730):1591
33. Kalter SS, Heberling RL, Barry JD, Tian PY (1995) Detection of Ebola-Reston (Filoviridae) virus antibody by dot-immunobinding assay. Lab Anim Sci 45(5):523–525
34. Ksiazek TG, Rollin PE, Jahrling PB, Johnson E, Dalgard DW, Peters CJ (1992) Enzyme immunosorbent assay for Ebola virus antigens in tissues of infected primates. J Clin Microbiol 30(4):947–950
35. Merzlikin NV, Chepurnov AA, Istomina NN, Ofitserov VI, Vorob'eva MS (1995) Development and application of an immunoenzyme test system for diagnosing Ebola fever. Vopr Virusol 40(1):31–35
36. Borisevich IV, Mikhailov VV, Potryvaeva NV, Malinkin Iu N, Kirillov AP, Krasnianskii VP, Markov VI, Makhlai AA, Lebedinskaia EV (1996) Development of the immunoenzyme test-system for detection of Ebola virus antigen. Vopr Virusol 41(5):232–234
37. Volchkov VE, Becker S, Volchkova VA, Ternovoj VA, Kotov AN, Netesov SV, Klenk HD (1995) GP mRNA of Ebola virus is edited by the Ebola virus polymerase and by T7 and vaccinia virus polymerases. Virology 214(2):421–430
38. Sanchez A, Trappier SG, Mahy BW, Peters CJ, Nichol ST (1996) The virion glycoproteins of Ebola viruses are encoded in two reading frames and are expressed through transcriptional editing. Proc Natl Acad Sci U S A 93(8):3602–3607
39. Mehedi M, Falzarano D, Seebach J, Hu X, Carpenter MS, Schnittler HJ, Feldmann H (2011) A new Ebola virus nonstructural glycoprotein expressed through RNA editing. J Virol 85(11):5406–5414. doi:10.1128/JVI.02190-10
40. Volchkova VA, Klenk HD, Volchkov VE (1999) Delta-peptide is the carboxy-terminal cleavage fragment of the nonstructural small glycoprotein sGP of Ebola virus. Virology 265(1):164–171
41. Volchkov VE, Volchkova VA, Slenczka W, Klenk HD, Feldmann H (1998) Release of viral glycoproteins during Ebola virus infection. Virology 245(1):110–119
42. Dolnik O, Volchkova V, Garten W, Carbonnelle C, Becker S, Kahnt J, Stroher U, Klenk HD, Volchkov V (2004) Ectodomain shedding of the glycoprotein GP of Ebola virus. EMBO J 23(10):2175–2184
43. Volchkova VA, Dolnik O, Martinez MJ, Reynard O, Volchkov VE (2015) RNA editing of the GP gene of Ebola virus is an important pathogenicity factor. J Infect Dis 212(Suppl 2):S226–S233. doi:10.1093/infdis/jiv309
44. Alazard-Dany N, Volchkova V, Reynard O, Carbonnelle C, Dolnik O, Ottmann M, Khromykh A, Volchkov VE (2006) Ebola virus glycoprotein GP is not cytotoxic when expressed constitutively at a moderate level. J Gen Virol 87(Pt 5):1247–1257
45. Yang Z, Delgado R, Xu L, Todd RF, Nabel EG, Sanchez A, Nabel GJ (1998) Distinct cellular interactions of secreted and transmembrane Ebola virus glycoproteins. Science 279(5353):1034–1037
46. Kindzelskii AL, Yang Z, Nabel GJ, Todd RF 3rd, Petty HR (2000) Ebola virus secretory glycoprotein (sGP) diminishes Fc gamma RIIIB-to-CR3 proximity on neutrophils. J Immunol 164(2):953–958
47. Mohan GS, Li W, Ye L, Compans RW, Yang C (2012) Antigenic subversion: a novel mechanism of host immune evasion by Ebola virus.

PLoS Pathog 8(12):e1003065. doi:10.1371/journal.ppat.1003065
48. Wahl-Jensen VM, Afanasieva TA, Seebach J, Stroher U, Feldmann H, Schnittler HJ (2005) Effects of Ebola virus glycoproteins on endothelial cell activation and barrier function. J Virol 79(16):10442–10450
49. Radoshitzky SR, Warfield KL, Chi X, Dong L, Kota K, Bradfute SB, Gearhart JD, Retterer C, Kranzusch PJ, Misasi JN, Hogenbirk MA, Wahl-Jensen V, Volchkov VE, Cunningham JM, Jahrling PB, Aman MJ, Bavari S, Farzan M, Kuhn JH (2011) Ebolavirus delta-peptide immunoadhesins inhibit marburgvirus and ebolavirus cell entry. J Virol 85(17): 8502–8513. doi:10.1128/JVI.02600-10
50. Escudero-Perez B, Volchkova VA, Dolnik O, Lawrence P, Volchkov VE (2014) Shed GP of Ebola virus triggers immune activation and increased vascular permeability. PLoS Pathog 10(11):e1004509. doi:10.1371/journal.ppat.1004509
51. Dolnik O, Volchkova VA, Escudero-Perez B, Lawrence P, Klenk HD, Volchkov VE (2015) Shedding of Ebola virus surface glycoprotein is a mechanism of self-regulation of cellular cytotoxicity and has a direct effect on virus infectivity. J Infect Dis 212(Suppl 2):S322–S328. doi:10.1093/infdis/jiv268
52. Ruiz-Arguello MB, Goni FM, Pereira FB, Nieva JL (1998) Phosphatidylinositol-dependent membrane fusion induced by a putative fusogenic sequence of Ebola virus. J Virol 72(3):1775–1781
53. Weissenhorn W, Calder LJ, Wharton SA, Skehel JJ, Wiley DC (1998) The central structural feature of the membrane fusion protein subunit from the Ebola virus glycoprotein is a long triple-stranded coiled coil. Proc Natl Acad Sci U S A 95(11):6032–6036
54. Sanchez A, Yang ZY, Xu L, Nabel GJ, Crews T, Peters CJ (1998) Biochemical analysis of the secreted and virion glycoproteins of Ebola virus. J Virol 72(8):6442–6447
55. Weissenhorn W, Carfi A, Lee KH, Skehel JJ, Wiley DC (1998) Crystal structure of the Ebola virus membrane fusion subunit, GP2, from the envelope glycoprotein ectodomain. Mol Cell 2(5):605–616
56. Volchkov VE, Feldmann H, Volchkova VA, Klenk HD (1998) Processing of the Ebola virus glycoprotein by the proprotein convertase furin. Proc Natl Acad Sci U S A 95(10):5762–5767
57. Wool-Lewis RJ, Bates P (1999) Endoproteolytic processing of the ebola virus envelope glycoprotein: cleavage is not required for function. J Virol 73(2):1419–1426
58. Neumann G, Feldmann H, Watanabe S, Lukashevich I, Kawaoka Y (2002) Reverse genetics demonstrates that proteolytic processing of the Ebola virus glycoprotein is not essential for replication in cell culture. J Virol 76(1):406–410
59. Neumann G, Geisbert TW, Ebihara H, Geisbert JB, Daddario-DiCaprio KM, Feldmann H, Kawaoka Y (2007) Proteolytic processing of the Ebola virus glycoprotein is not critical for Ebola virus replication in nonhuman primates. J Virol 81(6):2995–2998. doi:10.1128/JVI.02486-06
60. Yang ZY, Duckers HJ, Sullivan NJ, Sanchez A, Nabel EG, Nabel GJ (2000) Identification of the Ebola virus glycoprotein as the main viral determinant of vascular cell cytotoxicity and injury. Nat Med 6(8):886–889
61. Groseth A, Marzi A, Hoenen T, Herwig A, Gardner D, Becker S, Ebihara H, Feldmann H (2012) The Ebola virus glycoprotein contributes to but is not sufficient for virulence in vivo. PLoS Pathog 8(8):e1002847. doi:10.1371/journal.ppat.1002847
62. Hoenen T, Marzi A, Scott DP, Feldmann F, Callison J, Safronetz D, Ebihara H, Feldmann H (2015) Soluble glycoprotein is not required for Ebola virus virulence in guinea pigs. J Infect Dis 212(Suppl 2):S242–S246. doi:10.1093/infdis/jiv111
63. Chan SY, Empig CJ, Welte FJ, Speck RF, Schmaljohn A, Kreisberg JF, Goldsmith MA (2001) Folate receptor-alpha is a cofactor for cellular entry by Marburg and Ebola viruses. Cell 106(1):117–126
64. Alvarez CP, Lasala F, Carrillo J, Muniz O, Corbi AL, Delgado R (2002) C-type lectins DC-SIGN and L-SIGN mediate cellular entry by Ebola virus in cis and in trans. J Virol 76(13):6841–6844
65. Simmons G, Reeves JD, Grogan CC, Vandenberghe LH, Baribaud F, Whitbeck JC, Burke E, Buchmeier MJ, Soilleux EJ, Riley JL, Doms RW, Bates P, Pohlmann S (2003) DC-SIGN and DC-SIGNR bind ebola glycoproteins and enhance infection of macrophages and endothelial cells. Virology 305(1):115–123
66. Takada A, Fujioka K, Tsuiji M, Morikawa A, Higashi N, Ebihara H, Kobasa D, Feldmann H, Irimura T, Kawaoka Y (2004) Human macrophage C-type lectin specific for galactose and N-acetylgalactosamine promotes filovirus entry. J Virol 78(6):2943–2947
67. Kondratowicz AS, Lennemann NJ, Sinn PL, Davey RA, Hunt CL, Moller-Tank S, Meyerholz DK, Rennert P, Mullins RF, Brindley M, Sandersfeld LM, Quinn K,

Weller M, McCray PB Jr, Chiorini J, Maury W (2011) T-cell immunoglobulin and mucin domain 1 (TIM-1) is a receptor for Zaire Ebolavirus and Lake Victoria Marburgvirus. Proc Natl Acad Sci U S A 108(20):8426–8431. doi:10.1073/pnas.1019030108
68. Carette JE, Raaben M, Wong AC, Herbert AS, Obernosterer G, Mulherkar N, Kuehne AI, Kranzusch PJ, Griffin AM, Ruthel G, Dal Cin P, Dye JM, Whelan SP, Chandran K, Brummelkamp TR (2011) Ebola virus entry requires the cholesterol transporter Niemann-Pick C1. Nature 477(7364):340–343. doi:10.1038/nature10348
69. Cote M, Misasi J, Ren T, Bruchez A, Lee K, Filone CM, Hensley L, Li Q, Ory D, Chandran K, Cunningham J (2011) Small molecule inhibitors reveal Niemann-Pick C1 is essential for Ebola virus infection. Nature 477(7364):344–348. doi:10.1038/nature10380
70. Bhattacharyya S, Hope TJ, Young JA (2011) Differential requirements for clathrin endocytic pathway components in cellular entry by Ebola and Marburg glycoprotein pseudovirions. Virology 419(1):1–9. doi:10.1016/j.virol.2011.07.018
71. Bhattacharyya S, Warfield KL, Ruthel G, Bavari S, Aman MJ, Hope TJ (2010) Ebola virus uses clathrin-mediated endocytosis as an entry pathway. Virology 401(1):18–28. doi:10.1016/j.virol.2010.02.015
72. Empig CJ, Goldsmith MA (2002) Association of the caveola vesicular system with cellular entry by filoviruses. J Virol 76(10):5266–5270
73. Quinn K, Brindley MA, Weller ML, Kaludov N, Kondratowicz A, Hunt CL, Sinn PL, McCray PB Jr, Stein CS, Davidson BL, Flick R, Mandell R, Staplin W, Maury W, Chiorini JA (2009) Rho GTPases modulate entry of Ebola virus and vesicular stomatitis virus pseudotyped vectors. J Virol 83(19):10176–10186. doi:10.1128/JVI.00422-09
74. Saeed MF, Kolokoltsov AA, Albrecht T, Davey RA (2010) Cellular entry of ebola virus involves uptake by a macropinocytosis-like mechanism and subsequent trafficking through early and late endosomes. PLoS Pathog 6(9):e1001110. doi:10.1371/journal.ppat.1001110
75. Nanbo A, Imai M, Watanabe S, Noda T, Takahashi K, Neumann G, Halfmann P, Kawaoka Y (2010) Ebolavirus is internalized into host cells via macropinocytosis in a viral glycoprotein-dependent manner. PLoS Pathog 6(9):e1001121. doi:10.1371/journal.ppat.1001121
76. Mulherkar N, Raaben M, de la Torre JC, Whelan SP, Chandran K (2011) The Ebola virus glycoprotein mediates entry via a non-classical dynamin-dependent macropinocytic pathway. Virology 419(2):72–83. doi:10.1016/j.virol.2011.08.009
77. Aleksandrowicz P, Marzi A, Biedenkopf N, Beimforde N, Becker S, Hoenen T, Feldmann H, Schnittler HJ (2011) Ebola virus enters host cells by macropinocytosis and clathrin-mediated endocytosis. J Infect Dis 204(Suppl 3):S957–S967. doi:10.1093/infdis/jir326
78. Sanchez A (2007) Analysis of filovirus entry into vero e6 cells, using inhibitors of endocytosis, endosomal acidification, structural integrity, and cathepsin (B and L) activity. J Infect Dis 196(Suppl 2):S251–S258. doi:10.1086/520597
79. Hunt CL, Kolokoltsov AA, Davey RA, Maury W (2011) The Tyro3 receptor kinase Axl enhances macropinocytosis of Zaire ebolavirus. J Virol 85(1):334–347. doi:10.1128/JVI.01278-09
80. Chandran K, Sullivan NJ, Felbor U, Whelan SP, Cunningham JM (2005) Endosomal proteolysis of the Ebola virus glycoprotein is necessary for infection. Science 308(5728):1643–1645
81. Schornberg K, Matsuyama S, Kabsch K, Delos S, Bouton A, White J (2006) Role of endosomal cathepsins in entry mediated by the Ebola virus glycoprotein. J Virol 80(8):4174–4178
82. Martinez O, Johnson J, Manicassamy B, Rong L, Olinger GG, Hensley LE, Basler CF (2010) Zaire Ebola virus entry into human dendritic cells is insensitive to cathepsin L inhibition. Cell Microbiol 12(2):148–157. doi:10.1111/j.1462-5822.2009.01385.x
83. Misasi J, Chandran K, Yang JY, Considine B, Filone CM, Cote M, Sullivan N, Fabozzi G, Hensley L, Cunningham J (2012) Filoviruses require endosomal cysteine proteases for entry but exhibit distinct protease preferences. J Virol 86(6):3284–3292. doi:10.1128/JVI.06346-11
84. Gnirss K, Kuhl A, Karsten C, Glowacka I, Bertram S, Kaup F, Hofmann H, Pohlmann S (2012) Cathepsins B and L activate Ebola but not Marburg virus glycoproteins for efficient entry into cell lines and macrophages independent of TMPRSS2 expression. Virology 424(1):3–10. doi:10.1016/j.virol.2011.11.031
85. Marzi A, Reinheckel T, Feldmann H (2012) Cathepsin B & L are not required for ebola virus replication. PLoS Negl Trop Dis 6(12):e1923. doi:10.1371/journal.pntd.0001923
86. Brecher M, Schornberg KL, Delos SE, Fusco ML, Saphire EO, White JM (2012) Cathepsin cleavage potentiates the Ebola virus glycoprotein to undergo a subsequent fusion-relevant conformational change. J Virol 86(1):364–372. doi:10.1128/JVI.05708-11

87. Muhlberger E, Weik M, Volchkov VE, Klenk HD, Becker S (1999) Comparison of the transcription and replication strategies of Marburg virus and Ebola virus by using artificial replication systems. J Virol 73(3): 2333–2342
88. Weik M, Modrof J, Klenk HD, Becker S, Muhlberger E (2002) Ebola virus VP30-mediated transcription is regulated by RNA secondary structure formation. J Virol 76(17):8532–8539
89. Biedenkopf N, Hartlieb B, Hoenen T, Becker S (2013) Phosphorylation of Ebola virus VP30 influences the composition of the viral nucleocapsid complex: impact on viral transcription and replication. J Biol Chem 288(16):11165–11174. doi:10.1074/jbc.M113.461285
90. Biedenkopf N, Lier C, Becker S (2016) Dynamic phosphorylation of VP30 is essential for Ebola virus life cycle. J Virol 90(10):4914–4925. doi:10.1128/JVI.03257-15
91. Halfmann P, Kim JH, Ebihara H, Noda T, Neumann G, Feldmann H, Kawaoka Y (2008) Generation of biologically contained Ebola viruses. Proc Natl Acad Sci U S A 105(4):1129–1133. doi:10.1073/pnas.0708057105
92. Mehedi M, Hoenen T, Robertson S, Ricklefs S, Dolan MA, Taylor T, Falzarano D, Ebihara H, Porcella SF, Feldmann H (2013) Ebola virus RNA editing depends on the primary editing site sequence and an upstream secondary structure. PLoS Pathog 9(10):e1003677. doi:10.1371/journal.ppat.1003677
93. Becker S, Rinne C, Hofsass U, Klenk HD, Muhlberger E (1998) Interactions of Marburg virus nucleocapsid proteins. Virology 249(2):406–417
94. Han Z, Boshra H, Sunyer JO, Zwiers SH, Paragas J, Harty RN (2003) Biochemical and functional characterization of the Ebola virus VP24 protein: implications for a role in virus assembly and budding. J Virol 77(3): 1793–1800
95. Bharat TA, Noda T, Riches JD, Kraehling V, Kolesnikova L, Becker S, Kawaoka Y, Briggs JA (2012) Structural dissection of Ebola virus and its assembly determinants using cryo-electron tomography. Proc Natl Acad Sci U S A 109(11):4275–4280. doi:10.1073/pnas.1120453109
96. Watt A, Moukambi F, Banadyga L, Groseth A, Callison J, Herwig A, Ebihara H, Feldmann H, Hoenen T (2014) A novel life cycle modeling system for Ebola virus shows a genome length-dependent role of VP24 in virus infectivity. J Virol 88(18):10511–10524. doi:10.1128/JVI.01272-14
97. Hoenen T, Jung S, Herwig A, Groseth A, Becker S (2010) Both matrix proteins of Ebola virus contribute to the regulation of viral genome replication and transcription. Virology 403(1):56–66. doi:10.1016/j.virol.2010.04.002
98. Watanabe S, Noda T, Halfmann P, Jasenosky L, Kawaoka Y (2007) Ebola virus (EBOV) VP24 inhibits transcription and replication of the EBOV genome. J Infect Dis 196(Suppl 2):S284–S290. doi:10.1086/520582
99. Hoenen T, Groseth A, Kolesnikova L, Theriault S, Ebihara H, Hartlieb B, Bamberg S, Stroher U, Feldmann H, Becker S (2006) Infection of naive target cells with virus-like partices – implications for the function of Ebola virus VP24. J Virol 80(14):7260–7264
100. Hoenen T, Shabman RS, Groseth A, Herwig A, Weber M, Schudt G, Dolnik O, Basler CF, Becker S, Feldmann H (2012) Inclusion bodies are a site of ebolavirus replication. J Virol 86(21):11779–11788. doi:10.1128/JVI.01525-12
101. Nanbo A, Watanabe S, Halfmann P, Kawaoka Y (2013) The spatio-temporal distribution dynamics of Ebola virus proteins and RNA in infected cells. Sci Rep 3:1206. doi:10.1038/srep01206
102. Brauburger K, Boehmann Y, Tsuda Y, Hoenen T, Olejnik J, Schumann M, Ebihara H, Muhlberger E (2014) Analysis of the highly diverse gene borders in Ebola virus reveals a distinct mechanism of transcriptional regulation. J Virol 88(21):12558–12571. doi:10.1128/JVI.01863-14
103. Watanabe S, Watanabe T, Noda T, Takada A, Feldmann H, Jasenosky LD, Kawaoka Y (2004) Production of novel ebola virus-like particles from cDNAs: an alternative to ebola virus generation by reverse genetics. J Virol 78(2):999–1005
104. Schudt G, Dolnik O, Kolesnikova L, Biedenkopf N, Herwig A, Becker S (2015) Transport of Ebolavirus nucleocapsids is dependent on actin polymerization: live-cell imaging analysis of Ebolavirus-infected cells. J Infect Dis 212(Suppl 2):S160–S166. doi:10.1093/infdis/jiv083
105. Adu-Gyamfi E, Digman MA, Gratton E, Stahelin RV (2012) Single-particle tracking demonstrates that actin coordinates the movement of the Ebola virus matrix protein. Biophys J 103(9):L41–L43. doi:10.1016/j.bpj.2012.09.026
106. Ruigrok RW, Schoehn G, Dessen A, Forest E, Volchkov V, Dolnik O, Klenk HD, Weissenhorn W (2000) Structural characterization and membrane binding properties of the matrix protein VP40 of Ebola virus. J Mol Biol 300(1):103–112
107. Timmins J, Scianimanico S, Schoehn G, Weissenhorn W (2001) Vesicular release of

Ebola virus matrix protein VP40. Virology 283(1):1–6
108. Jasenosky LD, Neumann G, Lukashevich I, Kawaoka Y (2001) Ebola virus VP40-induced particle formation and association with the lipid bilayer. J Virol 75(11):5205–5214
109. Noda T, Sagara H, Suzuki E, Takada A, Kida H, Kawaoka Y (2002) Ebola virus VP40 drives the formation of virus-like filamentous particles along with GP. J Virol 76(10):4855–4865
110. Harty RN, Brown ME, Wang G, Huibregtse J, Hayes FP (2000) A PPxY motif within the VP40 protein of Ebola virus interacts physically and functionally with a ubiquitin ligase: implications for filovirus budding. Proc Natl Acad Sci U S A 97(25):13871–13876
111. Licata JM, Simpson-Holley M, Wright NT, Han Z, Paragas J, Harty RN (2003) Overlapping motifs (PTAP and PPEY) within the Ebola virus VP40 protein function independently as late budding domains: involvement of host proteins TSG101 and VPS-4. J Virol 77(3):1812–1819
112. Timmins J, Schoehn G, Ricard-Blum S, Scianimanico S, Vernet T, Ruigrok RW, Weissenhorn W (2003) Ebola virus matrix protein VP40 interaction with human cellular factors Tsg101 and Nedd4. J Mol Biol 326(2):493–502
113. Yasuda J, Nakao M, Kawaoka Y, Shida H (2003) Nedd4 regulates egress of Ebola virus-like particles from host cells. J Virol 77(18):9987–9992
114. Han Z, Sagum CA, Bedford MT, Sidhu SS, Sudol M, Harty RN (2016) ITCH E3 ubiquitin ligase interacts with Ebola virus VP40 to regulate budding. J Virol 90(20):9163–9171. doi:10.1128/JVI.01078-16
115. Neumann G, Ebihara H, Takada A, Noda T, Kobasa D, Jasenosky LD, Watanabe S, Kim JH, Feldmann H, Kawaoka Y (2005) Ebola virus VP40 late domains are not essential for viral replication in cell culture. J Virol 79(16):10300–10307
116. Han Z, Madara JJ, Liu Y, Liu W, Ruthel G, Freedman BD, Harty RN (2015) ALIX rescues budding of a double PTAP/PPEY L-domain deletion mutant of Ebola VP40: a role for ALIX in Ebola virus egress. J Infect Dis 212(Suppl 2):S138–S145. doi:10.1093/infdis/jiu838
117. Dessen A, Volchkov V, Dolnik O, Klenk HD, Weissenhorn W (2000) Crystal structure of the matrix protein VP40 from Ebola virus. EMBO J 19(16):4228–4236
118. Dessen A, Forest E, Volchkov V, Dolnik O, Klenk HD, Weissenhorn W (2000) Crystallization and preliminary X-ray analysis of the matrix protein from Ebola virus. Acta Crystallogr D Biol Crystallogr 56(Pt 6):758–760
119. Scianimanico S, Schoehn G, Timmins J, Ruigrok RH, Klenk HD, Weissenhorn W (2000) Membrane association induces a conformational change in the Ebola virus matrix protein. EMBO J 19(24):6732–6741
120. Hoenen T, Biedenkopf N, Zielecki F, Jung S, Groseth A, Feldmann H, Becker S (2010) Oligomerization of Ebola virus VP40 is essential for particle morphogenesis and regulation of viral transcription. J Virol 84(14):7053–7063. doi:10.1128/JVI.00737-10
121. Soni SP, Adu-Gyamfi E, Yong SS, Jee CS, Stahelin RV (2013) The Ebola virus matrix protein deeply penetrates the plasma membrane: an important step in viral egress. Biophys J 104(9):1940–1949. doi:10.1016/j.bpj.2013.03.021
122. Soni SP, Stahelin RV (2014) The Ebola virus matrix protein VP40 selectively induces vesiculation from phosphatidylserine-enriched membranes. J Biol Chem 289(48):33590–33597. doi:10.1074/jbc.M114.586396
123. Harcourt BH, Sanchez A, Offermann MK (1998) Ebola virus inhibits induction of genes by double-stranded RNA in endothelial cells. Virology 252(1):179–188
124. Harcourt BH, Sanchez A, Offermann MK (1999) Ebola virus selectively inhibits responses to interferons, but not to interleukin-1beta, in endothelial cells. J Virol 73(4):3491–3496
125. Basler CF, Wang X, Muhlberger E, Volchkov V, Paragas J, Klenk HD, Garcia-Sastre A, Palese P (2000) The Ebola virus VP35 protein functions as a type I IFN antagonist. Proc Natl Acad Sci U S A 97(22):12289–12294
126. Basler CF, Mikulasova A, Martinez-Sobrido L, Paragas J, Muhlberger E, Bray M, Klenk HD, Palese P, Garcia-Sastre A (2003) The Ebola virus VP35 protein inhibits activation of interferon regulatory factor 3. J Virol 77(14):7945–7956
127. Reid SP, Cardenas WB, Basler CF (2005) Homo-oligomerization facilitates the interferon-antagonist activity of the ebolavirus VP35 protein. Virology 341(2):179–189
128. Hartman AL, Towner JS, Nichol ST (2004) A C-terminal basic amino acid motif of Zaire ebolavirus VP35 is essential for type I interferon antagonism and displays high identity with the RNA-binding domain of another interferon antagonist, the NS1 protein of influenza A virus. Virology 328(2):177–184
129. Cardenas WB, Loo YM, Gale M Jr, Hartman AL, Kimberlin CR, Martinez-Sobrido L, Saphire EO, Basler CF (2006) Ebola virus VP35 protein binds double-stranded RNA and inhibits alpha/beta interferon

production induced by RIG-I signaling. J Virol 80(11):5168–5178
130. Prins KC, Cardenas WB, Basler CF (2009) Ebola virus protein VP35 impairs the function of interferon regulatory factor-activating kinases IKKepsilon and TBK-1. J Virol 83(7):3069–3077. doi:10.1128/JVI.01875-08
131. Bale S, Julien JP, Bornholdt ZA, Krois AS, Wilson IA, Saphire EO (2013) Ebolavirus VP35 coats the backbone of double-stranded RNA for interferon antagonism. J Virol 87(18):10385–10388. doi:10.1128/JVI.01452-13
132. Bray M (2001) The role of the Type I interferon response in the resistance of mice to filovirus infection. J Gen Virol 82(Pt 6):1365–1373
133. Hartman AL, Dover JE, Towner JS, Nichol ST (2006) Reverse genetic generation of recombinant Zaire Ebola viruses containing disrupted IRF-3 inhibitory domains results in attenuated virus growth in vitro and higher levels of IRF-3 activation without inhibiting viral transcription or replication. J Virol 80(13):6430–6440. doi:10.1128/JVI.00044-06
134. Hartman AL, Bird HB, Towner JS, Antoniadou Z, Zaki S, Nichol ST (2008) Inhibition of IRF-3 activation by VP35 is critical for the high level of virulence of Ebola virus. J Virol 82(6):2699–2704
135. Prins KC, Delpeut S, Leung DW, Reynard O, Volchkova VA, Reid SP, Ramanan P, Cardenas WB, Amarasinghe GK, Volchkov VE, Basler CF (2010) Mutations abrogating VP35 interaction with double-stranded RNA render Ebola virus avirulent in guinea pigs. J Virol 84(6):3004–3015. doi:10.1128/JVI.02459-09
136. Kash JC, Muhlberger E, Carter V, Grosch M, Perwitasari O, Proll SC, Thomas MJ, Weber F, Klenk HD, Katze MG (2006) Global suppression of the host antiviral response by Ebola- and Marburgviruses: increased antagonism of the type I interferon response is associated with enhanced virulence. J Virol 80(6):3009–3020
137. Guito JC, Albarino CG, Chakrabarti AK, Towner JS (2016) Novel activities by ebolavirus and marburgvirus interferon antagonists revealed using a standardized in vitro reporter system. Virology 501:147–165. doi:10.1016/j.virol.2016.11.015
138. Reid SP, Leung LW, Hartman AL, Martinez O, Shaw ML, Carbonnelle C, Volchkov VE, Nichol ST, Basler CF (2006) Ebola virus VP24 binds karyopherin alpha 1 and blocks STAT1 nuclear accumulation. J Virol 80(11):5156–5167
139. Reid SP, Valmas C, Martinez O, Sanchez FM, Basler CF (2007) Ebola virus VP24 proteins inhibit the interaction of NPI-1 subfamily karyopherin alpha proteins with activated STAT1. J Virol 81(24):13469–13477. doi:10.1128/JVI.01097-07
140. Xu W, Edwards MR, Borek DM, Feagins AR, Mittal A, Alinger JB, Berry KN, Yen B, Hamilton J, Brett TJ, Pappu RV, Leung DW, Basler CF, Amarasinghe GK (2014) Ebola virus VP24 targets a unique NLS binding site on karyopherin alpha 5 to selectively compete with nuclear import of phosphorylated STAT1. Cell Host Microbe 16(2):187–200. doi:10.1016/j.chom.2014.07.008
141. Zhang AP, Bornholdt ZA, Liu T, Abelson DM, Lee DE, Li S, Woods VL Jr, Saphire EO (2012) The Ebola virus interferon antagonist VP24 directly binds STAT1 and has a novel, pyramidal fold. PLoS Pathog 8(2):e1002550. doi:10.1371/journal.ppat.1002550
142. Halfmann P, Neumann G, Kawaoka Y (2011) The Ebolavirus VP24 protein blocks phosphorylation of p38 mitogen-activated protein kinase. J Infect Dis 204(Suppl 3):S953–S956. doi:10.1093/infdis/jir325
143. Ishida H, Ohkawa K, Hosui A, Hiramatsu N, Kanto T, Ueda K, Takehara T, Hayashi N (2004) Involvement of p38 signaling pathway in interferon-alpha-mediated antiviral activity toward hepatitis C virus. Biochem Biophys Res Commun 321(3):722–727. doi:10.1016/j.bbrc.2004.07.015
144. Goh KC, Haque SJ, Williams BR (1999) p38 MAP kinase is required for STAT1 serine phosphorylation and transcriptional activation induced by interferons. EMBO J 18(20):5601–5608. doi:10.1093/emboj/18.20.5601
145. Kaletsky RL, Francica JR, Agrawal-Gamse C, Bates P (2009) Tetherin-mediated restriction of filovirus budding is antagonized by the Ebola glycoprotein. Proc Natl Acad Sci U S A 106(8):2886–2891. doi:10.1073/pnas.0811014106
146. Lopez LA, Yang SJ, Hauser H, Exline CM, Haworth KG, Oldenburg J, Cannon PM (2010) Ebola virus glycoprotein counteracts BST-2/Tetherin restriction in a sequence-independent manner that does not require tetherin surface removal. J Virol 84(14):7243–7255. doi:10.1128/JVI.02636-09
147. Kuhl A, Banning C, Marzi A, Votteler J, Steffen I, Bertram S, Glowacka I, Konrad A, Sturzl M, Guo JT, Schubert U, Feldmann H, Behrens G, Schindler M, Pohlmann S (2011) The Ebola virus glycoprotein and HIV-1 Vpu employ different strategies to counteract the antiviral factor tetherin. J Infect Dis 204(Suppl 3):S850–S860. doi:10.1093/infdis/jir378
148. Gnirss K, Fiedler M, Kramer-Kuhl A, Bolduan S, Mittler E, Becker S, Schindler M, Pohlmann

S (2014) Analysis of determinants in filovirus glycoproteins required for tetherin antagonism. Virus 6(4):1654–1671. doi:10.3390/v6041654

149. Brinkmann C, Nehlmeier I, Walendy-Gnirss K, Nehls J, Gonzalez Hernandez M, Hoffmann M, Qiu X, Takada A, Schindler M, Pohlmann S (2016) Tetherin antagonism by the Ebola virus glycoprotein requires an intact receptor-binding domain and can be blocked by GP1-specific antibodies. J Virol. doi:10.1128/JVI.01563-16

150. Gustin JK, Bai Y, Moses AV, Douglas JL (2015) Ebola virus glycoprotein promotes enhanced viral egress by preventing Ebola VP40 from associating with the host restriction factor BST2/tetherin. J Infect Dis 212(Suppl 2):S181–S190. doi:10.1093/infdis/jiv125

151. Feng Z, Cerveny M, Yan Z, He B (2007) The VP35 protein of Ebola virus inhibits the antiviral effect mediated by double-stranded RNA-dependent protein kinase PKR. J Virol 81(1):182–192

152. Schumann M, Gantke T, Muhlberger E (2009) Ebola virus VP35 antagonizes PKR activity through its C-terminal interferon inhibitory domain. J Virol 83(17):8993–8997. doi:10.1128/JVI.00523-09

153. Haasnoot J, de Vries W, Geutjes EJ, Prins M, de Haan P, Berkhout B (2007) The Ebola virus VP35 protein is a suppressor of RNA silencing. PLoS Pathog 3(6):e86. doi:10.1371/journal.ppat.0030086

154. Nelson EV, Schmidt KM, Deflube LR, Doganay S, Banadyga L, Olejnik J, Hume AJ, Ryabchikova E, Ebihara H, Kedersha N, Ha T, Muhlberger E (2016) Ebola virus does not induce stress granule formation during infection and sequesters stress granule proteins within viral inclusions. J Virol 90(16):7268–7284. doi:10.1128/JVI.00459-16

Chapter 3

Ebolavirus: An Overview of Molecular and Clinical Pathogenesis

Veronica Vine, Dana P. Scott, and Heinz Feldmann

Abstract

Ebolaviruses cause severe, often fatal hemorrhagic fever in Central, East, and West Africa. Until recently, they have been viewed as rare but highly pathogenic infections with regional, but limited, global public health impact. This view has changed with the emergence of the first epidemic of Ebola hemorrhagic fever in West Africa. In this chapter we provide an introduction of the pathogenesis of ebolaviruses as well as a description of clinical disease features. We also describe the current animal models used in ebolavirus research, detailing each model's unique strengths and weaknesses. We focus on Ebola virus representing the type species *Zaire ebolavirus* of the genus *Ebolavirus*, as most work relates to this pathogen.

Key words Ebolavirus, Filovirus, Clinical parameters, Pathophysiology, Pathogenesis, Animal models

1 Impacts

Ebolaviruses are known as rare but highly virulent causative agents of Ebola hemorrhagic fever (EHF). Since their discovery in 1976, research has focused on the biology and pathogenesis as well as the ecology and epidemiology of the viruses. Countermeasure development was intensified with the listing of the viruses as potential bioterrorism agents, yet, in general, progress has been hampered due to the need for high containment facilities and a general lack of interest in these rare infections with limited public health impact. This changed, however, with the recent West African EHF outbreak.

In December of 2013, a 2-year-old child living in Guéckédou prefecture fell ill with fever, black stool, and vomiting and died 4 days later. From that single child five more individuals fell ill exhibiting similar symptoms [1]. In March of 2014, these cases were confirmed to be the result of an ebolavirus infection which escalated into the largest outbreak and first epidemic ever recorded with 28,601 confirmed, probable, and suspected cases and 11,300 deaths, suggesting a case fatality rate of around 40% and spanning

Thomas Hoenen and Allison Groseth (eds.), *Ebolaviruses: Methods and Protocols*, Methods in Molecular Biology, vol. 1628, DOI 10.1007/978-1-4939-7116-9_3, © Springer Science+Business Media LLC 2017

the countries of Guinea, Liberia, Sierra Leone, Mali, Nigeria, Senegal, Italy, Spain, the UK, and the USA. While at the time of writing this outbreak seems to have been largely contained, it does highlight the destructive potential of this virus and the urgent need for both effective therapeutics and prevention measures [2].

2 Clinical Pathogenesis

2.1 Transmission

Transmission of EBOV occurs through direct contact with blood, secretions, or tissues from patients or infected animals. Virus has been isolated in patients from blood, saliva, urine, breast milk, semen, and, most recently, the aqueous humor of the eye. Infectious EBOV has been shown to persist in fluids from immune-privileged sites for up to 98 days after the onset of symptoms; 15 days for breast milk, 98 days for aqueous humor, and 82 days for semen [3–7]. Viable virus has also been isolated from urine for up to 26 days after the onset of symptoms [8].

2.2 Clinical Signs and Symptoms

After an asymptomatic incubation period of 2–21 days, Ebola hemorrhagic fever, recently also being referred to as Ebola virus disease (EVD), begins with nonspecific flu-like symptoms that predominately include fever, headache, weakness, dizziness, diarrhea, abdominal pain, and vomiting. While hemorrhage is often thought to be a hallmark sign of EBOV infection, it is actually only seen in fewer than half of patients and similar to the development of a maculopapular rash, which is sometimes considered another hallmark sign of infection, but is actually not commonly observed [9–12]. Additionally, EHF has also recently been linked to possible chronic neurological and ocular manifestations in survivors after the infection has been cleared. This can manifest in the form of headaches and potentially severe uveitis and vision loss possibly as a result of viral replication within the eye [7, 13, 14]. As illustrated by Fig. 1, upon entering the host EBOV evades the host immune system by impairment of dendritic cell (DC) function, interferon antagonism, and depletion of lymphocytes, like natural killer (NK) cells. As a consequence of infection macrophages are induced to secrete high levels of pro-inflammatory cytokines which further induce infected and noninfected macrophages to express cytokines and chemokines that stimulate fluid leakage into the interstitium and promote a pro-coagulant state, thus leading to hypovolemic shock and the development of disseminated intravascular coagulation (DIC). Death typically occurs 6–16 days after the onset of symptoms due to hypovolemic shock and multiorgan failure.

2.3 Impairment of the Immune Response

Monocytes, macrophages and dendritic cells have been shown to be the most important early target cells and are preferred by EBOV infection [15, 16]. After DC infection in the lymphoid organs and

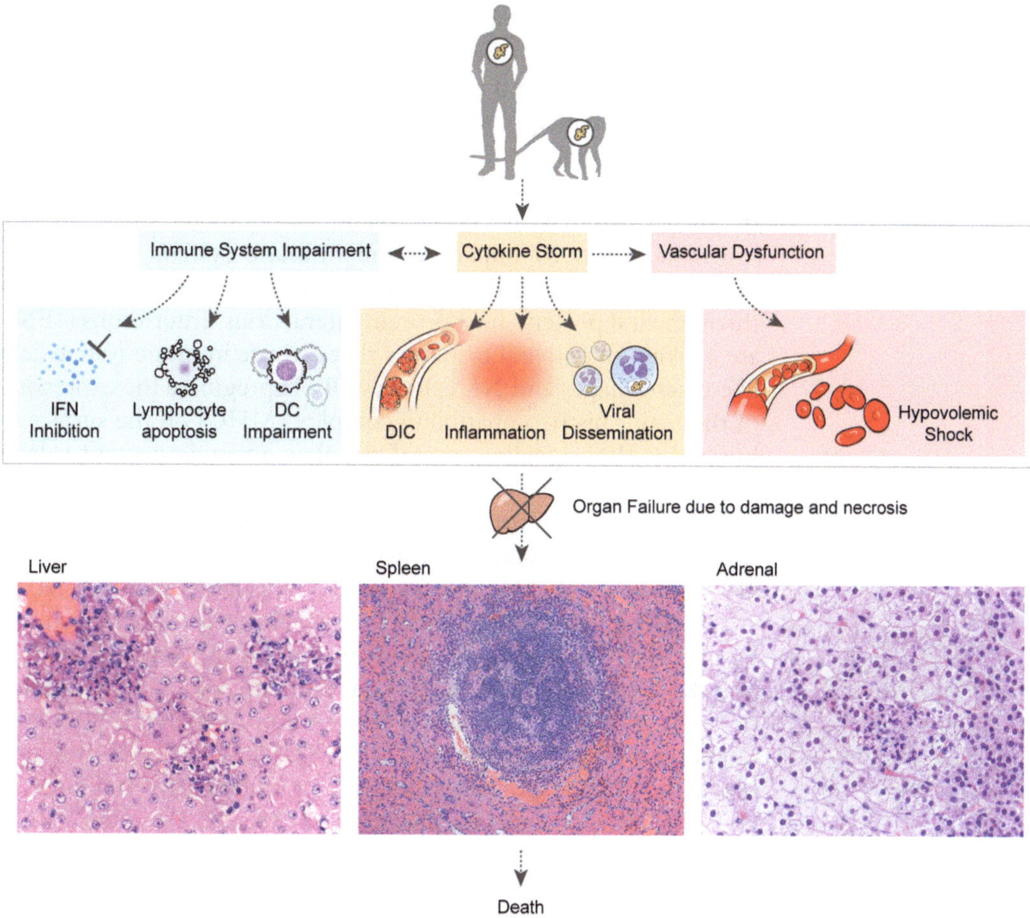

Fig. 1 Schematic representation of essential pathologic events leading to fatal outcome during EHF

subsequent infection of circulating macrophages, the virus can be further disseminated into the blood stream and to other organs such as the liver where hepatocytes and resident macrophages (Kupffer cells) can also be infected. Other cells shown to be susceptible to EBOV infection include endothelial cells, fibroblasts, hepatocytes, adrenal cortical cells, and several types of epithelial cells [9, 17, 18]. Targeting immune system cells (e.g., dendritic cells and macrophages) allows the virus to negatively impact the immune response by impairing and inhibiting their function, while also facilitating virus dissemination. EBOV infection efficiently suppresses DC maturation with infected DCs being unable to produce inflammatory cytokines and being impaired in their ability to activate T-cells [16, 19]. Studies in vitro suggest a role of VP35 in this impairment showing that VP35 is capable of suppressing cytokine expression and activation of T-cells in DCs activated by RIG-I-like receptor signaling [20]. EBOV further downregulates the immune system by the induction of bystander lymphocyte apoptosis which is most likely mediated via the extrinsic apoptotic pathway through FasL/FasR receptor binding and the upregulation of TNF-α production and possibly via the intrinsic pathway caused by viral induced damage of surrounding tissue [15, 21–27]. In contrast to DC infection, infection of macrophages results in activation and extensive cytokine production, termed the "cytokine storm," which may also play a role in fatal outcome.

In addition to impairing the immune system by infecting important effector cells, EBOV also causes immune disruption through viral protein–host protein interactions. Interferons (IFNs) are a group of antiviral cytokines that activate immune system cells like macrophages and NK cells as well as upregulate the expression of the major histocompatibility complex (MHC) on the surface of host cells. IFN signaling, therefore, plays a very important role in the innate immune response against viral infection. EBOV has developed several proteins specifically designed to downregulate this response and contribute to viral resilience. VP35 has been shown to block phosphorylation of IFN-regulatory factor 3 (IRF3) and IFN-regulatory factor 7 (IRF7), which act as transcription factors for IFN production, by acting as a decoy substrate for kinases, like IKKε and TBK1, involved in the pathway. VP35 is also capable of binding dsRNA to suppress recognition of the viral RNA and further inhibit IFN production [17, 18]. VP24 has been shown to block the IFN response by acting as a competitive inhibitor of pSTAT1 and preventing nuclear entry [18, 28, 29]. The EBOV GP also serves as a tetherin antagonist. Tetherin is an interferon-induced cellular response factor that blocks the release of many viruses, including EBOV, from infected cells. By inhibiting this factor, GP allows for budding and release of the virus [17, 30, 31]. In addition sGP, secreted by EBOV infected cells, has been shown to bind anti-GP antibodies and neutrophils and inhibit their activity [27, 32, 33].

2.4 Dysregulation of the Immune System: The Cytokine Storm

In addition to impairing the immune system by infection of effector cells and the actions of viral proteins, EBOV also stimulates immune system cells to release large amounts of pro-inflammatory cytokines and chemokines (the "cytokine storm") which can be detrimental to the host. Monocytes, macrophage precursors, infected with EBOV have been shown to secrete MCP-1, CXCL1 (gro-α), IL-6, IL-1β, IL-8, MIP-1α, RANTES and TNF-α. Similarly increased levels of IL-1β, IL-1RA, IL-6, IL-8, IL-15, IL-16, MIP-1a, MIP-1b, MCP-1, M-CSF, MIF, IP-10, CXCL1, and CCL11 (eotaxin) have been found in human cases [26, 27, 34, 35]. The upregulation of these chemokines results in further activation of the coagulation cascade and disruption of vascular integrity that increase the likelihood of fatal outcome. Additionally, this results in the mass recruitment of immune system cells which may further facilitate dissemination and survival of the virus by allowing more susceptible cells to congregate to areas of infection. This can also specifically result in the attraction of neutrophils which have been shown in vitro to be induced by EBOV to release even more inflammatory cytokines and chemokines that may contribute to vasodilation and increased vascular permeability [36]. Elevated levels of IL-8 have also been shown to antagonize IFN, thus its upregulation may also aid EBOV in suppressing the IFN response [21]. In this way the hosts own immune response to EBOV infection causes damage, while those actions essential for viral clearance are inhibited.

2.5 Vascular Dysfunction and DIC

The vascular dysfunction that accompanies EHF can lead to hypovolemic shock and is, therefore, an important contributor to mortality. It is thought that infection of macrophages induces the release of cytokines, like TNF-α, that allow for fluid leakage into the interstitium [23, 27, 37–40]. While EBOV has been shown to infect endothelial cells this seems to typically occur during late stage infection and is accompanied by little or no endothelial damage, suggesting this loss of vascular integrity is not initially due to the physical destruction of endothelial cells, although this activity may play a larger role as the disease progresses [27, 41]. Nitric oxide (NO) released by infected macrophages may also play a key role in the vascular dysfunction seen in EHF. NO is a potent endogenous vasodilator and is involved in the development of vasodilatory shock. Increased levels of NO in the blood have been found in human cases of EBOV infection as well as in NHP infection [17, 18].

The development of disseminated intravascular coagulation (DIC) has also been shown to be a contributor in fatal EHF outcome. DIC is the result of overuse of coagulation proteins, which create small blood clots (thrombi) and eventually results in the over consumption of those proteins rendering the body unable to clot properly and increasing the risk of hemorrhage. This manifests in the form of a low platelet count (thrombocytopenia) which is seen in most human and NHP cases of EHF and possible organ damage due to these small thrombi impeding blood flow [8, 10, 42–44]. EBOV has been shown in NHP infection to directly induce the expression of tissue factor (TF), the principle initiator of the coagulation cascade, on the surface of macrophages, which could result in the development of DIC seen in EBOV patients [45]. It is also thought that the expression of TF may be further upregulated by IL-6 secreted by infected macrophages [46].

3 Clinical Chemistry and Hematology

In light of the most recent EBOV outbreak much more clinical data has become available, allowing for a much clearer picture of the clinical progression of EHF, as well as the identification of possible predictors of fatal outcome. These abnormalities can be grouped into those related to hematology, clinical chemistry, and coagulation. Hematological abnormalities include early leukocytosis, due mainly to neutrophilia, followed by late lymphopenia while clinical chemistry abnormalities include elevated liver enzymes (ALT, AST, ALP, and/or GGT), decreased serum albumin, and increased serum creatinine and blood urea nitrogen (BUN). Coagulation abnormalities, which are thought to initiate the progression of DIC, include early elevation of fibrinogen levels followed by thrombocytopenia, as well as prolonged clotting times. Figure 2 illustrates the time course of these events paired alongside the clinical signs and symptoms observed based on human and NHP data.

An early increase in white blood cells (leukocytosis) is mainly due to an increase in the number of neutrophils, possibly caused by cytokine recruitment and/or damage to the blood vessel endothelium, and is followed by a decrease in their numbers due to

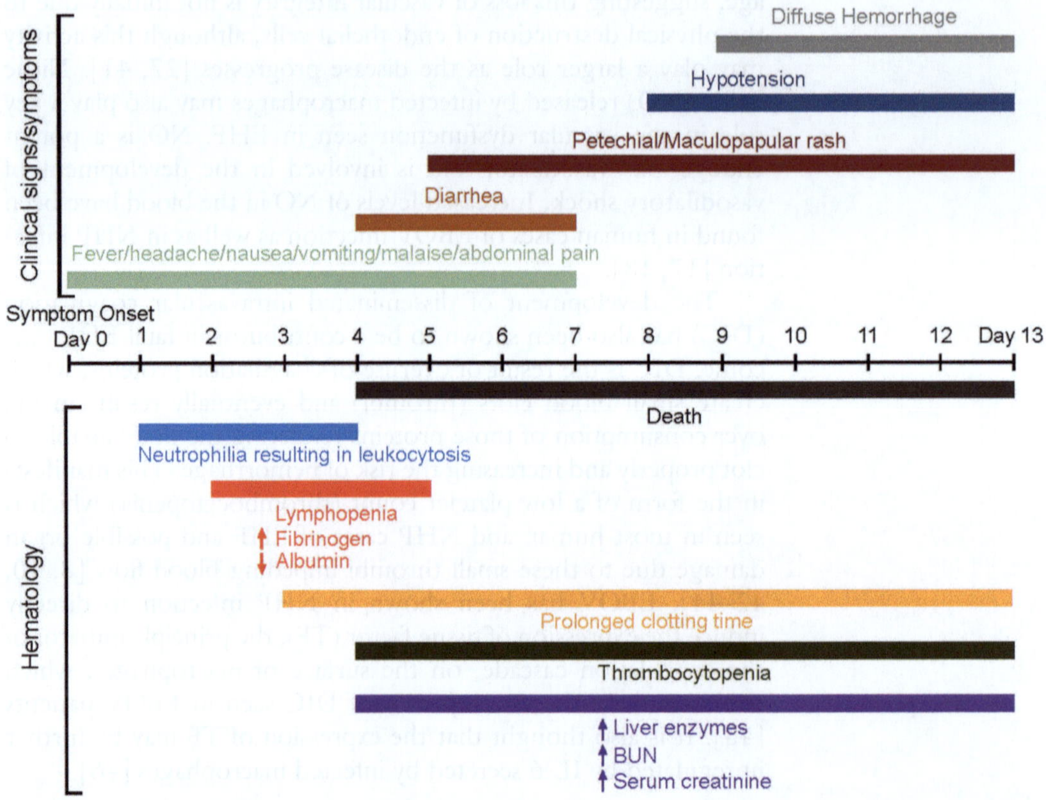

Fig. 2 Gantt chart depicting clinical and hematological progression of EHF beginning at symptom onset (day 0). Clinical signs and symptoms were derived primarily from human data available after the most recent outbreak in West Africa. Hematology data were derived primarily from NHP data, as available. Human data were complicated due to treatment regimens; however, an attempt was made to correlate recent human data obtained in the West African outbreak with available NHP data

lymphopenia, which is thought to be caused primarily by bystander lymphocyte apoptosis. This trend is seen in both NHPs as well as humans [8, 10, 42, 43, 47]. Liver enzyme tests like ALT and AST serve as a direct measurement of liver damage. Alanine aminotransferase (ALT) and aspartate aminotransferase (AST) are present within hepatocytes and are released into the blood in large quantities due to lysis of these cells. Alkaline phosphatase (ALP) and gamma-glutamyl transpeptidase (GGT) can be found bound to the hepatocyte cellular membrane and can be released due to direct damage, but their release can also be induced by the obstruction of bile ducts within the liver. Elevated levels of these enzymes can be found later in infection in both NHPs as well as human cases and have been shown to be strong predictors of fatal outcome [10, 11, 44, 47, 48]. Because albumin is synthesized by the liver, a decrease can also indicate liver dysfunction and damage, and this can indeed be seen in NHPs as well as human cases [44]. Similarly, elevation of both serum creatinine as well as BUN can indicate kidney dysfunction and are also considered predictors of fatal outcome in NHPs as well as humans [10, 11, 44, 47].

Due to both increased expression of TF, as well as cytokine production and possible endothelial damage, the coagulation cascade is initiated. As the disease progresses this results in increased fibrinogen levels, which leads to consumption of platelets and clotting factors, with a corresponding decreased platelet count (thrombocytopenia) and an increase in clotting time parameters, as seen in fatal cases [10, 11, 42, 44, 47–49]. This is followed by a sharp decline in fibrinogen levels as platelets and clotting factors are used up, and with the liver being unable to synthesize additional clotting factors, leading to a progression to DIC and eventually death.

4 Gross Pathology and Histology

EHF is a multisystem disease affecting numerous organs; however, the most extensive damage can be seen in the liver, spleen, lungs, gastrointestinal tract, lymph nodes, adrenal glands, and kidneys. Gross pathology includes enlarged, hemorrhagic lymph nodes, hepatomegaly, and splenomegaly, as well as hemorrhage and congestion of the gastrointestinal tract [15, 50]. Histology of the liver demonstrates hepatocyte necrosis with occasional mild inflammation as well as Kupffer cell hyperplasia with viral antigen being present primarily within Kupffer cells and hepatocytes, but also occasionally found in endothelial cells. Fibrin can also occasionally be found within the sinusoids of infected NHPs [45]. Lung pathology consists of congestion, focal intra-alveolar edema and hemorrhage with no significant inflammation. Viral antigen can be found in alveolar macrophages, endothelial cells, fibroblasts and other interstitial cells. The gastrointestinal tract appears congested with mild hemorrhagic lesions and antigen detection within macrophages and endothelial cells. Within the spleen and lymph nodes there is evidence of widespread, diffuse lymphoid necrosis and depletion. Ultrastructural evaluation of the lymph nodes and spleen also reveal the presence of apoptotic bodies within the follicular debris with histological evidence of antigen in dendritic cells, fibroblasts and other phagocytic cells. Within the spleen extensive fibrin deposition as well as necrosis of the red pulp with occasional hemorrhage of the marginal zone can also be seen. Necrosis has also been noted in the adrenal gland as well as acute tubular necrosis within the kidneys [9, 15, 18, 24, 26, 45].

5 Available Animal Models

5.1 Mouse

Mouse models of disease are quite popular as they are, compared to other animal models, easy to care for and house in large numbers, inexpensive, and their use create less ethical quandaries than other animal models; however, their physiological differences from humans make them a less valuable model for the assessment of pathogenesis and therapeutic efficacy. In order to develop EHF, immunocompetent mouse models must use mouse-adapted EBOV (MA-EBOV) as WT-EBOV does not cause disease. Additionally, the disease caused by MA-EBOV does not accurately replicate certain features of human disease, such as coagulation abnormalities and vascular permeability, thus it is unsurprising that this model often fails to predict countermeasure efficacy in NHPs [51, 52].

While there has recently been advancement of a humanized mouse model, this still may retain similar problems to wild-type laboratory mice strains when looking at pathogenesis or therapeutics. Humanized mice can potentially replicate features of human disease; however, they are more expensive, complicated to produce and replicate, and time consuming to acquire and also suffer from the same difference in physiology when not looking at the humanized tissues of the animal. On the other hand one of the strengths of this model is its ability to produce disease when using WT-EBOV rather than MA-EBOV, which separates it from all the other mouse models currently available [53]. Overall, while the mouse model can be useful as an initial first step in determining therapeutic effectiveness, because of the use of MA-EBOV in all except the humanized mouse model as well as the differences in disease features it is generally not considered as a stand-alone model.

5.2 Guinea Pig and Syrian Hamster

Similar to the mouse model, both the guinea pig and hamster models require the use of guinea pig adapted EBOV (GPA-EBOV) or MA-EBOV respectively, rather than wild-type to cause disease [54, 55]. Also similar to the mouse model, guinea pigs do not accurately recapitulate those features seen in human cases. Upon infection with GPA-EBOV guinea pigs, unlike mice, do develop coagulation abnormalities, though they are not as marked as those seen in humans; however, they do not exhibit bystander lymphocyte apoptosis while mice do. The correlation between countermeasure efficacy in guinea pigs and the outcome in NHPs has been shown to be slightly more predictive than the mouse model [56].

Though relatively new, the Syrian hamster model does seem to replicate many of the features of human EHF; however, this still requires an adapted version of the virus (MA-EBOV). Upon infection with MA-EBOV these hamsters develop severe coagulopathy, lymphocyte apoptosis, cytokine dysregulation, and histopathological lesions similar to those seen in NHPs. However, because this model for EBOV is so new it remains to be seen how effective it will be at correlating predicted countermeasure efficacy in NHPs and humans [51, 54]. Similar to the mouse model it seems that the guinea pig and hamster models both can be useful as an initial first step in determining therapeutic effectiveness, although because of the hamster model's ability to more accurately replicate important features of human disease it may prove more useful in the future than the other two in this area.

5.3 Nonhuman Primates (NHP)

Because of their close physiological relationship with humans, NHPs are considered the gold standard of animal models. Infection of NHPs with WT-EBOV produces similar clinical signs to what is seen in humans as well as similar hematology and clinical chemistry features, such as lymphopenia, thrombocytopenia and elevated liver enzymes with cytokine dysregulation, DIC and vascular dysfunction [42, 57]. The most commonly used NHPs are the rhesus and cynomolgus macaques as they replicate human features of the disease more accurately than other NHP species, apart from baboons which are more logistically challenging due to their size and present safety concerns when working in BSL-4 facilities. However, the common marmoset has also been shown to be a very useful model for filovirus infection. Similar to the macaque models, the marmoset model also accurately recapitulates human features of EHF while also providing fewer logistical and safety concerns due to their small size [58, 59]. Unfortunately, their small size can also serve as a hindrance due to the limited volume of blood and other bodily fluids that can be obtained during an exam. NHPs generally serve as a final step before evaluating vaccine and therapeutic efficacy in humans.

6 Hope for the Future

As described above, EBOV produces a severe and often fatal hemorrhagic disease for which there are currently no licensed vaccines or therapeutics. Normally, treatment for EHF is mainly supportive involving fluid and electrolyte replacement; however, due to the recent outbreak in 2014 the urgency for the development of vaccines and therapeutics has increased. There are currently several vaccine candidates undergoing Phase I–III clinical trials as well as several therapeutic measures, including the transfusion of convalescent plasma and monoclonal antibodies [60]. However, more research is needed for EBOV as well as other filoviruses in order to prevent future outbreaks with a similar devastating capacity to that seen in 2014.

Acknowledgments

This research was supported by the Intramural Research Program of the NIH, NIAID. The authors would like to thank Patrick Hanley (NIAID, NIH) for discussions and advice regarding the clinical chemistry and hematology section as well as Anita Mora (NIAID, NIH) for all her help with designing the figures.

References

1. Baize S, Pannetier D, Oestereich L et al (2014) Emergence of Zaire Ebola virus disease in Guinea. N Engl J Med 371(15):1418–1425. doi:10.1056/NEJMoa1404505
2. WHO (2015) Ebola situation report — 30 December 2015. http://apps.who.int/ebola/current-situation/ebola-situation-report-30-december-2015. Accessed 4 Jan 2016
3. Bausch DG, Towner JS, Dowell SF et al (2007) Assessment of the risk of Ebola virus transmission from bodily fluids and fomites. J Infect Dis 196:S142–S147. doi:10.1086/520545
4. Christie A, Davies-Wayne GJ, Cordier-Lasalle T et al (2015) Possible sexual transmission of Ebola virus — Liberia, 2015. MMWR 64(17): 479–481
5. Judson S, Prescott J, Munster V (2015) Understanding Ebola virus transmission. Viruses 7:511–521. doi:10.3390/v7020511
6. Mate SE, Kugelman JR, Nyenswah TG et al (2015) Molecular evidence of sexual transmission of Ebola virus. N Engl J Med. doi:10.1056/NEJMoa1509773. Brief Report
7. Varkey JB, Shantha JG, Crozier I et al (2015) Persistence of Ebola virus in ocular fluid during convalescence. N Engl J Med 372(25):2423–2427. doi:10.1056/NEJMoa1500306
8. Kreuels B, Wichmann D, Emmerich P et al (2014) A case of severe Ebola virus infection complicated by Gram-negative septicemia. N Engl J Med 371(25):2394–2401. doi:10.1056/NEJMoa1411677
9. Martines RB, Ng DL, Greer PW, Rollin PE, Zaki SR (2014) Tissue and cellular tropism, pathology and pathogenesis of Ebola and Marburg viruses. J Pathol 235:153–174. doi:10.1002/path.4456
10. Hunt L, Gupta-Wright A, Simms V et al (2015) Clinical presentation, biochemical, and haematological parameters and their association with outcome in patients with Ebola virus disease: an observational cohort study. Lancet Infect Dis 15:1292–1299. doi:10.1016/S1473-3099(15)00144-9
11. Schieffelin JS, Shaffer JG, Goba A et al (2014) Clinical illness and outcomes in patients with Ebola in Sierra Leone. N Engl J Med 371(22): 2092–2100. doi:10.1056/NEJMoa1411680
12. WHO Ebola Response Team (2014) Ebola virus disease in West Africa — the first 9 months of the epidemic and forward projections. N Engl J Med 371(16):1481–1495. doi:10.1056/NEJMoa1411100
13. Alves DA, Honko AN, Kortepeter MG et al (2016) Necrotizing scleritis, conjunctivitis, and other pathologic findings in the left eye and brain of an Ebola virus-infected rhesus macaque (*Macaca mulatta*) with apparent recovery and a delayed time of death. J Infect Dis 213:57–60. doi:10.1093/infdis/jiv357
14. WHO (2014) Sierra Leone: helping the Ebola survivors turn the page. http://www.who.int/features/2014/post-ebola-syndrome/en. Accessed 4 Jan 2016

15. Geisbert TW, Hensley LE, Larsen T et al (2003) Pathogenesis of Ebola hemorrhagic fever in cynomolgus macaques: evidence that dendritic cells are early and sustained targets of infection. Am J Pathol 163(6):2347–2370
16. Mahanty S, Hutchinson K, Agarwal S et al (2003) Cutting edge: impairment of dendritic cells and adaptive immunity by Ebola and Lassa viruses. J Immunol 170:2797–2801. doi:10.4049/jimmunol.170.6.2797
17. Feldmann H, Geisbert TW (2011) Ebola haemorrhagic fever. Lancet 377:849–862. doi:10.1016/S0140-6736(10)60667-8
18. Messaoudi I, Amarasinghe GK, Basler CF (2015) Filovirus pathogenesis and immune evasion: insights from Ebola virus and Marburg virus. Nat Rev Microbiol 13:663–676. doi:10.1038/nrmicro3524
19. Bosio CM, Aman MJ, Grogan C et al (2003) Ebola and Marburg viruses replicate in monocyte-derived dendritic cells without inducing the production of cytokines and full maturation. J Infect Dis 188:1630–1638
20. Yen B, Mulder LCF, Martinez O, Basler CF (2014) Molecular basis for Ebolavirus VP35 suppression of human dendritic cell maturation. J Virol 88(21):12500–12510. doi:10.1128/JVI.02163-14
21. Bixler SL, Goff AJ (2015) The role of cytokines and chemokines in filovirus infection. Viruses 7:5489–5507. doi:10.3390/v7102892
22. Bradfute SB, Swanson PE, Smith MA et al (2010) Mechanisms and consequences of Ebolavirus-induced lymphocyte apoptosis. J Immunol 184:327–335. doi:10.4049/jimmunol.0901231
23. Bradley JR (2008) TNF-mediated inflammatory disease. J Pathol 214:149–160. doi:10.1002/path.2287
24. Geisbert TW, Hensley LE, Gibb TR, Steele KE, Jaax NK et al (2000) Apoptosis induced in vitro and in vivo during infection by Ebola and Marburg viruses. Lab Invest 80:171–186
25. McElroy AK, Akondy RS, Davis CW et al (2015) Human Ebola virus infection results in substantial immune activation. Proc Natl Acad Sci U S A 112(15):4719–4724. doi:10.1073/pnas.1502619112
26. Wauquier N, Becquart P, Padilla C, Baize S, Leroy EM (2010) Human fatal Zaire Ebola virus infection is associated with an aberrant innate immunity and with massive lymphocyte apoptosis. PLoS Negl Trop Dis 4(10):e837. doi:10.1371/journal.pntd.0000837
27. Zampieri CA, Sullivan NJ, Nabel GJ (2007) Immunopathology of highly virulent pathogens: insights from Ebola virus. Nat Immunol 8(11):1159–1164. doi:10.1038/ni1519
28. Dunham EC, Banadyga L, Groseth A et al (2015) Assessing the contribution of interferon antagonism to the virulence of West African Ebola viruses. Nat Commun 8000:1–6. doi:10.1038/ncomms9000
29. Reid SP, Valmas C, Martinez O, Sanchez FM, Basler CF (2007) Ebola virus VP24 proteins inhibit the interaction of NPI-1 subfamily karyopherin α proteins with activated STAT1. J Virol 81:13469–13477
30. Kaletsky RL, Francica JR, Agrawal-Gamse C, Bates P (2009) Tetherin-mediated restriction of filovirus budding is antagonized by the Ebola glycoprotein. Proc Natl Acad Sci U S A 106(8): 2886–2891. doi:10.1073/pnas.0811014106
31. Vande Burgt NH, Kaletsky RL, Bates P (2015) Requirements within the Ebola viral glycoprotein for tetherin antagonism. Viruses 7:5587–5602. doi:10.3390/v7102888
32. Mohan GS, Li W, Ye L, Compans RW, Yang C (2012) Antigenic subversion: a novel mechanism of host immune evasion by Ebola virus. PLoS Pathog 8(12):e1003065. doi:10.1371/journal.ppat.1003065
33. Yang Z, Delgado R, Xu L et al (1998) Distinct cellular interactions of secreted and transmembrane Ebola virus glycoproteins. Science 279:1034–1037
34. Gupta M, Mahanty S, Ahmed R, Rollin PE (2001) Monocyte-derived human macrophages and peripheral blood mononuclear cells infected with Ebola virus secrete MIP-1a and TNF-a and inhibit poly-IC-induced IFN-a in vitro. Virology 284:20–25. doi:10.1006/viro.2001.0836
35. Stroher U, West E, Bugany H et al (2001) Infection and activation of monocytes by Marburg and Ebola viruses. J Virol 75(22):11025–11033. doi:10.1128/JVI.75.22.11025-11033.2001
36. Mohamadzadeh M, Coberley SS, Olinger GG et al (2006) Activation of triggering receptor expressed on myeloid cells-1 on human neutrophils by Marburg and Ebola viruses. J Virol 80(14):7235–7244. doi:10.1128/JVI.00543-06
37. Escudero-Perez B, Volchkova VA, Dolnik O, Lawrence P, Volchkov VE (2014) Shed GP of Ebola virus triggers immune activation and increased vascular permeability. PLoS Pathog 10(11):e1004509. doi:10.1371/journal.ppat.1004509
38. Hensley LE, Young HA, Jahrling PB, Geisbert TW (2002) Proinflammatory response during Ebola virus infection of primate models: possible involvement of the tumor necrosis factor receptor superfamily. Immunol Lett 80: 169–179

39. Wahl-Jensen V, Kurz S, Feldmann F, Buehler LK, Kindrachuk J et al (2011) Ebola virion attachment and entry into human macrophages profoundly effects early cellular gene expression. PLoS Negl Trop Dis 5(10):e1359. doi:10.1371/journal.pntd.0001359
40. Wahl-Jensen VM, Afanasieva TA, Seebach J et al (2005) Effects of Ebola virus glycoproteins on endothelial cell activation and barrier function. J Virol 79(16):10442–10450. doi:10.1128/JVI.79.16.10442-10450.2005
41. Geisbert TW, Hensley LE, Larsen T et al (2003) Pathogenesis of Ebola hemorrhagic fever in primate models: evidence that hemorrhage is not a direct effect of virus-induced cytolysis of endothelial cells. Am J Pathol 163(6):2347–2370
42. Ebihara H, Rockx B, Marzi A et al (2011) Host response dynamics following lethal infection of rhesus macaques with Zaire ebolavirus. J Infect Dis 204:S991–S999. doi:10.1093/infdis/jir336
43. Schibler M, Vetter P, Cherpillod P et al (2015) Clinical features and viral kinetics in a rapidly cured patient with Ebola virus disease: a case report. Lancet Infect Dis 15:1034–1040. doi:10.1016/S1473-3099(15)00229-7
44. Wolf T, Kann G, Becker S et al (2015) Severe Ebola virus disease with vascular leakage and multiorgan failure: treatment of a patient in intensive care. Lancet 385:1428–1435. doi:10.1016/S0140-6736(14)62384-9
45. Geisbert TW, Young HA, Jahrling PB, Davis KJ, Kagan E, Hensley LE (2003) Mechanisms underlying coagulation abnormalities in Ebola hemorrhagic fever: overexpression of tissue factor in primate monocytes/macrophages is a key event. J Infect Dis 188:1618–1629. doi:10.1086/379724
46. Neumann FJ, Ott I, Marx N et al (1997) Effect of human recombinant interleukin-6 and interleukin-8 on monocyte procoagulant activity. Arterioscler Thromb Vasc Biol 17:3399–3405. doi:10.1161/01.ATV.17.12.3399
47. Kortepeter MG, Lawler JV, Honko A et al (2011) Real-time monitoring of cardiovascular function in rhesus macaques infected with Zaire ebolavirus. J Infect Dis 204:S1000–S1010. doi:10.1093/infdis/jir337
48. Lyon GM, Mehta AK, Varkey JB et al (2015) Clinical care of two patients with Ebola virus disease in the United States. N Engl J Med 371(25):2402–2409. doi:10.1056/NEJMoa1409838
49. Martins K, Cooper C, Warren T et al (2015) Characterization of clinical and immunological parameters during Ebola virus infection of rhesus macaques. Viral Immunol 28(1):32–41. doi:10.1089/vim.2014.0085
50. Baskerville A, Bowen ETW, Platt GS et al (1978) The pathology of experimental Ebola virus infection in monkeys. J Pathol 125: 131–138
51. Wahl-Jensen V, Bollinger L, Safronetz D et al (2012) Use of the Syrian hamster as a new model of Ebola virus disease and other viral hemorrhagic fevers. Viruses 4:3754–3784. doi:10.3390/v4123754
52. Bray M, Davis K, Geisbert T et al (1999) A mouse model for evaluation of prophylaxis and therapy of Ebola hemorrhagic fever. J Infect Dis 179:S248–S258
53. Bird BH, Spengler JR, Chakrabarti AK et al (2015) Humanized mouse model of Ebola virus disease mimics the immune responses in human disease. J Infect Dis 213:703–711. doi:10.1093/infdis/jiv538
54. Ebihara H, Zivcec M, Gardner D et al (2013) A Syrian golden hamster model recapitulating ebola hemorrhagic fever. J Infect Dis 207:306–318. doi:10.1093/infdis/jis626
55. Volchkov VE, Chepurnov AA, Volchkova VA, Ternovoj VA, Klenk HD (2000) Molecular characterization of guinea pig-adapted variants of Ebola virus. Virology 277:147–155. doi:10.1006/viro.2000.0572
56. Cross RW, Fenton KA, Geisbert JB et al (2015) Modeling the disease course of Zaire ebolavirus infection in the outbred guinea pig. J Infect Dis 212:S305–S315. doi:10.1093/infdis/jiv237
57. Geisbert TW, Strong JE, Feldmann H (2015) Considerations in the use of nonhuman primate models of Ebola virus and Marburg virus infection. J Infect Dis 212:S91–S97. doi:10.1093/infdis/jiv284
58. Carrion R Jr, Patterson JL (2012) An animal model that reflects human disease: the common marmoset (Callithrix jacchus). Curr Opin Virol 2(3):357–362. doi:10.1016/j.coviro.2012.02.007
59. Carrion R Jr, Youngtae R, Hoosien K et al (2011) A small nonhuman primate model for filovirus-induced disease. Virology 420:117–124. doi:10.1016/j.virol.2011.08.022
60. WHO (2015) Essential medicines and health products. http://www.who.int/medicines/emp_ebola_q_as/en/. Accessed 4 Jan 2016

Part II

Studying Ebolaviruses Under Biosafety Level 2 Conditions

Chapter 4

Production of Filovirus Glycoprotein-Pseudotyped Vesicular Stomatitis Virus for Study of Filovirus Entry Mechanisms

Rachel B. Brouillette and Wendy Maury

Abstract

Members of the family *Filoviridae* are filamentous, enveloped, and nonsegmented negative-stranded RNA viruses that can cause severe hemorrhagic disease in humans and nonhuman primates with high mortality rates. Current efforts to analyze the structure and biology of these viruses as well as the development of antivirals have been hindered by the necessity of biosafety level 4 containment (BSL4). Here, we outline how to produce and work with Ebola virus glycoprotein bearing vesicular stomatitis virus (VSV) pseudovirions. These pseudovirions can be safely used to evaluate early steps of the filovirus life cycle without need for BSL4 containment. Virus gene expression in the transduced cells is easy to assess since the pseudovirions encode a reporter gene in place of the VSV G glycoprotein gene. Adoption of VSV for use as a pseudovirion system for filovirus GP has significantly expanded access for researchers to study specific aspects of the viral life cycle outside of BSL4 containment and has allowed substantial growth of filovirus research.

Key words Filovirus, Pseudovirus, Pseudotyped-vesicular stomatitis virus, Entry

1 Introduction

Members of the family *Filoviridae* are filamentous, enveloped, and nonsegmented negative-stranded RNA viruses [1, 2] of which there are three genera: *Ebolavirus*, *Cuevavirus*, and *Marburgvirus*. Filoviruses can cause severe hemorrhagic disease in humans and nonhuman primates with high mortality rates [1, 2]. As a consequence of the absence of countermeasures against these viruses, their lethality, and their transmissibility between individuals, filoviruses are listed as a Category A pathogen by the NIH. Current efforts to analyze the structure and biology of these viruses as well as the development of antivirals have been hindered by the necessity of biosafety level 4 (BSL4) containment. After it was observed that VSV could efficiently incorporate foreign viral glycoproteins on its particles [3], VSV pseudovirions bearing filovirus glycoprotein

Thomas Hoenen and Allison Groseth (eds.), *Ebolaviruses: Methods and Protocols*, Methods in Molecular Biology, vol. 1628, DOI 10.1007/978-1-4939-7116-9_4,

(GP) were developed as a surrogate to investigate filoviral biology under more accessible BSL2 conditions [4–6].

VSV is a member of the family *Rhabdoviridae*, and like filoviruses, is within the taxonomic order of *Mononegavirales* and therefore has a similar genome structure. Although the concept of viral glycoprotein pseudotyping was discovered to occur during coinfections between VSV and other viruses in the 1980s [7], these initial studies included within the VSV genome an intact G gene encoding the native VSV glycoprotein. Thus the use of VSV pseudovirions to investigate the biology of the pseudotyped glycoprotein did not occur for more than another decade [3]. Subsequent studies used VSV genomes that deleted the G gene, thus allowing the study of the biology of the pseudotyped GP [4]. Early studies showed that Ebola virus GP conferred cellular tropism to VSV corresponding to the host range tropism of Ebola virus [4]. VSV pseudovirions have proven useful for the study of a variety of aspects of the biology of virion-associated glycoproteins from a number of high containment, bio-defense viral pathogens, including arenaviruses, paramyxoviruses, filoviruses, and bunyaviruses [8–15]. Additionally, VSV preparations are convenient for production of pseudoparticles; VSV grows in many animal and some insect cells and can be propagated in large quantities [16]. Adoption of VSV for use as a pseudovirion system for filovirus GP has significantly expanded access for researchers to study the virus outside of BSL4 containment and has allowed studies to emerge that demonstrate a range of findings from the basic structure of the Ebola glycoprotein, to entry, to its post internalization complex processing, to mechanisms of antibody-mediated neutralization and more.

The following protocol, as summarized in Fig. 1, describes the production of recombinant VSV containing a reporter gene, green fluorescent protein (GFP), instead of the VSV G protein gene, which thus are not infectious unless a gene expressing a viral glycoprotein responsible for receptor binding and membrane fusion is provided in *trans* (VSVΔG-GFP) [4].

2 Materials

2.1 Cell Culture Reagents

1. Dulbecco's Modified Eagle's Medium (DMEM) supplemented with 10% fetal bovine serum (FBS) and 1% penicillin–streptomycin (P/S) (*see* **Notes 1–3**).
2. DMEM supplemented with 1.5% FBS and 1% P/S.
3. 1× phosphate buffered saline (PBS) (*see* **Note 4**).
4. An inverted light microscope.

2.2 Cells

1. HEK293T cells (*Homo sapiens* embryonic kidney epithelial cells containing the SV40 T antigen; ATCC CRL-3216).
2. Vero cells (*Cercopithecus aethiops* kidney epithelial cells; ATCC CCL-81).

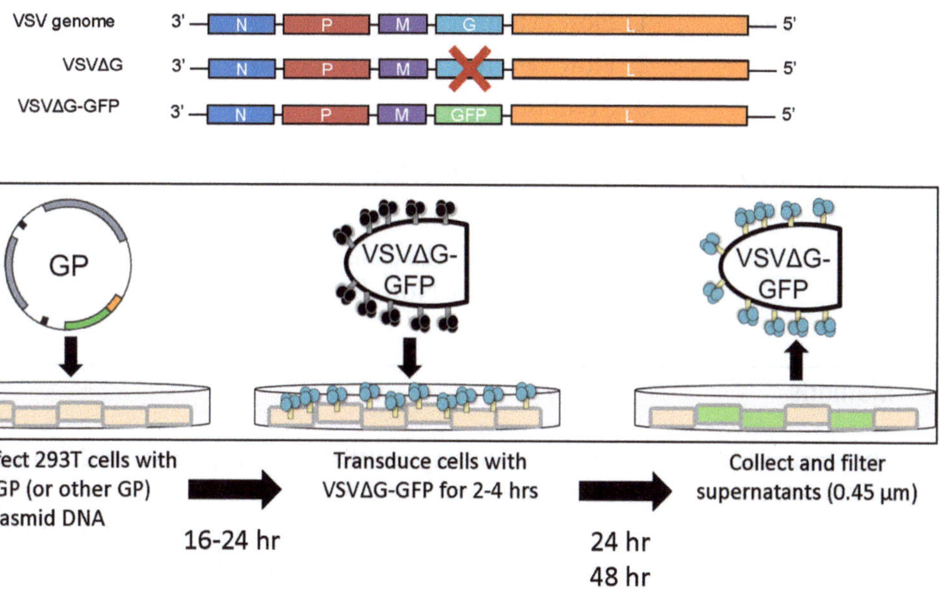

Fig. 1 Outline of VSV pseudovirion production. (**a**) Schematic of wild-type VSV genome, the VSV genome deleted for the G gene and insertion of the reporter gene, GFP, in place of G. (**b**) Production of EBOV GP pseudotyped VSV

2.3 Transfection Reagents

1. Polyethylenimine (PEI) solution (1 mg/mL) (*see* **Note 5**).
2. 150 mM NaCl solution.
3. DNA plasmid containing a filovirus glycoprotein gene driven by a mammalian enhancer/promoter, such as the CMV promoter. Using this protocol, we have generated VSV pseudovirions containing Ebola virus, Sudan virus, Bundibugyo, or Marburg virus GP (*see* **Note 6**).

2.4 Transduction Reagent

1. VSVΔG-GFP pseudotyped with a viral glycoprotein that readily mediates entry into HEK293T cells (*see* **Note 7**). VSVΔG-stocks encoding a reporter gene (GFP, dsRed or luciferase) are available from Kerafast (http://www.kerafast.com/c-310-delta-g-vsv-pseudotyping-system.aspx).

2.5 Concentration/ Purification of Viral Particles (Optional)

1. A high-speed centrifuge that can achieve ≥5400 × *g*.
2. 250 or 500 mL sterile polypropylene bottles (depending on centrifuge rotor size).
3. An ultracentrifuge that can achieve ≥80,000 × *g*.
4. 3 mL sterile polycarbonate ultracentrifuge tubes for Beckman SW SW60Ti rotor or 30 mL sterile ultracentrifuge tube for Beckman SW32Ti rotor (or similar equipment if using an alternative ultracentrifuge).
5. PBS.
6. 20% sucrose in PBS (sterilized through a ≤0.45 μm filter).

2.6 Quantification of Viral Particles

1. Flow cytometer.
2. Flow cytometer analysis software.
3. Accutase.

2.7 Normalization of Pseudovirus Stocks

1. 5× lysis buffer (0.125% NP40 in PBS).
2. 96-Well dot blot apparatus.
3. Primary/secondary antibodies and blocking buffer for immunostaining (*see* **Note 8**).

3 Methods

3.1 Preparation of 1 mg/mL PEI Solution

1. Dissolve 100 mg of PEI powder in 80–90 mL water on a stir plate with a magnetic bar.
2. Adjust the pH to 7.0 with HCl (the solution clarifies as the correct pH is approached; full dissolution may take several hours).
3. Add water until a final volume of 100 mL is reached.
4. Sterilize the solution through a 0.22 μm filter and dispense into aliquots.
5. Store aliquots at −20 °C or −80 °C for long term storage; an in-use stock can be kept at 4 °C for up to several months (*see* **Note 5**).

3.2 Transfection of HEK293T Cells

1. In the afternoon, seed tissue culture plates to achieve 70–80% confluency 24 h after seeding in DMEM supplemented with 10% FBS and 1% P/S. *See* Table 1 for cell numbers.
2. Twenty four hours following seeding, transfect cells with filovirus glycoprotein plasmid DNA (*see* **Notes 9–11**).
3. Prepare two tubes for transfection (*see* Table 1):

   Tube 1—Mix 25 μL of NaCl with every 1 μg of DNA.

   Tube 2—Mix 22 μL of NaCl with 3 μL of PEI for every 1 μg of DNA in Tube 1.
4. Combine the two tubes, vortex well (10–15 s), and incubate at room temperature for 7–20 min.
5. Add the total volume of DNA/PEI/NaCl solution dropwise to cells and place tissue culture plate(s) back in 37 °C incubator with 5% CO_2 until transduction.

3.3 Transduction with VSVΔG-GFP

1. Add a stock of GP pseudotyped VSVΔG at an MOI of ~1–3 to cells 16–24 h post transfection (there is no need to change the medium). The MOI of the seeding virus added will need to be empirically determined for your conditions and seeding stock.

The stock we typically use is a Lassa virus GPC-pseudotyped VSVΔG containing GFP (*see* **Note 7**).

2. Maintain the seeding stock on cells at 37 °C for 2–4 h. As filovirus GP-pseudotyped VSV particles are produced as early as 6 h following transduction, be sure to continue with **step 3** before then.
3. Remove media and the wash cells with PBS twice to remove unbound input virus before replacing with DMEM supplemented with 1.5% FBS and 1% P/S.
4. Return plate(s) to the 37 °C incubator 5% CO_2.

3.4 Collection of Filovirus GP-Pseudotyped VSVΔG-GFP

1. Collect medium in sterile syringes at 24 h following transduction and filter them through 0.45 μm filters into sterile containers.
2. If making small stocks of pseudovirus, a single collection at 24 h may be sufficient. Store in ~0.5 mL aliquots and discard used culture plates. However, if making large stocks of pseudovirus, add fresh DMEM to cells and repeat collection at 48 h following transduction.
3. Store collected stocks at −80 °C or continue directly to ultracentrifugation, if desired (*see* **Notes 12** and **13**).

3.5 Concentration and Purification via Ultracentrifugation Through a Sucrose Cushion (Optional) (See Note 13)

1. If collecting ≥30 mL of supernatant pseudovirus, first combine all collections into a single sterile bottle and spin for at least 16 h at 5400 × *g* in a high speed centrifuge at 4 °C to pellet the viral particles. Resuspend the pellet in ~1 mL sterile PBS. If concentrating <30 mL of supernatant pseudovirus, begin at **step 2**, using the instructions for the SW32Ti rotor and the larger centrifuge tubes.
2. Add 0.5 mL of sterile 20% sucrose PBS to the bottom of a 3 mL polycarbonate ultracentrifuge tube if using the SW60Ti rotor; or add 3 mL of sterile 20% sucrose PBS to the bottom of a 30 mL ultracentrifuge tube if using the SW32Ti rotor (*see* **Note 14**).
3. Without penetrating the sucrose cushion, carefully layer either the supernatant collection or the resuspended virus pellet into the tube. Completely fill the tube, using PBS as necessary, to prevent collapse of the tube during centrifugation.
4. Balance the rotor according to the ultracentrifuge manufacturer's instructions and spin at 80,000 × *g* for 2 h at 4 °C.
5. Discard liquid from the tubes and wipe out any remaining liquid with a sterile swab. Place the pellets on ice, add sterile PBS (to a desired volume) and resuspend the pellet. It is best to allow the pellet to loosen by incubating on ice for at least 30 min prior to attempting to resuspend. With highly

Table 1

Quantity of reagents needed for transfection performed with different sized plates

Tissue culture plate	# Cells to seed	Transfection tube 1		Transfection tube 2		Total volume (μL)
		μg DNA	μL NaCl	μL PEI	μL NaCl	
6-Well	5.00E+05	2 per well	50 per well	6 per well	44 per well	100 per well
10 cm	3.50E+06	16	400	48	352	800
15 cm	1.00E+07	32	800	96	704	1600

concentrated stocks, resuspension of virion clumps can take several hours and may require scraping with a pipette tip and/or extensive vortexing. Highly concentrated stocks will have a milky appearance.

6. Distribute resuspension into aliquots and store at −80 °C for future use.

3.6 Quantification of Pseudovirus Stocks

1. To determine the pseudovirion titer (transducing units/mL) in Vero cells, seed 50,000 Vero cells per well in a 48-well format in DMEM supplemented with 10% FBS and 1% P/S in the afternoon.
2. On the following day, make several dilutions of your pseudovirus stock in DMEM supplemented with 1.5% FBS and 1% P/S. Remove medium from Vero cells and add 250 μL of each concentration of pseudovirus to duplicate or triplicate wells. In our lab, to titer a typically concentrated and purified virus stock, a series of several dilutions (two-, five-, or ten-fold) is assessed. *See* Fig. 2 for a typical detailed dilution curve.
3. Incubate plates in a 5% CO_2 incubator at 37 °C for 18–24 h.
4. Remove media and detach cells with 100 μL 37 °C Accutase per well. Transfer cells into 4 mL polystyrene round-bottom tubes appropriate for use on a flow cytometer.
5. Quantify the number of GFP-expressing cells using a flow cytometer. Using values in the linear range of your dilution curve, calculate transducing units/mL after analysis with the flow cytometer software: [proportion of GFP-positive cells] × [75,000 cells]/[volume of pseudovirus in mL] as shown in Fig. 2 (*see* **Note 15**).

3.7 Normalization of Pseudovirus Stocks

For many experiments, equivalent numbers of virions from different stocks need to be compared, and, to do this, you need to normalize the number of virions added. For these types of studies, we normalize by comparing the amount of VSV matrix protein in each stock.

1. Mix an aliquot containing an equivalent volume of each pseudovirus stock with PBS and a sufficient quantity of 5× lysis buffer to achieve a 1× lysis buffer in the final concentration. For stocks concentrated by ultracentrifugation, as little as 1–6 μL of stock can be used for normalization studies. The total sample volume then should be brought to 160 μL with PBS and 40 μL of 5× lysis buffer added to achieve a final sample size of 200 μL. For unconcentrated supernatant virus stocks, 160 μL of stock can be mixed directly with 5× lysis buffer. Incubate virions in the lysis buffer for at least 2 min.
2. Make several two-fold dilutions of your lysed virions in PBS.
3. Apply 100 μL of each dilution of each pseudovirus directly to a nitrocellulose membrane pre-wetted with PBS through a 96-well dot blot apparatus attached to a vacuum pump. Wash each well with 1× buffer several times.
4. Proceed with immunoblotting according to a standard protocol. Primary antibodies targeting either the filovirus GP or the VSV matrix protein are appropriate. For detection of VSV matrix protein, monoclonal antibody 23H12 (1 μg/mL) works well and is available from Kerafast.com. Antisera and monoclonal antibodies (0.1–0.5 μg/mL) for the detection of Ebola virus glycoprotein by immunoblotting are available from IBT Bioservices (*see* **Notes 8** and **16**).
5. Quantify densities of dots and calculate the amount of particles relative to a standard stock as shown in Fig. 3. Immune detection of VSV matrix protein in our lab is performed using secondary

Percent GFP+	Proportion GFP+	Total # cells	# GFP+ cells	μL Pseudovirus	Transducing units/mL	Average
66.1	0.661	75,000	49575	0.03541	1.40E+09	2.35E+09
47.0	0.470	75,000	35250	0.01517	2.32E+09	
28.8	0.288	75,000	21600	0.00650	3.32E+09	

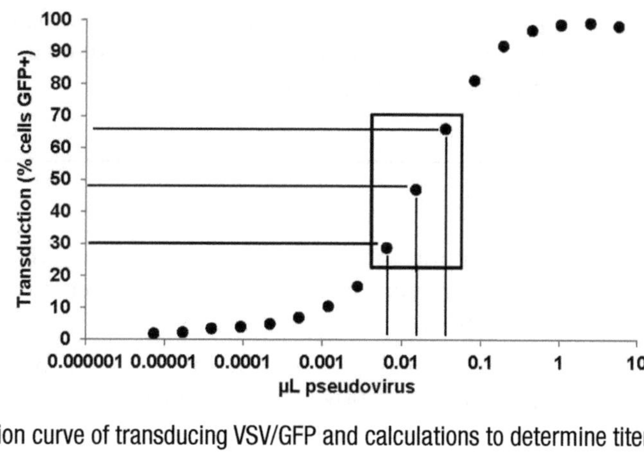

Fig. 2 Typical dilution curve of transducing VSV/GFP and calculations to determine titers of stocks

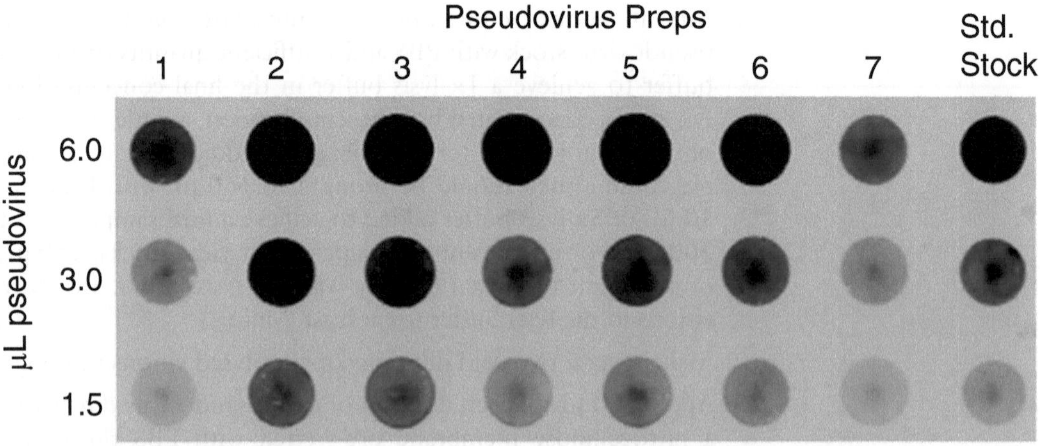

	Signal Intensities							
µL virus	Virus 1	Virus 2	Virus 3	Virus 4	Virus 5	Virus 6	Virus 7	Std Stock
6	4280	15798	13633	8539	10739	9279	3121	6823
3	1432	6740	6026	3246	4425	3600	1189	3240
1.5	505	2823	2586	1205	1764	1345	500	805
	Relative to Std Stock							
	Virus 1	Virus 2	Virus 3	Virus 4	Virus 5	Virus 6	Virus 7	Std Stock
6	1.59	0.43	0.50	0.80	0.64	0.74	2.19	1.00
3	2.26	0.48	0.54	1.00	0.73	0.90	2.72	1.00
1.5	1.59	0.29	0.31	0.67	0.46	0.60	1.61	1.00
Average	1.82	0.40	0.45	0.82	0.61	0.74	2.71	1.00

Fig. 3 Analysis and normalization of concentrated virus stocks in dot blots. Shown are eight different virus stocks (including a standard stock) that are to be normalized for the quantity of VSV capsid. A solution of 6 µL of each stock was assessed as well as two 2-fold dilutions. In the example provided, 1.82 µL of Virus 1 should be added for every 1.00 µL of Standard Stock virus in order to be considered roughly equivalent in particle numbers

antibodies conjugated to far red fluorophores and imaged using a LiCor Odyssey imaging system. However, other quantitative imaging approaches also work.

6. For further experiments, the amount of a new pseudovirion preparation used should then be adjusted based on its relative reactivity in dot blot compared to the standard stock. A sample calculation is presented in Fig. 3.

4 Notes

1. It is important to use premium FBS, as HEK293T cells may not transfect or transduce well under suboptimal growth conditions.
2. Stock penicillin–streptomycin (P/S) can be purchased commercially as a 100× solution, which contains 10,000 units/mL penicillin and 10,000 µg/mL streptomycin.

3. All reagents used in transfections, transductions, and general cell culture must be sterile.
4. To prevent potential bacterial contamination, you may supplement PBS with 1% P/S for washing cells. For 1 L 20× PBS: 160 g NaCl, 4 g KCl, 28.8 g Na_2PO_4, 4.8 g KH_2PO_4; adjust to pH 7.4 after dilution to 1× concentration.
5. Sterile PEI solutions may be frozen at −20 °C for long term storage. Upon thawing, 1 mg/mL concentrations of PEI tend to precipitate out of solution. To resuspend precipitant, heat PEI solution in a 37 °C water bath for 30 min to an hour and vortex well. The PEI transfection method provided here was optimized for our lab's conditions. Protocols for PEI transfection vary widely from lab to lab, so we encourage you to troubleshoot accordingly if transfection efficiency is low (less than 60–70% of an HEK293T culture).
6. VSV recruits its viral glycoprotein at the plasma membrane. Those viral glycoproteins that do not traffic to the plasma membrane, such as flaviviruses, pseudotype VSV poorly. For those viral glycoproteins, other systems such as transcription- and replication-competent virus-like particle (trVLP) approaches should be used to study entry.
7. The VSVΔG-GFP stock used to generate your new stocks of VSVΔG-GFP bearing filovirus GP needs to robustly transduce HEK293T cells. Viral glycoproteins that mediate strong transduction of HEK293T cells include Lassa virus GPC, VSV G, or Venezuelan Equine Encephalitis Virus GP. If you wish to use your previously generated filovirus GP-VSVΔG-eGFP stock as your seed stock, *see* **Note 11**.
8. 5× lysis buffer contains low concentrations of a mild detergent (NP40) that does not affect the detection of the native conformation of viral glycoproteins; therefore conformationally dependent antibodies will work in the dot blot immunostaining protocol.
9. To achieve the goal of a high titer stock with this transfection/transduction protocol, optimal levels of transfection are needed. Ideally, close to 100% of the HEK 293 T cells should be expressing the transfected viral glycoprotein. Introduction of pseudotyped VSVΔG stock into an untransfected cell will generate defective VSVΔG that do not bear a viral glycoprotein and, consequently, produce only defective viral particles.
10. It is not required to refresh the media before PEI transfection of HEK293T cells, but if you choose to do so, make sure the new media is warmed in 37 °C at 5% CO_2 for at least 30 min before transfection.
11. Filovirus glycoproteins on the surface of the VSV pseudovirions do not mediate robust entry into HEK293T cells.

This allows only one efficient cycle of pseudovirus production. To amplify production of filovirus GP pseudotyped VSV, you may co-transfect the HEK293T cells with a plasmid expressing TIM-1cDNA (9:1 ratio of GP expressing to TIM-1 expressing plasmid). TIM-1 is a cell surface, phosphatidylserine receptor for filoviruses. Including a TIM-1 expressing plasmid in the transfection protocol enhances filovirus pseudovirion production by allowing multi-cycle amplification.

12. After collection of supernatants, cells can be incubated with a sterile 5 mM EDTA/PBS solution for 5 min to release any cell-associated particles. You may pool this collection with the supernatant collection; however, it is only advised to do so if you plan to concentrate and purify by ultracentrifugation, as EDTA can affect downstream applications. EDTA also has the potential to lift cells from the plate, so in the event that too many cells accompanied your collection of supernatant pseudovirus spin the collection at 170 × *g* for 3–5 min to pellet the cells and debris before filtering.
13. Pseudovirus can be used directly as supernatant collections. Preparations can also be concentrated by overnight centrifugation in a high speed centrifuge (at least 16 h at 5400 × *g* at 4 °C) and/or concentrated and purified via ultracentrifugation (80,000 × *g* for 2 h at 4 °C). The method will depend on your application of interest.
14. The ultracentrifugation protocol is based on using the Beckman Coulter Optima™ L-90K (Class S) model using rotors SW60Ti or SW32Ti. If using alternative equipment, modifications should be made accordingly.
15. While we find that titering the viral stock based on flow cytometric analysis of transduced cells provides the most accurate titer determination, an alternative titering approach is to serially dilute your stock in an end point dilution assay in a 96-well format. For these assays, stocks are usually diluted in three- or fivefold serial dilutions and as many as eight replicates are assessed at each dilution. End point dilution titers are evaluated at day 5 after infection by assessing GFP expression in each well using an inverted fluorescent microscope.
16. Primary antibodies or antisera used for the dot blot normalization assays must not react with nonspecific background proteins. In a similar manner, use of affinity purified secondary antibodies will prevent detection of irrelevant background proteins.

Acknowledgments

This work was supported by the National Institutes of Health R21 AI123616 (W.M.).

References

1. Feldmann H, Klenk HD, Sanchez A (1993) Molecular biology and evolution of filoviruses. Arch Virol Suppl 7:81–100
2. Peters CJ, Sanchez A, Rollin PE, Ksiazek TG, Murphy FA (1996) Filoviruses. Fields virology. Lippincott-Raven, Philadelphia, PA
3. Schnell MJ, Buonocore L, Kretzschmar E, Johnson E, Rose JK (1996) Foreign glycoproteins expressed from recombinant vesicular stomatitis viruses are incorporated efficiently into virus particles. Proc Natl Acad Sci U S A 93(21):11359–11365
4. Takada A, Robison C, Goto H, Sanchez A, Murti KG, Whitt MA, Kawaoka Y (1997) A system for functional analysis of Ebola virus glycoprotein. Proc Natl Acad Sci U S A 94(26):14764–14769
5. Ito H, Watanabe S, Sanchez A, Whitt MA, Kawaoka Y (1999) Mutational analysis of the putative fusion domain of Ebola virus glycoprotein. J Virol 73(10):8907–8912
6. Ito H, Watanabe S, Takada A, Kawaoka Y (2001) Ebola virus glycoprotein: proteolytic processing, acylation, cell tropism, and detection of neutralizing antibodies. J Virol 75(3):1576–1580. doi:10.1128/jvi.75.3.1576-1580.2001
7. Zavada J (1982) The pseudotypic paradox. J Gen Virol 63(Pt 1):15–24. doi:10.1099/0022-1317-63-1-15
8. Suda Y, Fukushi S, Tani H, Murakami S, Saijo M, Horimoto T, Shimojima M (2016) Analysis of the entry mechanism of Crimean-Congo hemorrhagic fever virus, using a vesicular stomatitis virus pseudotyping system. Arch Virol. doi:10.1007/s00705-016-2803-1
9. Shtanko O, Nikitina RA, Altuntas CZ, Chepurnov AA, Davey RA (2014) Crimean-Congo hemorrhagic fever virus entry into host cells occurs through the multivesicular body and requires ESCRT regulators. PLoS Pathog 10(9):e1004390. doi:10.1371/journal.ppat.1004390
10. Fukushi S, Tani H, Yoshikawa T, Saijo M, Morikawa S (2012) Serological assays based on recombinant viral proteins for the diagnosis of arenavirus hemorrhagic fevers. Virus 4(10):2097–2114. doi:10.3390/v4102097
11. Ray N, Whidby J, Stewart S, Hooper JW, Bertolotti-Ciarlet A (2010) Study of Andes virus entry and neutralization using a pseudovirion system. J Virol Methods 163(2):416–423. doi:10.1016/j.jviromet.2009.11.004
12. Moller-Tank S, Kondratowicz AS, Davey RA, Rennert PD, Maury W (2013) Role of the phosphatidylserine receptor TIM-1 in enveloped-virus entry. J Virol 87(15):8327–8341. doi:10.1128/JVI.01025-13
13. Tamin A, Harcourt BH, Lo MK, Roth JA, Wolf MC, Lee B, Weingartl H, Audonnet JC, Bellini WJ, Rota PA (2009) Development of a neutralization assay for Nipah virus using pseudotype particles. J Virol Methods 160(1–2):1–6. doi:10.1016/j.jviromet.2009.02.025
14. Ogino M, Ebihara H, Lee BH, Araki K, Lundkvist A, Kawaoka Y, Yoshimatsu K, Arikawa J (2003) Use of vesicular stomatitis virus pseudotypes bearing hantaan or seoul virus envelope proteins in a rapid and safe neutralization test. Clin Diagn Lab Immunol 10(1):154–160
15. Tani H, Iha K, Shimojima M, Fukushi S, Taniguchi S, Yoshikawa T, Kawaoka Y, Nakasone N, Ninomiya H, Saijo M, Morikawa S (2014) Analysis of Lujo virus cell entry using pseudotype vesicular stomatitis virus. J Virol 88(13):7317–7330. doi:10.1128/JVI.00512-14
16. Lawson ND, Stillman EA, Whitt MA, Rose JK (1995) Recombinant vesicular stomatitis viruses from DNA. Proc Natl Acad Sci U S A 92(10):4477–4481

Chapter 5

Lentiviral Vectors Pseudotyped with Filoviral Glycoproteins

Patrick L. Sinn, Jeremy E. Coffin, Natarajan Ayithan, Kathleen H. Holt, and Wendy Maury

Abstract

Pseudotyping lentivirus-based vectors is a strategy used to study conferred vector tropism and mechanisms of envelope glycoprotein function. Lentiviruses and filoviruses both assemble at the plasma membrane and have homotrimeric structural envelope glycoproteins that mediate both receptor binding and fusion. Such similarities help foster efficient pseudotyping. Importantly, filovirus glycoprotein pseudotyping of lentiviral vectors allows investigators to study virus entry at substantially less restrictive levels of biosafety containment than that required for wild-type filovirus work (biosafety level-2 vs. biosafety level-4, respectively). Standard lentiviral vector production involves transient transfection of viral component expression plasmids into producer cells, supernatant collection, and centrifuge concentration. Because the envelope glycoprotein expression plasmid is provided *in trans*, wild type or variant filoviral glycoproteins from marburgvirus or ebolavirus species may be used for pseudotyping and compared side-by-side. In this chapter we discuss the manufacture of pseudotyped lentiviral vector with an emphasis on small-scale laboratory grade production.

Key words Viral vector, Pseudotyping, Production, Purification, Titering, Lentivirus, Filovirus, Ebolavirus, Marburgvirus, Glycoprotein

1 Introduction

Several features of lentiviral vectors make them useful tools for basic science and preclinical research applications, including their large packaging capacity, efficient gene transfer capabilities, and persistent transgene expression. Lentiviral vectors also integrate into both dividing and nondividing cells, expanding the range of cells that can be targeted [1]. Pseudotyping is the act of replacing the native envelope protein with glycoproteins from other enveloped viruses [2, 3]. By pseudotyping lentiviral vectors, a wide range of cell types may be transduced at a biosafety level of containment available to most laboratories (*see* **Note 1**). The vesicular stomatitis virus glycoprotein (VSV-G) is by far the most common envelope glycoprotein used for lentiviral pseudotyping; however, VSV-G may not be a suitable choice for many scientific questions.

Thomas Hoenen and Allison Groseth (eds.), *Ebolaviruses: Methods and Protocols*, Methods in Molecular Biology, vol. 1628, DOI 10.1007/978-1-4939-7116-9_5, © Springer Science+Business Media LLC 2017

A wide variety of heterologous viral envelope glycoproteins have successfully been used to pseudotype lentiviral vectors; including those from vesiculoviruses, lyssaviruses [4, 5], arenaviruses [6, 7], hepadnaviruses [8], flaviviruses [9], paramyxoviruses [10], orthomyxoviruses [11], baculovirus [12, 13], alphaviruses [14, 15], and filoviruses [11, 16–18].

Pseudotyping lentivirus is a proven strategy to study filovirus receptor binding, entry, and endosomal escape outside of BSL-4 containment [19–21]. Pseudotyping lentivirus with filoviral glycoproteins effectively directs the tropism to the central nervous system or the apical surface of airway epithelial cells [16, 17, 22, 23], providing evidence that the filovirus glycoprotein targets cells within these organs. Lentiviral vectors can be effectively pseudotyped with wild-type Ebola virus (EBOV) or Marburg virus (MARV) glycoproteins, which is of use to many filoviral-related research applications. However, if the goal is maximum pseudotyping efficiency and high titers, modifications to the glycoprotein peptide sequence have been shown to be beneficial [24, 25]. Pseudotyping efficiency may be improved by either a directed evolution approach [26] or simply truncating the glycoprotein C-terminal tail [27, 28]. Specifically for pseudotyping lentiviral vectors with the EBOV envelope glycoprotein, we and others observed that efficiency improved when the heavily O-glycosylated mucin domain was deleted [17, 24]. Presumably, streamlined post-translational processing results in better surface display of the glycoprotein, which leads to more efficient incorporation into the budding virions.

In this chapter we outline a simple method for producing an EBOV GP pseudotyped human immunodeficiency virus (HIV)-based lentiviral vector. The approach does not require exotic equipment or materials and can easily be adapted to other lentiviral vectors such as simian or feline immunodeficiency virus-based lentiviral vectors. Furthermore, these methods will easily translate to pseudotyping with other envelope glycoproteins, such as MARV GP, VSV-G and baculovirus GP64 [29]. Many commercially available titering kits for lentiviral vectors detect VSV-G, thus they are not suitable for titering filoviral pseudotyped lentiviral vectors. Here, we outline three alternate methods for vector titration.

2 Materials

2.1 Transfection Reagents

1. 2× HEPES buffered saline (HBS): For 1 L: 11.9 g HEPES, 16.4 g NaCl, 0.21 g Na_2HPO_4. Bring up to 1 L with sterile water. Bring pH to 7.1 with 1 N NaOH. Filter through a 0.22 μm bottle top filter. Aliquot and store at −20 °C.

2. 2.5 M $CaCl_2$: 187.3 g $CaCl_2$ dihydrate. Bring volume up to 500 mL with sterile water. Filter through a 0.22 μm bottle top filter. Store solution at 4 °C or aliquot and store at −20 °C.
3. Serum-free Dulbecco's Modified Eagle Medium (DMEM) with 1% penicillin/streptomycin (Pen/Strep).
4. DMEM with 2% Nu-Serum IV Culture Supplement (Corning) and 1% Pen/Strep.
5. DMEM with 10% FBS, 1% Pen/Strep for cell maintenance.
6. A filovirus glycoprotein-expressing plasmid and third generation lentiviral vector production plasmids (*see* **Note 2**).
7. Resuspension Lactose/PBS buffer (40 mg/mL): 4 g D-lactose monohydrate. Bring up to 100 mL with 1× PBS solution (Gibco). Filter through a 0.22 μm bottle top filter. Aliquot and store at 4 °C.

2.2 Cultured Cells

1. Producer Cells: HEK293FT–Human Embryonic Kidney Cells with T antigen (Invitrogen).
2. Titering Cells: HT1080 cells (ATCC CCL-121) are derived from a human fibrosarcoma and are typically highly permissive to lentiviral transduction. Thus, this cell line is useful for titration of lentiviral particles. Vero cells are commonly used for titering many viruses but should be avoided for titering HIV-based vector due to potential simian TRIM5α-mediated restriction [30].

2.3 Flow Cytometry Titering Reagents

1. DMEM with 10% FBS, 1% Pen/Strep.
2. DMEM with 2% FBS, 1% Pen/Strep.
3. 8 mg/mL Polybrene (hexadimethrine bromide) stock (Sigma), *optional* (long-term storage at −20 °C).
4. 5 mL polystyrene tubes with caps, 12 × 75 mm (e.g., Evergreen Scientific) or BD Falcon 5 mL polystyrene round-bottom tubes, 12 × 77 mm (e.g., BD Biosciences).
5. Accumax cell detachment solution (Millipore) (storage at 4 °C with a recommended shelf life of 6 weeks, or the product can be frozen in small aliquots and stored at −20 °C indefinitely).
6. Falcon nylon cell strainer, 70 μm.
7. Propidium iodide (PI): 50 μg/mL in 1× PBS (store at 4 °C, protect from light).

2.4 Real-Time PCR Titering Reagents

1. 12-well tissue culture plate.
2. 1.5 mL microcentrifuge tubes, sterile.
3. DMEM (e.g., Cellgro) with 2% FBS (e.g., Atlas) and 1% Pen/Strep (e.g., Cellgro).

4. Positive control—Lentivirus expressing eGFP (HIV-eGFP) with a known titer.
5. 8 mg/mL Polybrene (hexadimethrine bromide) stock (Sigma)—*optional.*
6. 10 mg/mL RNase A stock.
7. Genomic DNA Purification Kit (Promega).
8. Isopropanol.
9. 70% Ethanol.
10. RNase/DNase Free Water.
11. MicroAmp Optical 384-well Reaction Plate with Barcode (ABI).
12. MicroAmp Optical Adhesive Film (ABI).
13. TaqMan Universal Master Mix II, No UNG (ABI).
14. RNase/DNase-free water.
15. Primers and probe: HIVforward 5′-CGA CTG GTG AGT ACG CCA AA-3′; HIVreverse 5′-CGC ACC CAT CTC TCTCCT TCT-3′; HIVprobe 5′–/FAM/ATT TTG ACT AGC GGA GGC/Black Hole/–3′ (*see* **Note 3**).

2.5 p24 Assay

1. MagPlex Microspheres Bead Region 42 (Luminex Corporation).
2. p24 monoclonal antibody (ImmunoDiagnostics).
3. 0.1 M monobasic sodium phosphate (pH 6.2).
4. Sulfo-*N*-hydroxysuccinimide (Sulfo-NHS, 50 mg/mL).
5. 1-ethyl-3-[3-dimethylaminopropyl] carbodiimide Hydrochloride (EDC, 50 mg/mL).
6. 1× PBS (pH 7.4).
7. Microsphere Resuspension Buffer: 1× PBS (pH 7.4), 0.1% BSA and 0.02% Tween-20.
8. Luminex Wash Buffer: 1× PBS with 0.01% Tween-20 and 2% 1 M Tris–HCl (pH 8).
9. Normal Goat Serum (e.g., Thermo Fisher Scientific).
10. Normal Mouse Serum (e.g., Thermo Fisher Scientific).
11. Luminex Assay Buffer: (Luminex Wash buffer + 1% Normal Goat Serum + 1% Normal Mouse Serum).
12. HIV-1 p24 IIIB (Baculo) standard (NIH AIDS Reagent Program).
13. RD1-labelled anti-p24 KC57 detectionantibody (Beckman Coulter).
14. 96-Well Black Flat Bottom Microplate (e.g., Greiner Bio-One).

15. Microplate Sealer (e.g., Fisher Scientific).
16. Gyrotory Shaker G2 (New Brunswick), or similar.
17. Bio-Plex Pro II Wash Station (Bio-Rad).
18. Bio-Plex 200 System (Bio-Rad).

2.6 Equipment

1. 500 mL centrifuge bottles (e.g., Beckman Coulter).
2. Avanti J-25 centrifuge with JA-10 rotor (Beckmann Coulter), or similar.
3. 0.5 mL sterile microfuge tubes.
4. Biological Safety Cabinet (*see* **Note 1**).
5. BioPlex Multiplex System.

3 Methods

3.1 Lentiviral Vector Production by Four-Plasmid Transfection

1. Split HEK 293FT cells 1–4 days prior to transfecting cells in 150 mm dishes with DMEM (10% FBS) medium (*see* **Note 4**). Seed 1.3×10^7 cells per 150 mm plate the day prior to transfection, 2.5×10^6 cells per 150 mm plate 3 days prior to transfection, or 1.5×10^6 cells per 150 mm plate 4 days prior to transfection. Incubate the cells at 37 °C until ready for use.
2. Add the appropriate volume of RNase/DNase-free sterile H_2O (Table 1) to a 50 mL conical tube. Note that for all values in Table 1, it is good practice to multiple by an additional 10% to account for pipetting error.

Table 1
Reagent amounts for transfection of producer cells to generate pseudotyped lentiviral particles

		Amount per 150 mm plate	Amount for n 150 mm plate
TUBE 1:	2× HBS	2000 μL	2000 μL × n
TUBE 2:	Viral glycoprotein plasmid	7.5 μg	7.5 μg × n
	pLP1 (expresses gag/pol)	22.5 μg	22.5 μg × n
	pLP2 (expresses rev)	7.5 μg	7.5 μg × n
	Plasmid containing transgene-of-interest within a packagable lentiviral genome	22.5 μg	22.5 μg × n
	2.5 M $CaCl_2$	200 μL	200 μL × n
	Sterile water	Bring to 2000 μL	Bring to 2000 μL × n

3. Add plasmids to H_2O tube (*see* **Note 2**).
4. Add room temperature 2.5 M $CaCl_2$ to H_2O/Plasmid tube and mix by inversion.
5. Combine tubes 1 and 2 by adding the required amount of 2× HBS to Tube 2 (plasmid + H_2O + $CaCl_2$ mixture) in a dropwise manner and mix gently by inversion. After the addition of the HBS solution, the DNA should visibly precipitate (the solution will appear cloudy).
6. Add the DNA–calcium phosphate solution to serum-free DMEM (15 mL/plate) and mix well by inversion.
7. Aspirate the culture medium from the HEK 293FT (producer) cells.
8. Gently add 15 mL of the DMEM–calcium phosphate solution to each 150 mm plate without disturbing the cell layer.
9. Incubate plates at 37 °C with 5% CO_2 for a minimum of 4 h or a maximum of 6 h.
10. Remove the transfection medium and replace with 15 mL of DMEM with 2% Nu-Serum, 1% Pen/Strep.
11. At 24 h post-transfection, collect the supernatant and store at 4 °C. Replace 15 mL of fresh medium to each plate.
12. Repeat the collection at 40 h post-transfection and again at 64 h post-transfection.
13. Combine the collected supernatants and pass the entire collected volume through a 0.22 μm bottle top filter.
14. Viral supernatant can be temporarily stored at 4 °C for up to 1 week or for longer periods of time at −80 °C until ready to concentrate. Care should be taken to limit samples to a single freeze/thaw.

3.2 Low-Speed Centrifuge Concentration

1. If necessary, thaw vector supernatant at room temperature or at 4 °C.
2. In a biological safety cabinet, transfer vector supernatant into sterile 500 mL centrifuge bottles.
3. Spin the supernatant at 9000 × *g* overnight at 4 °C.
4. The next day, remove the bottles from the centrifuge and visually confirm there is a pellet before removing the medium.
5. In a biological safety cabinet, aspirate medium without disturbing the pellet.
6. Invert the bottle onto a clean paper napkin and let remaining medium drain.
7. Aspirate any additional medium attached to the bottle walls and around pellet to avoid including any of the medium with your resuspension.

8. Resuspend the pellet in 40 mg/mL of Lactose/PBS buffer (120 μL/plate).
9. Store the concentrated lentiviral vector at 4 °C for a short period (30 min to 1 h) to allow the pellet to completely dissolve in the buffer.
10. Transfer resuspended lentiviral vector to a 0.5 mL microcentrifuge tube.
11. Invert the tube a few times to mix and then spin for 10 s at maximum speed in a benchtop microcentrifuge (~15,000–20,000 × *g*) to remove any protein debris.
12. Dispense 100 μL aliquots of lentiviral vector into sterile 0.5 mL tubes. Also, dispense at least one 50 μL aliquot for titering. Store for short term at 4 °C, and for long term at −80 °C.
13. Centrifugation also concentrates cellular debris and culture medium components such as bovine serum albumin (BSA). For potential strategies to further purify laboratory grade lentiviral vectors, *see* **Note 5**.

3.3 Vector Titration Using Flow Cytometry

This method is only suitable for the analysis of lentiviral vectors containing fluorescent transgenes (GFP, mCherry, etc.).

1. Split HT1080 cells the day prior to infecting cells. Plate HT1080 cells in a 6-well plate at 7 × 10^5 cells/well with DMEM (10% FBS) medium. Incubate cells for ~24 h at 37 °C. The cells should be ~80–90% confluent (or ~2.0 × 10^6 cells/well) before proceeding.
2. Prepare DMEM (2% FBS) with optional 1 μL/mL Polybrene. Mix well.
3. Perform serial dilutions of the concentrated lentiviral vector (100-fold to 10,000-fold) in DMEM with 2% FBS–Polybrene in 2.0 mL microcentrifuge tubes. Mix thoroughly between tubes.
4. Aspirate medium and plate 1 mL of lentiviral vector dilutions. You must be very gentle to avoid disrupting the cells. To negative control wells, add DMEM (2% FCS) medium only.
5. Use one or two additional control wells to trypsinize, harvest and count the number of cells. The 6-well plates have a surface area of approximately 951 mm^2 per well, although this may vary slightly depending on the supplier, and you should end up with ~2300 cells/mm^2. A visual screening and evaluation of the cell layer can be very deceptive.
6. Incubate cells for ~65 h at 37 °C.
7. Label 5 mL polystyrene tubes and add 20 μL per tube of propidium iodide.

8. Aspirate medium from the wells and add 0.5–1.0 mL of Accumax per well.
9. Pipet dislodged cells into tubes only immediately prior to analysis. This prevents the formation of clumps that will clog the machine. Vortexing should be avoided if possible because it causes cell lysis. Cell strainers are preferable.
10. Perform flow cytometry for the fluorescent reporter transgene using the appropriate filters and according to the manufacturer's instructions.
11. The titer in Transducing Units (TU)/mL is calculated as (% fluorescence positive cells/100) × (dilution factor) × total cell number/well.

3.4 Vector Titration Using Quantitative Real-Time PCR

This method is appropriate for the analysis of HIV-based lentiviral vectors containing any type of transgene (fluorescent, luminescent, enzymatic, etc.) as long as their sequence contains the sequences corresponding to the primer and probe sequences specified in Subheading 2.4, **item 15**.

1. Seed HT1080 (titering cells) on a 12 well plate at a density of 2.5×10^5 cells per well. Incubate at 37 °C with 5% CO_2 overnight.
2. The next day, add 1 μL of Polybrene per 1 mL of DMEM, for a final concentration of 8 μg/mL. Addition of Polybrene can be considered optional but may result in more reproducible titers.
3. Make serial dilutions (100- and 1000-fold) for each sample; include an HIV-eGFP control and leave one well of cells untransduced (i.e., DMEM without lentivirus) as a negative control.
4. Apply 500 μL of each dilution per well.
5. Incubate at 37 °C with 5% CO_2 for 3 days.
6. Harvest cells and extract DNA using the Genomic DNA Purification Kit (Promega) according to the manufacturer's instructions.
7. Resuspend DNA in 100 μL of water and store at 4 °C.
8. Prepare lentivirus vector samples in triplicate on a 384-well plate.
9. Prepare a linearized plasmid standard curve ranging from 10^{10} to 10^5 copies/μL.
10. Dilute primers and probe to 100 pmol/μL (*see* sequences in Subheading 2.4, **item 15** above).
11. Make a PCR master mix using TaqMan Universal Master Mix II (Table 2) and aliquot 6.4 μL in each well. Note that for all values

Table 2
Preparation of quantitative real-time PCR reactions for lentiviral RNAdetection

	1 Reaction (μL)	Amount for n reactions (μL × n)
Water	1.16	1.16
TaqMan Universal Master Mix II	5.0	5.0
100 μM probe	0.08	0.08
100 μM Primer 1	0.08	0.08
100 μM Primer 2	0.08	0.08
Total volume	6.4	6.4

in Table 2, it is good practice to multiple by an additional 10% to account for pipetting error.

12. Add 3.6 μL of sample per well or 3.6 μL water for the *No Template Control*.
13. Spin plate briefly to pull down samples.
14. Run the samples using the following program: (1) 50 °C 2 min, (2) 95 °C 10 min, (3) 95 °C 15 s, (4) 60 °C 1 min, (5) repeat **steps 3** and **4** for 35 cycles.
15. Proceed with data analysis and titer calculations (*see* **Note 6**).

3.5 Vector Quantitation Using a Bioplex-Based p24 ELISA

This method is appropriate for the analysis of HIV-based lentiviral vectors containing any type of transgene (fluorescent, luminescent, enzymatic, etc.).

1. Prior to performing a HIV p24 (capsid) detection assay, magnetic beads region 42 (MagPlex Carboxymethylated Microspheres) should be coupled with anti-p24 mAb: Pellet magnetic microsphere beads (1.25×10^7) in a low protein-binding microcentrifuge tube by spinning them at 8000 × *g* for 2 min, and resuspend them in 160 μL of 0.1 M monobasic sodium phosphate (pH 6.2).
2. Activate beads by the addition of 20 μL of Sulfo-NHS and 20 μL of EDC (freshly prepared in water) for 20 min at RT.
3. Wash the beads twice in 1 mL of 1× PBS (pH 7.4) by centrifugation at 8000 × *g* for 2 min. Resuspend beads in 250 μL 1× PBS (pH 7.4).
4. Add 100 μL of anti-p24 mAb (1 mg/mL) and bring to a final volume to 1 mL with 1× PBS (pH 7.4).
5. Incubate for 2 h at room temperature with gentle mixing on a Gyrotory shaker G2 or thermomixer.

6. Wash the beads twice in 1× PBS and resuspend in 1 mL of Microsphere Resuspension Buffer.
7. Determine the bead concentration by counting on an automated cell counter or a hemocytometer and store at 4 °C for p24 assay.
8. Prepare Luminex Wash Buffer as described in Subheading 2.5, **item 8**.
9. Prepare Luminex Assay Buffer as described in Subheading 2.5, **item 11**.
10. Mix anti-p24 mAb coupled beads in Luminex assay buffer at a concentration of ~1200 beads/sample. Beads can be counted on a hemocytometer or an automated cell counter.
11. Add 50 μL of bead suspension (~1200 beads) in duplicate well for each sample in a 96-well black flat bottom microplate.
12. Prepare p24 standards (1.5–10,000 pg/mL, threefold dilutions) and dilute samples in Luminex Buffer. Also include a negative control where all reagents are added except the antigen.
13. Add 50 μL p24 standards and samples/well in duplicates and gently mix the plate containing the magnetic beads at the bottom.
14. Seal the 96-well plate with adhesive film and cover the plates with aluminum foil.
15. Shake the plate (~50 rpm) at room temperature for 2 h.
16. Wash the plate three to four times with Luminex wash buffer in a 96-well magnetic plate washer with 2 min/wash.
17. Dilute the detection antibody (RD1-labelled anti-p24 KC57) in Luminex Assay Buffer to a concentration of 0.5 μg/mL.
18. Add 100 μL of detection anti-p24 antibody to the magnetic beads and gently mix the plate.
19. Reseal the 96-well plate with adhesive film and cover the plates with aluminum foil.
20. Shake the plate (~50 rpm) at room temperature for 1 h.
21. Repeat the wash step (same as **step 10** above).
22. Add 60 μL Luminex Assay Buffer to each well and gently mix without introducing air bubbles.
23. The beads are analyzed for region of difference one (RD1) using the Bio-Plex Multiplex Systems instrument (controlled by the Bioplex manager software 4.11). Assay results should be based on at least 50 beads/sample in order to obtain reproducible data.
24. The amount of HIV-p24 levels is calculated by the Luminex System itself which is directly proportional to the fluorescence intensity.

4 Notes

1. All work with lentiviral vectors should be performed inside a certified Class II, biosafety cabinet. Consult the local Institutional Biosafety Committee (IBC) for information regarding additional regulations. At most institutions, biosafety level-2 containment is sufficient.
2. Plasmid transfection of one 150 mm plate is performed by calcium phosphate precipitation of 7.5 μg viral glycoprotein (envelope) plasmid (i.e., EBOV GP), 22.5 μg gag/pol packaging plasmid (i.e., LP1), 7.5 μg rev plasmid (i.e., LP2), and 22.5 μg lentiviral vector genome plasmid containing your transgene of choice (e.g., HIV CMV eGFP). We found that a 1:3:1:3 ratio yielded optimal titers; however, results may vary. Pure, endotoxin free plasmids typically yield the best vector product. Mammalian expression vectors containing the glycoprotein of the species *Zaire ebolavirus* or the Marburg virus Musoke isolate glycoprotein are available from BEI Resources (www.beiresources.org) (Catalog #s:NR-19814 and NR-19815, respectively). Multiple HIV vectors are available through addgene.com, which carry reporter genes such as mCherry, GFP, and firefly luciferase. Empty vectors allow for construction of vectors carrying other genes of interest if desired.
3. These primer and probe sequences listed here are commonly used with standard lentiviral vector constructs. Verify that these are complementary to your lentiviral vector before use.
4. For consistent results, ensure that producer cells are at a low passage number (<27) and are not over confluent.
5. Adverse inflammatory and immunogenic reactions resulting from in vivo delivery of lentiviral vectors are largely attributable to these components in the centrifuge-concentrated vector preparations [31]. Clearly, the removal of unnecessary foreign proteins from vector preparations is an important step in improving the safety and efficacy of in vivo gene transfer. Several methods have been proposed to partially purify pseudotyped lentiviral preparations. A strong anion exchange HPLC column was effective in purifying functional vector particles, resulting in decreased cell toxicity [32]. Weak anion exchange hollow fiber has also been used to purify pseudotyped HIV vector [33]. Mustang Q strong anion exchange capsules have been used in the purification of large-scale viral preparations [34]. These processes are based on the known electrostatic charge of the vector envelope. There are several alternative approaches for depleting stocks of contaminants. Stocks can be ultracentrifuged through a 20% sucrose (in 1× PBS)

cushion (80,000 × *g* for 2 h in 4 °C). Alternately, size exclusion chromatography has also been used to purify large amounts of pseudotyped lentiviral particles [35]. Tangential flow filtration is also an effective and widely used technique to purify lentiviral preparations [36]; however, the equipment and materials can be cost prohibitive for small-scale laboratory grade applications.

6. The RNA copy number (*x*-axis) and Cycle Threshold value (C_T; *y*-axis) are used to create a standard curve and the formula for a semilogarithmic curve ($y = be^{mx}$) is generated. The sample C_T values are fitted to the curve to determine the HIV RNA copy number. Each sample is run in triplicate and averaged together. Each sample of concentrated vector was infected at 100- and 1000-fold dilutions; therefore the calculated copy number of the appropriate sample is multiplied by 100 or 1000, respectively. Because 500 μL of each dilution was added per well, the copy number is multiplied by 2 to convert it to RNA copy number per milliliter. The values obtained for RNA copy number from each dilution are averaged together to determine the real-time PCR titer.

   Each plate is run with a sample of genomic DNA from HT1080 cells infected with HIV-eGFP concentrated vectors. The flow cytometry titer of HIV-eGFP is determined by FACS (as described in Subheading 3.3) and the real time PCR titers (as determined as described in Subheading 3.5) are normalized to the flow cytometry titer using the following formula:

$$\left(\text{control } HIV\text{-eGFP titer}\left(\text{flow cytometry}\right) / \text{control } HIV\text{-eGFP titer}\left(\text{real-time}\right)\right) \times \text{Sample titer}\left(\text{real-time}\right) = \text{Normalized sample titer}\left(TU / mL\right)$$

Acknowledgments

This work was supported by the National Institutes of Health R01 HL105821 (PLS) and R21 AI123616 (WM). The University of Iowa Viral Vector Core is partially supported by the Cystic Fibrosis Foundation and the National Institutes of Health: P30 DK054759, P01 HL051670, and P30 CA086862.

References

1. Naldini L, Blomer U, Gallay P, Ory D, Mulligan R, Gage FH, Verma IM, Trono D (1996) In vivo gene delivery and stable transduction of nondividing cells by a lentiviral vector. Science 272:263–267
2. Sanders DA (2002) No false start for novel pseudotyped vectors. Curr Opin Biotechnol 13(5):437–442
3. Cronin J, Zhang XY, Reiser J (2005) Altering the tropism of lentiviral vectors through pseudotyping. Curr Gene Ther 5(4):387–398
4. Sena-Esteves M, Tebbets JC, Steffens S, Crombleholme T, Flake AW (2004) Optimized large-scale production of high titer lentivirus vector pseudotypes. J Virol Methods 122(2): 131–139. doi:10.1016/j.jviromet.2004.08.017

5. Hislop JN, Islam TA, Eleftheriadou I, Carpentier DC, Trabalza A, Parkinson M, Schiavo G, Mazarakis ND (2014) Rabies virus envelope glycoprotein targets lentiviral vectors to the axonal retrograde pathway in motor neurons. J Biol Chem 289(23):16148–16163. doi:10.1074/jbc.M114.549980
6. Zhang C, Hu B, Xiao L, Liu Y, Wang P (2014) Pseudotyping lentiviral vectors with lymphocytic choriomeningitis virus glycoproteins for transduction of dendritic cells and in vivo immunization. Hum Gene Ther Methods 25(6):328–338. doi:10.1089/hgtb.2014.105
7. Beyer WR, Westphal M, Ostertag W, von Laer D (2002) Oncoretrovirus and lentivirus vectors pseudotyped with lymphocytic choriomeningitis virus glycoprotein: generation, concentration, and broad host range. J Virol 76(3):1488–1495
8. Sung VM, Lai MM (2002) Murine retroviral pseudotype virus containing hepatitis B virus large and small surface antigens confers specific tropism for primary human hepatocytes: a potential liver-specific targeting system. J Virol 76(2):912–917
9. Hsu M, Zhang J, Flint M, Logvinoff C, Cheng-Mayer C, Rice CM, McKeating JA (2003) Hepatitis C virus glycoproteins mediate pH-dependent cell entry of pseudotyped retroviral particles. Proc Natl Acad Sci U S A 100(12): 7271–7276. doi:10.1073/pnas.0832180100
10. Funke S, Schneider IC, Glaser S, Muhlebach MD, Moritz T, Cattaneo R, Cichutek K, Buchholz CJ (2009) Pseudotyping lentiviral vectors with the wild-type measles virus glycoproteins improves titer and selectivity. Gene Ther 16(5):700–705. doi:10.1038/gt.2009.11
11. Cheng H, Koning K, O'Hearn A, Wang M, Rumschlag-Booms E, Varhegyi E, Rong L (2015) A parallel genome-wide RNAi screening strategy to identify host proteins important for entry of Marburg virus and H5N1 influenza virus. Virol J 12:194. doi:10.1186/s12985-015-0420-3
12. Kumar M, Bradow BP, Zimmerberg J (2003) Large-scale production of pseudotyped lentiviral vectors using baculovirus gp64. Hum Gene Ther 14:67–77
13. Sinn PL, Cooney AL, Oakland M, Dylla DE, Wallen TJ, Pezzulo AA, Chang EH, McCray PB Jr (2012) Lentiviral vector gene transfer to porcine airways. Mol Ther Nucleic Acids 1:e56. doi:10.1038/mtna.2012.47
14. Kang Y, Stein CS, Heth JA, Sinn PL, Penisten AK, Staber PD, Ratliff KL, Shen H, Barker CK, Martins I, Sharkey CM, Sanders DA, McCray PB Jr, Davidson BL (2002) In vivo gene transfer using a nonprimate lentiviral vector pseudotyped with ross river virus glycoproteins. J Virol 76:9378–9388
15. Kolokoltsov AA, Weaver SC, Davey RA (2005) Efficient functional pseudotyping of oncoretroviral and lentiviral vectors by Venezuelan equine encephalitis virus envelope proteins. J Virol 79(2):756–763. doi:10.1128/JVI.79.2.756-763.2005
16. Kobinger GP, Weiner DJ, Yu QC, Wilson JM (2001) Filovirus-pseudotyped lentiviral vector can efficiently and stably transduce airway epithelia in vivo. Nat Biotechnol 19(3):225–230. doi:10.1038/85664
17. Sinn PL, Hickey MA, Staber PD, Dylla DE, Jeffers SA, Davidson BL, Sanders DA, McCray PB Jr (2003) Lentivirus vectors pseudotyped with filoviral envelope glycoproteins transduce airway epithelia from the apical surface independently of folate receptor alpha. J Virol 77(10):5902–5910
18. Limberis MP, Bell CL, Heath J, Wilson JM (2010) Activation of transgene-specific T cells following lentivirus-mediated gene delivery to mouse lung. Mol Ther 18(1):143–150. doi:10.1038/mt.2009.190
19. Kondratowicz AS, Lennemann NJ, Sinn PL, Davey RA, Hunt CL, Moller-Tank S, Meyerholz DK, Rennert P, Mullins RF, Brindley M, Sandersfeld LM, Quinn K, Weller M, McCray PB Jr, Chiorini J, Maury W (2011) T-cell immunoglobulin and mucin domain 1 (TIM-1) is a receptor for Zaire Ebolavirus and Lake Victoria Marburgvirus. Proc Natl Acad Sci U S A 108(20):8426–8431. doi:10.1073/pnas.1019030108
20. Brindley MA, Hunt CL, Kondratowicz AS, Bowman J, Sinn PL, McCray PB Jr, Quinn K, Weller ML, Chiorini JA, Maury W (2011) Tyrosine kinase receptor Axl enhances entry of Zaire ebolavirus without direct interactions with the viral glycoprotein. Virology 415(2): 83–94. doi:10.1016/j.virol.2011.04.002
21. Alvarez CP, Lasala F, Carrillo J, Muniz O, Corbi AL, Delgado R (2002) C-type lectins DC-SIGN and L-SIGN mediate cellular entry by Ebola virus in cis and in trans. J Virol 76(13):6841–6844
22. Watson DJ, Kobinger GP, Passini MA, Wilson JM, Wolfe JH (2002) Targeted transduction patterns in the mouse brain by lentivirus vectors pseudotyped with VSV, Ebola, Mokola, LCMV, or MuLV envelope proteins. Mol Ther 5(5 Pt 1):528–537. doi:10.1006/mthe.2002.0584
23. Medina MF, Kobinger GP, Rux J, Gasmi M, Looney DJ, Bates P, Wilson JM (2003)

Lentiviral vectors pseudotyped with minimal filovirus envelopes increased gene transfer in murine lung. Mol Ther 8(5):777–789

24. Jeffers SA, Sanders DA, Sanchez A (2002) Covalent modifications of the ebola virus glycoprotein. J Virol 76(24):12463–12472
25. Sandrin V, Boson B, Salmon P, Gay W, Negre D, Le Grand R, Trono D, Cosset FL (2002) Lentiviral vectors pseudotyped with a modified RD114 envelope glycoprotein show increased stability in sera and augmented transduction of primary lymphocytes and CD34+ cells derived from human and nonhuman primates. Blood 100:823–832
26. Merten CA, Stitz J, Braun G, Poeschla EM, Cichutek K, Buchholz CJ (2005) Directed evolution of retrovirus envelope protein cytoplasmic tails guided by functional incorporation into lentivirus particles. J Virol 79(2):834–840. doi:10.1128/JVI.79.2.834-840.2005
27. Mammano F, Salvatori F, Indraccolo S, De Rossi A, Chieco-Bianchi L, Gottlinger HG (1997) Truncation of the human immunodeficiency virus type 1 envelope glycoprotein allows efficient pseudotyping of Moloney murine leukemia virus particles and gene transfer into CD4+ cells. J Virol 71(4):3341–3345
28. Trabalza A, Eleftheriadou I, Sgourou A, Liao TY, Patsali P, Lee H, Mazarakis ND (2014) Enhanced central nervous system transduction with lentiviral vectors pseudotyped with RVG/HIV-1gp41 chimeric envelope glycoproteins. J Virol 88(5):2877–2890. doi:10.1128/JVI.03376-13
29. Kang Y, Xia L, Tran DT, Stein CS, Hickey M, Davidson BL, McCray PB Jr (2005) Persistent expression of factor VIII in vivo following nonprimate letiviral gene transfer. Blood 106:1552–1558
30. Stremlau M, Owens CM, Perron MJ, Kiessling M, Autissier P, Sodroski J (2004) The cytoplasmic body component TRIM5alpha restricts HIV-1 infection in old world monkeys. Nature 427(6977):848–853. doi:10.1038/nature02343
31. Tuschong L, Soenen SL, Blaese RM, Candotti F, Muul LM (2002) Immune response to fetal calf serum by two adenosine deaminase-deficient patients after T cell gene therapy. Hum Gene Ther 13(13):1605–1610
32. Yamada K, McCarty DM, Madden VJ, Walsh CE (2003) Lentivirus vector purification using anion exchange HPLC leads to improved gene transfer. BioTechniques 34(5):1074–1078, 1080
33. Matsuoka H, Miyake K, Shimada T (1998) Improved methods of HIV vector mediated gene transfer. Int J Hematol 67(3):267–273
34. Slepushkin V, Chang N, Cohen R, Gan Y, Jiang B, Deausen E, Berlinger D, Binder G, Andre K, Humeau L, Dropulic B (2003) Large-scale purification of a lentiviral vector by size exclusion chromatography or mustang Q exchange capsule. Bioprocess J 2:89–95
35. Transfiguracion J, Jaalouk DE, Ghani K, Galipeau J, Kamen A (2003) Size-exclusion chromatography purification of high-titer vesicular stomatitis virus G glycoprotein-pseudotyped retrovectors for cell and gene therapy applications. Hum Gene Ther 14(12):1139–1153
36. Cooper AR, Patel S, Senadheera S, Plath K, Kohn DB, Hollis RP (2011) Highly efficient large-scale lentiviral vector concentration by tandem tangential flow filtration. J Virol Methods 177(1):1–9. doi:10.1016/j.jviromet.2011.06.019

Chapter 6

Modeling Ebola Virus Genome Replication and Transcription with Minigenome Systems

Tessa Cressey, Kristina Brauburger, and Elke Mühlberger

Abstract

In this chapter, we describe the minigenome system for Ebola virus (EBOV), which reconstitutes EBOV polymerase activity in cells and can be used to model viral genome replication and transcription. This protocol comprises all steps including cell culture, plasmid preparation, transfection, and luciferase reporter assay readout.

Key words *Zaire ebolavirus*, Ebola virus, Filoviruses, Minigenome, Replication, Transcription, Replication promoter, Reporter gene, Firefly luciferase assay, β-galactosidase enzyme assay

1 Introduction

Research on Ebola virus (EBOV) has been limited, in part, due to the necessity for working with this virus under the highest biosafety level conditions, BSL-4. In this regard, minigenome systems, such as the one developed for EBOV, are extremely useful, allowing researchers to study aspects of the EBOV replication cycle under BSL-2 conditions [1]. The nonsegmented negative-sense RNA genome of EBOV is about 19 kb in length and contains short promoter sequences at its 3′ and 5′ ends that are crucial *cis*-acting signals for genome replication and viral gene transcription [2]. Seven viral genes are arranged in a linear order on the EBOV genome, each flanked by a transcription start signal sequence (gene start) and a transcription termination/polyadenylation signal sequence (gene end) (reviewed in [3, 4]). A monocistronic minigenome is a shortened version of the viral genome which contains only a single gene (Fig. 1a) along with the 3′ and 5′ ends of the genome, which are required for replication and transcription. To construct a monocistronic minigenome, all viral genes are removed and replaced by a single (nonviral) reporter gene flanked by the virus-specific gene start and gene end sequences (reviewed in [5, 6]).

Thomas Hoenen and Allison Groseth (eds.), *Ebolaviruses: Methods and Protocols*, Methods in Molecular Biology, vol. 1628, DOI 10.1007/978-1-4939-7116-9_6, © Springer Science+Business Media LLC 2017

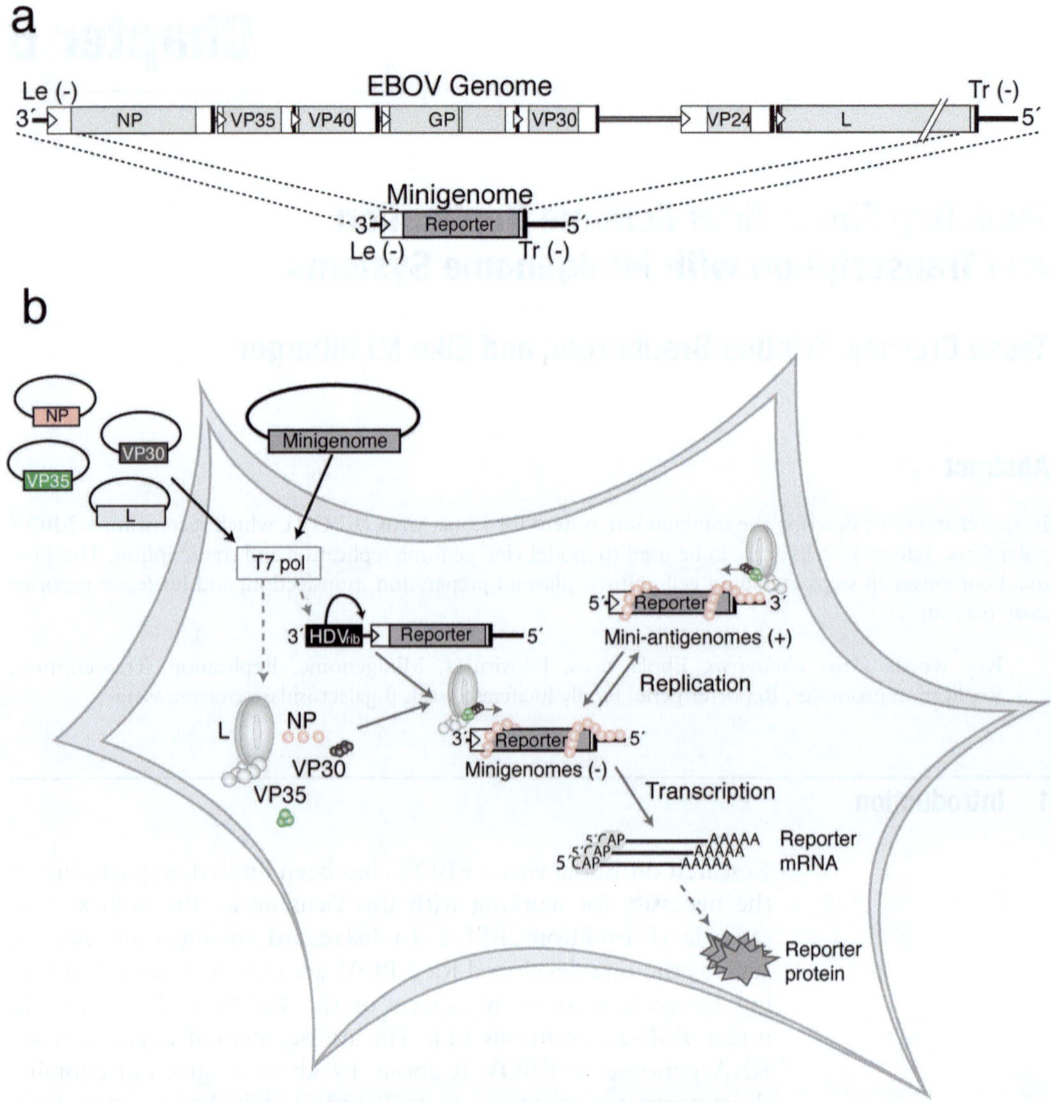

Fig. 1 Scheme of Ebola virus minigenome structure and the minigenome system. (**a**) Ebola virus (EBOV) genome (*top*) and minigenome (*bottom*) structure in negative-sense (3′–5′) orientation. Leader (Le) and trailer (Tr) sequences are shown as *black lines*. The open reading frames encoding the seven genes are shown in *light gray* and the untranslated regions are illustrated in *white*. Transcription start and stop signals that mark the beginning and end of each gene are depicted as *white triangles* or *black bars*, respectively. Intergenic regions that separate the different genes are shown as *dark gray lines*. At gene borders missing an intergenic region, the transcription signals of consecutive genes overlap directly (e.g., VP35/VP40). The editing site of the GP gene is shown as a *white bar*. The regions of the EBOV genome contained within the minigenome are emphasized by *dashed lines*. Most of the full-length genome is replaced by a reporter gene. (**b**) Scheme showing the minigenome system. Cells expressing the T7 RNA polymerase (T7 pol) are transfected with expression plasmids encoding the minigenome and the four EBOV proteins required for transcription and replication, NP, VP35, VP30 and L, under the control of the T7 RNA polymerase promoter. Transcription by the T7 RNA polymerase generates negative-sense minigenomes containing the minigenome sequence (as shown in (**a**)) and the hepatitis delta virus ribozyme sequence (HDV_{rib}) attached to the 3′ end of the minigenome. Once transcribed, the HDV_{rib} will autocatalytically cleave itself off, creating a defined 3′ minigenome end.

Thus, the minigenome contains all *cis*-acting signals necessary for the EBOV polymerase to direct minigenome replication and reporter gene transcription. Several reporter genes have been used in the EBOV minigenome system [7–12]. We focus on the usage of a monocistronic minigenome with firefly luciferase as a reporter gene [11]. In our system, the EBOV minigenome sequence, including the reporter gene, was inserted into the vector p2,0 [13] between the T7 RNA polymerase promoter and the hepatitis delta virus ribozyme sequence [1]. When cells constitutively expressing the T7 RNA polymerase are transfected with this plasmid, the negative-sense, single-stranded minigenome RNA will be transcribed by the T7 RNA polymerase. Precise 3′ minigenome ends (negative-sense orientation) are generated by the autocatalytic cleavage activity of the ribozyme [13]. Four EBOV proteins are then required for replication and transcription of the minigenome: NP, VP35, VP30, and L (reviewed in [3, 4]). In the system described here, these four viral genes are each inserted into separate plasmids (pTM1 backbone; [14]) and are co-transfected into cells alongside the plasmid encoding the minigenome. T7 RNA polymerase will transcribe the viral genes from the plasmids and cellular machinery will translate the mRNA, producing the viral proteins NP, VP35, VP30, and L. Like the RNA genome, the minigenome is encapsidated by the nucleoprotein, NP (reviewed in [3, 4, 15]). NP associates with the viral polymerase, which is formed by a complex of two proteins: the large (L) catalytic protein and its cofactor VP35 [1]. VP30 is a transcription enhancement factor [16, 17]. The viral polymerase complex recognizes the encapsidated minigenome as a template for replication and transcription. In order to replicate the minigenome, the viral polymerase initiates RNA synthesis at the 3′ end of the minigenome to produce the positive-sense, single-stranded replicative intermediate (mini-antigenome). The mini-antigenome is used as a template by the viral polymerase to synthesize additional copies of the negative-sense minigenome, thus modeling genome replication. These products can be detected by Northern blot analysis. In addition, the polymerase will use the minigenomes as a template to transcribe the reporter gene. The polymerase recognizes the gene start signal and initiates transcription of the reporter gene. It will then elongate the mRNA until it reaches the gene end signal,

Fig. 1 (continued) These negative-sense minigenomes (–) can then be recognized as a template by the EBOV polymerase complex. Replication of the minigenomes leads to the generation of positive-sense mini-antigenomes (+). Replication from the 3′ complementary trailer region subsequently generates more negative-sense minigenomes. Mini-antigenomes can be detected by Northern hybridization analysis. Transcription of the minigenomes by the viral polymerase leads to production of reporter mRNA. Subsequent translation by the cellular machinery generates reporter protein (e.g., luciferase) that can be detected by reporter gene assay

where it will terminate transcription and polyadenylate the mRNA [18]. These mRNAs are translated by the cellular host machinery (as viral mRNAs would be during the course of viral infection), producing the reporter protein (Fig. 1b). Firefly luciferase can readily be measured to indirectly quantify the extent of transcription performed by the viral polymerase in a luciferase reporter assay [11]. Replication of the minigenome leads to the amplification of template RNA used for transcription and therefore enhances transcriptional activity. Alternatively, the reporter mRNA can be detected directly by Northern blot analysis (see Chapter 11). The system described here functions by transfecting five plasmids (encoding L, NP, VP35, VP30, and the minigenome) into BSR T7/5 cells, which express T7 RNA polymerase [19].

2 Materials

2.1 Passaging Cells

1. Supplemented Glasgow's Minimal Essential Media (MEM), 10% fetal bovine serum (FBS): 500 mL Glasgow's MEM, 50 mL FBS, 10 mL MEM amino acids solution, 5 mL 200 mM L-glutamine.
2. BSR T7/5 cells [19].
3. Phosphate Buffered Saline Mg^{2+}- and Ca^{2+}-free (PBS -Mg^{2+},-Ca^{2+}): 0.154 M sodium chloride; 5 mM sodium phosphate, dibasic anhydrous; 1 mM potassium phosphate, monobasic anhydrous.
4. 0.025% trypsin–0.01% EDTA solution.
5. Geneticin, 50 mg/mL.
6. Sterile 15 mL polypropylene tubes.

2.2 Plating Cells for Transfection

1. Supplemented Glasgow's MEM, 10% FBS.
2. BSR T7/5 cells [19].
3. 6-well tissue culture plate.
4. Neubauer cell counting chamber.

2.3 Transfection

1. EBOV support plasmids: pTM1-L_{EBOV}, pTM1-NP_{EBOV}, pTM1-$VP35_{EBOV}$, pTM1-$VP30_{EBOV}$ [11]. The support plasmids are available through Addgene (https://www.addgene.org/Elke_Muhlberger/).
2. pTM1-empty plasmid [14].
3. EBOV monocistronic minigenome plasmid 3E5E firefly luciferase [11]. The minigenome plasmid is available through Addgene (https://www.addgene.org/Elke_Muhlberger/).
4. pCAGGS-mCherry plasmid.

5. pSV-β-galactosidase control vector (Promega).
6. NanoDrop Spectrophotometer.
7. 5 mL sterile round-bottom polystyrene tubes with snap cover.
8. OptiMEM Reduced Serum Medium.
9. Lipofectamine LTX with PLUS Reagent (ThermoFisher Scientific).
10. PBS -Mg^{2+},-Ca^{2+}.
11. Supplemented Glasgow's MEM, 10% FBS.

2.4 Harvesting Cells for Enzyme Assays

1. Inverted microscope with appropriate filter for mCherry fluorescence (560 nm excitation, 630 nm emission).
2. PBS -Mg^{2+},-Ca^{2+}.
3. Luciferase Assay System (Promega).
4. Tabletop centrifuge.
5. Sterile microcentrifuge tubes.
6. Ultrapure, deionized H_2O.

2.5 Luciferase Assay

1. 96-well plates, white.
2. Luciferase Assay System (Promega).
3. Plate reader for luminescence with injector. We use the Luciferase Reporter Assay Program of the LUMIStar Omega Luminometer (BMG Labtech).

2.6 β-Galactosidase Assay

1. β-Galactosidase Enzyme Assay System with Reporter Lysis Buffer (Promega).
2. 96-well plates, clear.
3. Plate reader for measuring absorbance at 420 nm.

3 Methods

3.1 Passaging Cells

Grow BSR T7/5 cells in a T75 flask at 37 °C, 5% CO_2 in a total volume of 15 mL medium. Cells should be split approximately every 3–4 days (*see* **Note 1**). Use Supplemented Glasgow's MEM with 10% FBS unless otherwise noted.

1. Prepare Supplemented Glasgow's MEM with 10% FBS.
2. Remove medium from cells using a serological pipette (*see* **Note 2**).
3. Add 10 mL PBS. Rock gently over cells. Remove with pipette. Repeat once for a total of two washes (*see* **Note 3**).
4. Add 1.5 mL trypsin/EDTA solution to cells. Rock to gently coat cells with trypsin/EDTA solution.

5. Place flask in incubator (37 °C, 5% CO_2) for approximately 1.5 min until cells are ready to detach (i.e., easy to dislodge by smacking).
6. Smack side of flask to dislodge cells.
7. Resuspend cells with 8.5 mL medium by pipetting until a single-cell suspension is obtained (*see* **Note 4**).
8. Remove cell suspension from flask (~10 mL total) and save for passaging and plating cells for transfections in a sterile 15 mL tube.
9. To passage cells, make the necessary dilution of cell suspension into fresh Supplemented Glasgow's MEM with 10% FBS (*see* **Note 5**).
10. Add 300 μL geneticin (50 mg/mL) to the flask every other passage (*see* **Note 6**).

3.2 Plating Cells for Transfection

1. Plate 4 × 10^5 BSR T7/5 cells per well of a 6-well plate in a volume of 2 mL per well of Supplemented Glasgow's MEM with 10% FBS. We recommend making a master mix based on the number of wells being used (*see* **Note 7**). In this case, add 2 mL of diluted cell master mix to each well in 6-well plates.
2. Incubate at 37 °C, 5% CO_2 until ready for transfection on day two (*see* **Note 8**).

3.3 Transfection

1. Prepare master mixes for transfection of cells with the EBOV support plasmids, the minigenome plasmid, and the β-galactosidase plasmid based on the number of wells to be transfected (*see* **Note 9**): pTM1-L_{EBOV}, pTM1-NP_{EBOV}, pTM1-$VP35_{EBOV}$, pTM1-$VP30_{EBOV}$, 3E5E firefly luciferase, and pSV-β-galactosidase (*see* **Note 10**). One master mix will include functional L: pTM1-L_{EBOV} (Table 1). The second master mix will replace pTM1-L_{EBOV} with an empty vector pTM1-empty (*see* **Note 11**) (Table 1). If this assay is intended to be quantitative, we recommend performing the transfection in triplicate (i.e., transfect 3 separate wells with the appropriate plasmids for each condition being tested). A β-galactosidase enzyme assay should be performed to normalize the data from the luciferase assay to account for minor experimental inconsistencies in transfection efficiency and handling/processing of the different samples.
2. Prepare rack with polystyrene tubes (one for each well to be transfected). Include one extra for transfecting one well with pCAGGS-mCherry as a control for the transfection efficiency.
3. Add 500 μL OptiMEM per tube.
4. Add the appropriately prepared plasmid master mix (either with or without pTM1-L_{EBOV}), except for the pCAGGS-mCherry transfection control tube.

Table 1
Composition of support plasmid master mixes with and without pTM1-L$_{EBOV}$

	Master mix with pTM1-L$_{EBOV}$	Master mix without pTM1-L$_{EBOV}$
pTM1-L$_{EBOV}$	100	–
pTM1-empty	–	100
pTM1-NP$_{EBOV}$	500	500
pTM1-VP35$_{EBOV}$	500	500
pTM1-VP30$_{EBOV}$	100	100
3E5E firefly luciferase	1500	1500
pSV-β-galactosidase	100	100

Amounts are in ng per one well

5. In the transfection control tube, add 0.5 μg pCAGGS-mCherry and 2.3 μg pTM1-empty (or other empty vector) so that the total DNA amount (ng) is the same as in tubes with support plasmid master mix and minigenome plasmid.
6. Add 2.5 μL PLUS reagent per tube.
7. Vortex each tube three times briefly (*see* **Note 12**).
8. Incubate for 5 min at room temperature.
9. Add 6.25 μL Lipofectamine LTX per tube.
10. Vortex each tube three times briefly (*see* **Note 12**).
11. Incubate for 30 min at room temperature.
12. Approximately 5 min before end of incubation time (i.e., after ~25 min), wash cells with 1 mL PBS (room temperature). Remove PBS (*see* **Note 13**).
13. Add 1 mL OptiMEM to each well.
14. At the end of the 30 min incubation, add transfection mix dropwise on top of the cells (*see* **Note 14**). Place 6-well plates in incubator: 37 °C, 5% CO_2.
15. Five hours post transfection, add 1 mL per well of Supplemented Glasgow's MEM (10% FBS) per well. Return to incubator: 37 °C, 5% CO_2.
16. Day three: one day post transfection, replace the medium with fresh 2 mL Supplemented Glasgow's MEM (10% FBS) (*see* **Note 15**). Return to incubator: 37 °C, 5% CO_2.

17. Day four: two days post transfection, cells are ready for readout: e.g., luciferase assay, β-galactosidase assay, RNA extraction, etc.

3.4 Harvesting Cells for Enzyme Assays

1. Prior to harvesting cells, check transfection efficiency by examining mCherry fluorescence in transfection control well (transfected with pCAGGS-mCherry) (Fig. 2a). Examine well using a microscope with a filter appropriate for mCherry fluorescence (560 nm excitation; 630 nm emission).
2. On the laboratory bench, remove and discard the medium from each well.
3. Wash cells with 1 mL PBS. Remove and discard PBS.
4. Add 600 μL 1× Reporter Lysis Buffer to each well (*see* **Note 16**).
5. Rock on shaker for 10 min at room temperature.
6. Transfer cell lysate to 1.5 mL microcentrifuge tubes (*see* **Note 17**).

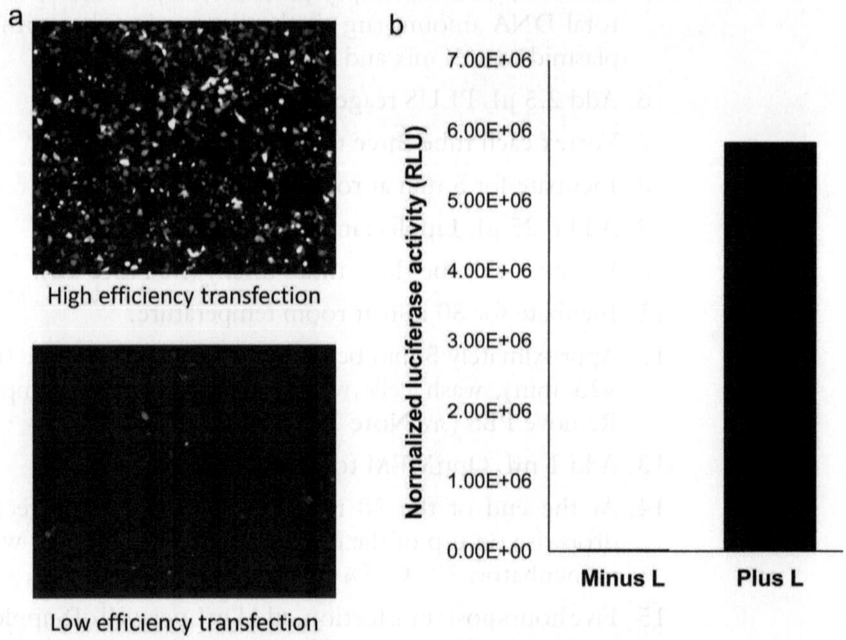

Fig. 2 Example results. (**a**) Example images of two wells of BSR T7/5 cells transfected with pCAGGS-mCherry 2 days post transfection taken using an mCherry filter (560 nm excitation; 630 nm emission). Transfection efficiency was high in one well (*top*) and low in the other well from a separate experiment (*bottom*). (**b**) Example results from a luciferase assay experiment with the EBOV minigenome system. Minus L: BSR T7/5 cells were transfected with the master mix containing pTM1-empty instead of pTM1-L_{EBOV} (Table 1, Subheading 3.3.). Plus L: BSR T7/5 cells were transfected with the master mix containing pTM1-L_{EBOV} (Table 1, Subheading 3.3.). Raw luciferase RLU (*blank subtracted*) were normalized using β-galactosidase activity (Table 2)

7. Vortex tubes at maximum speed for ~15 s.
8. Centrifuge at 960 × *g* for 5 min at room temperature (*see* **Note 18**).

3.5 Luciferase Assay

1. Pipet 50 μL of 1× Reporter Lysis Buffer into one well of 96 well plate as a negative control. Data obtained from this sample will serve as a blank in all calculations (*see* Subheading 3.7 Quantification).
2. Pipet 50 μL from each tube of cell lysates to be analyzed into the wells of white 96-well plate (*see* **Notes 19** and **20**).
3. Set up measurement conditions in the program for the plate reader and wash and prime the injectors with firefly luciferase reagent (*see* **Note 21**). Then read plates using an injection volume of 50 μL. We use a delay between injection and measurement of 0.2 s (inherent to the plate reader) and a measurement time of 23.5 s. Parameters may vary based on the equipment being used (*see* **Note 22** for our program conditions).

3.6 β-Galactosidase Assay

1. Thaw 2× Assay Buffer. Place on ice.
2. Pipet 50 μL 1× Reporter Lysis Buffer into a well of clear 96-well plate as a blank measurement.
3. Pipet 50 μL from each tube of cell lysates into wells of clear 96-well plate (*see* **Note 19**). These are the same cell lysates that have been prepared for use in luciferase assay.
4. Add 50 μL of 2× Assay Buffer to each well. Mix by pipetting.
5. Cover plate and incubate at 37 °C for 30 min. A yellow color should develop.
6. Add 150 μL 1 M sodium carbonate (supplied with the β-galactosidase Enzyme Assay System) to stop the reaction. Mix by pipetting. Avoid forming bubbles.
7. Read absorbance of each well at 420 nm in plate reader.

3.7 Quantification

1. Prepare a standard curve according to manufacturer's instructions (*see* **Note 23**).
2. Using the equation from this standard curve and the absorbance values determined for the samples in Subheading 3.5, determine the β-galactosidase activity (mU) for each sample (Table 2) and obtain a corrected β-galactosidase activity value for each sample by subtracting the blank value.
3. To normalize the luciferase assay, divide the relative light units (RLU) determined for each sample in the luciferase assay, minus the blank value, by the β-galactosidase activity (mU) for each sample (Table 2). These values are the normalized luciferase activity RLU for each sample and account for well-to-well variability in transfection efficiency and handling (Fig. 2b).

Table 2
Example normalization of luciferase RLU for samples from a luciferase assay

Sample from β-galactosidase assay	Absorbance (420 nm)	β-galactosidase (milliunits)	Raw luciferase RLU (blank subtracted)	Normalized luciferase activity RLU
Minus L	0.1937	1.618	3.95×10^4	2.44×10^4
Plus L	0.1770	1.570	9.16×10^6	5.83×10^6

Minus L refers to assays performed with cell lysates from cells transfected with master mix containing pTM1-empty instead of pTM1-L_{EBOV} (Table 1, Subheading 3.3.). Plus L refers to assays performed with cell lysates from cells transfected with master mix containing pTM1-L_{EBOV} (Table 1, Subheading 3.3.). Standard curve of absorbance (420 nm) versus concentration of β-galactosidase (milliunits) yielded line with equation: $y = 0.349x - 0.3709$ where, y = *absorbance (420 nm)* and x = *β-galactosidase (mU)*. Concentration of β-galactosidase (mU) calculated using absorbance and line equation from the standard curve. Normalized luciferase activity (RLU) is determined by dividing raw luciferase RLU (blank subtracted) by concentration of β-galactosidase (milliunits)

4 Notes

1. All cell culture work should be done in a biosafety cabinet with proper sterile technique. All medium should be stored at 4 °C. Media should not be cold when used on cells (either leave at room temperature until use or warm in 37 °C water bath prior to use).
2. Remove medium from flask. To maximize medium removal tilt flask and pipette from the flask corner.
3. Add PBS by pipetting against the large side of flask that cells are not adhered to. When removing PBS wash, tip flask slightly to maximize wash removal by pipetting from flask corner.
4. Resuspend cells by pipetting up and down several times. Be sure to pipet liquid against the side of the flask on which the cell monolayer was grown to maximize resuspension of cells.
5. Dilutions should be on the order of 1:12–1:35. Examine the cell monolayer in flask under a light microscope to determine confluency. Aim to passage cells when approximately confluent. Cells should be confluent two times per week. The efficacy of the system wanes with increasing passage number of the cells used.
6. BSR T7/5 cells stably express T7 RNA polymerase, which is encoded on a plasmid under control of the cytomegalovirus promoter and neomycin resistance gene. Selection with geneticin ensures cells will retain the plasmid [19].
7. If you aim to transfect 6 wells, make master mix with enough cells for 7 wells to account for pipetting errors. Load 10 μL cell suspension onto a Neubauer counting chamber. Count the number of cells in four big counting grid squares (each 1 × 1 mm)

and calculate the mean cell number per square. This is the number of cells in 0.1 μL of the cell suspension. Calculate the volume of cell suspension required per well as follows:

$$\textit{Volume of cell suspension per well} = \frac{\textit{Required number of cells} \times 0.1\,\mu L}{\textit{Mean\# of cells per grid}}$$

For example:

If the mean cell count per grid is 95 cells and one would like to seed 4 × 10^5 cells per well of a 6-well plate, the calculation will be:

$$\textit{Volume of cell suspension per well} = \frac{4 \times 10^5\,\textit{cells} \times 0.1\,\mu L}{95\,\textit{cells}}$$

$$= 421\,\mu L\ \textit{per well}\left(\textit{amount of cell suspension to add to each well to seed}\ 4 \times 10^5\ \textit{cells}\right)$$

For each well, mix calculated volume of cell suspension (421 μL in the example) with media to a final volume of 2 mL. Pipet 2 mL of cell suspension into each well of a 6-well plate.

8. Cells need to attach prior to transfection. Incubate plated cells overnight at 37 °C, 5% CO_2 and start transfection the following morning. Cells should be ~70% confluent on day two, at the time of transfection.
9. The amounts of each plasmid transfected and the ratios of the plasmids to each other are critical when using the minigenome system. We have recommended amounts to use in this protocol (Table 1). However, since plasmid quality and transfection efficiency may vary, initial titration experiments might be helpful to achieve optimal results.
10. Preparing master mixes of support and minigenome plasmids before transfection can be performed on the lab bench. We recommend frequently requantifying plasmid DNA concentrations using a NanoDrop Spectrophotometer.
11. Wells transfected with the master mix without pTM1-L_{EBOV} (instead pTM1-empty or other empty vector) will serve as negative controls. Add empty vector in the same amount (μg) as pTM1-L_{EBOV} to ensure each well is transfected with equal total DNA mass.
12. Vortexer should be set to "touch" mode. After adding reagent (PLUS or LTX), hold transfection tube at the top and press bottom to vortexer three times, very briefly (~1 s).
13. Remove PBS from wells by tilting 6-well plate and pipetting off liquid from the edge of the wells.
14. Pipet transfection mix onto cells gently. Use a P1000 pipet with sterile filter tips and slowly depress the plunger while the

tip is in the air over the medium in the well. Move pipet to different areas of well while adding transfection mix to add drops of mix all over the well ensuring even complex distribution.

15. This should be done as early as possible the next day. Remove medium by tipping 6-well plate and pipetting off liquid from the edge of the wells.

16. Prepare 1× Reporter Lysis Buffer from 5× Reporter Lysis Buffer stock in the Promega Luciferase Assay System kit by diluting with ultrapure deionized H_2O.

17. Before transferring the cell lysate to 1.5 mL microcentrifuge tubes, you may need to pipet the liquid up and down over the surface of the well to resuspend all cells from well in the lysis buffer. Pipet gently to minimize the production of bubbles from detergent in the buffer.

18. If needed, cell lysates can be stored at −20 °C, thawed, and analyzed at a later time. Be sure to recentrifuge at 960 × *g* for 5 min at room temperature prior to performing luciferase assay and/or β-galactosidase assay.

19. Be sure to pipet the 50 μL from the top of the 1.5 mL tube. Do not disturb the pellet at the bottom of the tube. If disturbed, recentrifuge and then pipet from the top.

20. You may need to dilute the lysate when performing luciferase assays (e.g., 1:10 dilution by adding 45 μL 1× Reporter Lysis Buffer and 5 μL cell lysate to the well in 96 well plate). The luciferase assay has an upper limit of linearity that may be reached when testing undiluted lysates. The range of linear detection may vary by instrument and should be verified for each instrument being used prior to sample analysis. If the assay is meant to be quantitative, analyze each sample at least twice by including duplicate wells with cell lysate in the 96-well plate for measurement.

21. Prepare firefly substrate as per manufacturer's instructions. This can be stored at −20 °C wrapped in aluminum foil for use in future experiments.

22. Program conditions for measurement using the Reporter Assay Program on the LUMIStar Omega Luminometer (BMG Labtech). Parameters will vary if other instruments/programs are used:

Basic parameters

(a) Positioning delay is 0.2 s

(b) No. of kinetic windows = 1

(c) Measurement start time = 0.0 s

(d) No. of intervals = 48

(e) Interval time = 0.5 s

(f) End Kinetic Window = 24
(g) Optic (right side) = Top Optic
(h) No. of multichromatics = 1
(i) Emission Filter = lens
(j) Gain = 3600
(k) Everything else is off or 0.00

Concentrations/Volumes/Shaking

(a) Volume of firefly luciferase reagent: 50 μL
(b) pump speed = 310 μL/s
(c) shaking = 0

23. We recommend constructing a standard curve for each assay performed. Determine the equation for the best fit curve of absorbance (420 nm) versus concentration of β-galactosidase.

Acknowledgment

This work was supported by the National Institute of Allergy and Infectious Diseases of the National Institutes of Health under award numbers U01AI082954, R03AI114293, and UC6-AI058618.

References

1. Mühlberger E, Weik M, Volchkov VE, Klenk H-D, Becker S (1999) Comparison of the transcription and replication strategies of marburg virus and Ebola virus by using artificial replication systems. J Virol 73(3):2333–2342
2. Weik M, Enterlein S, Schlenz K, Mühlberger E (2005) The Ebola virus genomic replication promoter is bipartite and follows the rule of six. J Virol 79(16):10660–10671
3. Mühlberger E (2007) Filovirus replication and transcription. Future Virol 2(2):205–215. doi:10.2217/17460794.2.2.205
4. Brauburger K, Deflubé LR, Mühlberger E (2015) Filovirus transcription and replication. In: Pattnaik AK, Whitt MA (eds) Biology and pathogenesis of Rhabdo- and Filoviruses. World Scientific Publishing Co. Pte. Ltd., Singapore, pp 515–5555
5. Conzelmann KK (2004) Reverse genetics of mononegavirales. Curr Top Microbiol Immunol 283:1–41
6. Hoenen T, Groseth A, de Kok-Mercado F, Kuhn JH, Wahl-Jensen V (2011) Minigenomes, transcription and replication competent virus-like particles and beyond: reverse genetics systems for filoviruses and other negative stranded hemorrhagic fever viruses. Antiviral Res 91(2): 195–208. doi:10.1016/j.antiviral.2011.06.003
7. Mühlberger E, Trommer S, Funke C, Volchkov V, Klenk H-D, Becker S (1996) Termini of all mRNA species of Marburg virus: sequence and secondary structure. Virology 223(2):376–380
8. Watanabe S, Watanabe T, Noda T, Takada A, Feldmann H, Jasenosky LD, Kawaoka Y (2004) Production of novel ebola virus-like particles from cDNAs: an alternative to ebola virus generation by reverse genetics. J Virol 78(2): 999–1005
9. Hoenen T, Groseth A, Kolesnikova L, Theriault S, Ebihara H, Hartlieb B, Bamberg S, Feldmann H, Stroher U, Becker S (2006) Infection of naive target cells with virus-like particles: implications for the function of ebola virus VP24. J Virol 80(14):7260–7264. doi:10.1128/JVI.00051-06
10. Jasenosky LD, Neumann G, Kawaoka Y (2010) Minigenome-based reporter system suitable for high-throughput screening of compounds able to inhibit Ebolavirus replication and/or transcription. Antimicrob Agents

Chemother 54(7):3007–3010. doi:10.1128/AAC.00138-10

11. Trunschke M, Conrad D, Enterlein S, Olejnik J, Brauburger K, Muhlberger E (2013) The L-VP35 and L-L interaction domains reside in the amino terminus of the Ebola virus L protein and are potential targets for antivirals. Virology 441(2):135–145. doi:10.1016/j.virol.2013.03.013
12. Uebelhoer LS, Albarino CG, McMullan LK, Chakrabarti AK, Vincent JP, Nichol ST, Towner JS (2014) High-throughput, luciferase-based reverse genetics systems for identifying inhibitors of Marburg and Ebola viruses. Antiviral Res 106:86–94. doi:10.1016/j.antiviral.2014.03.018
13. Pattnaik AK, Ball LA, LeGrone AW, Wertz GW (1992) Infectious defective interfering particles of VSV from transcripts of a cDNA clone. Cell 69(6):1011–1020
14. Moss B, Elroy-Stein O, Mizukami T, Alexander WA, Fuerst TR (1990) Product review. New mammalian expression vectors. Nature 348(6296):91–92. doi:10.1038/348091a0
15. Kuhn JH (2008) Filoviruses; a compendium of 40 years of epidemiological, clinical, and laboratory studies. Springer Verlag, Wien, Austria
16. Weik M, Modrof J, Klenk HD, Becker S, Muhlberger E (2002) Ebola virus VP30-mediated transcription is regulated by RNA secondary structure formation. J Virol 76(17):8532–8539
17. Martinez MJ, Biedenkopf N, Volchkova V, Hartlieb B, Alazard-Dany N, Reynard O, Becker S, Volchkov V (2008) Role of Ebola virus VP30 in transcription reinitiation. J Virol 82(24):12569–12573. doi:10.1128/JVI.01395-08. JVI.01395-08 [pii]
18. Whelan SP, Barr JN, Wertz GW (2004) Transcription and replication of nonsegmented negative-strand RNA viruses. Curr Top Microbiol Immunol 283:61–119
19. Buchholz UJ, Finke S, Conzelmann KK (1999) Generation of bovine respiratory syncytial virus (BRSV) from cDNA: BRSV NS2 is not essential for virus replication in tissue culture, and the human RSV leader region acts as a functional BRSV genome promoter. J Virol 73(1):251–259

Chapter 7

Quantification of RNA Content in Reconstituted Ebola Virus Nucleocapsids by Immunoprecipitation

Logan Banadyga and Hideki Ebihara

Abstract

Immunoprecipitations are commonly used to isolate proteins or protein complexes and assess protein-protein interactions; however, they can also be used to assess protein-RNA complexes. Here we describe an adapted RNA immunoprecipitation technique that permits the quantification of RNA content in Ebola virus nucleocapsids that have been reconstituted in vitro by transient transfection.

Key words Ebola virus, RNA, Nucleoprotein, VP35, VP24, Nucleocapsid, Immunoprecipitation, qRT-PCR

1 Introduction

Immunoprecipitation (IP) is a common molecular biology technique used to purify proteins from a particular sample, usually cell lysate. Typically, the protein of interest is bound by a specific antibody, which is, in turn, bound by protein A or G that has been pre-attached to agarose or magnetic beads. Pelleting the beads "precipitates" the protein-antibody complex and, in effect, purifies it from the general milieu of the cell lysate. Often, other proteins that interact with the protein of interest can be co-precipitated with the complex, and by subjecting the precipitated complexes to sodium dodecyl sulfate polyacrylamide gel electrophoresis (SDS-PAGE) and Western blotting, interactions between proteins can be assessed. A variation of this methodology can also be used to assess the ability of a protein to interact with RNA. By immunoprecipitating the protein of interest and then subjecting the precipitated complexes to quantitative real-time PCR (qRT-PCR), the amount of RNA that co-precipitates with the protein can be determined (Fig. 1). Here we describe an RNA IP procedure adapted from a variety of previous reports [1–4], to assess the RNA content of Ebola virus nucleocapsids reconstituted in vitro by transfection.

Thomas Hoenen and Allison Groseth (eds.), *Ebolaviruses: Methods and Protocols*, Methods in Molecular Biology, vol. 1628, DOI 10.1007/978-1-4939-7116-9_7, © Springer Science+Business Media LLC 2017

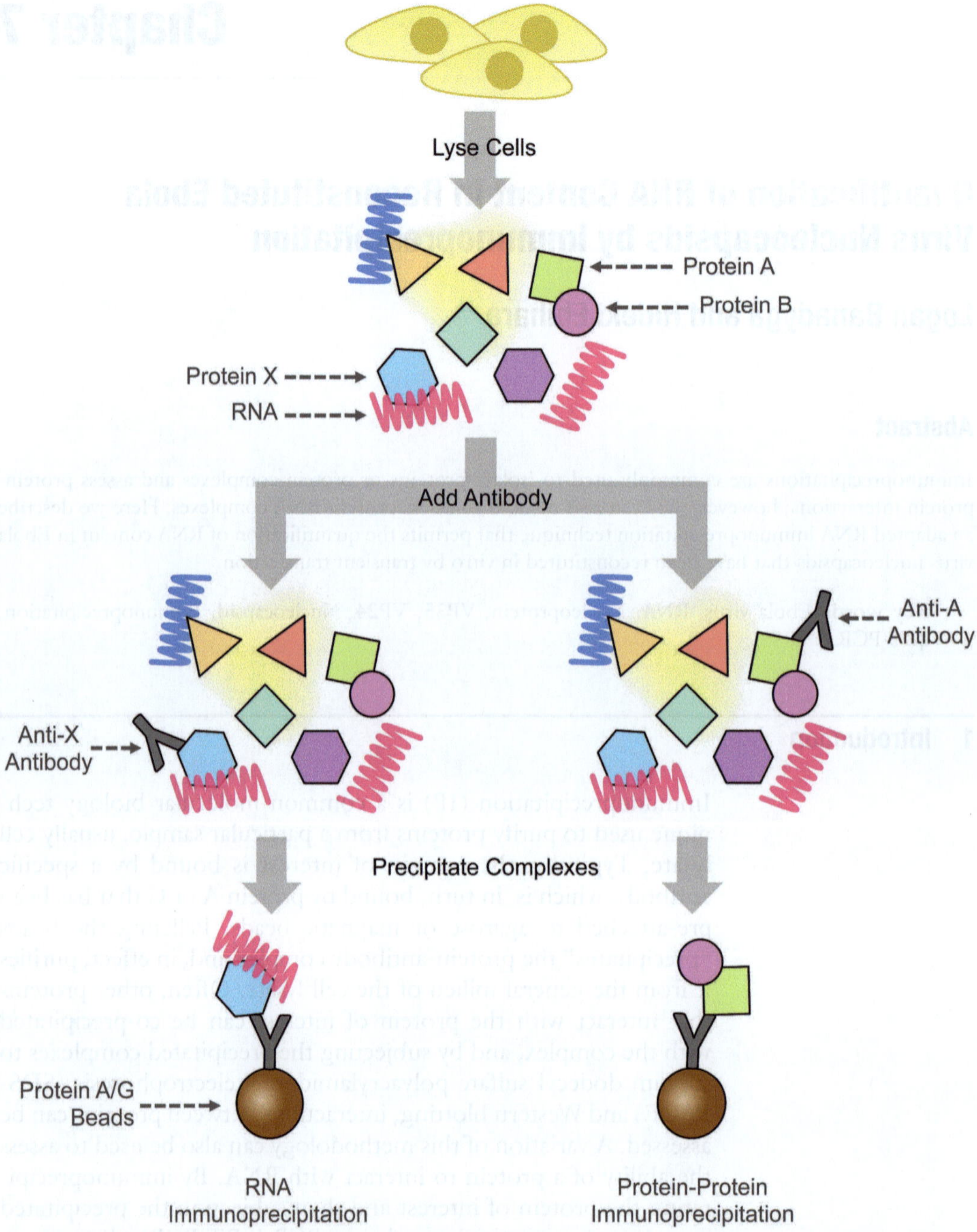

Fig. 1 A schematic outline of immunoprecipitations. Following cell lysis, samples are incubated with a specific antibody, which binds to its cognate antigen. Antibody is then bound to protein A or G attached to agarose or magnetic beads, and the complexes are precipitated by centrifugation or application of a magnet. In an RNA immunoprecipitation, RNA is precipitated along with an RNA-binding protein (Protein X) that is bound to antibody (Anti-X Antibody). In a protein-protein immunoprecipitation, a secondary protein (Protein B) is precipitated along with a primary protein (Protein A) that is bound to antibody (Anti-A Antibody)

The single-stranded, negative-sense RNA genome of Ebola virus is encapsidated by the nucleoprotein (NP) along with viral protein 35 (VP35) and VP24, which, together, are the minimum components required to assemble the viral nucleocapsid [5, 6]. Transfection of NP, VP35, and VP24 alone is sufficient to produce nucleocapsid-like structures, which incorporate cellular RNA [5, 6]. To assess the viral RNA content of Ebola virus nucleocapsids, we transfected cells with NP possessing a C-terminal FLAG-HA tag, VP35, and VP24. Viral RNA was produced by transfecting the T7 polymerase along with a T7-driven tetracistronic Ebola virus minigenome (MG) [7]. Immunoprecipitation of NP, using an anti-FLAG antibody, followed by quantitative real-time PCR (qRT-PCR) using primers directed against the transfected MG allowed us to quantify the viral RNA content of Ebola virus nucleocapsids. This protocol will be useful for investigating the contribution that various Ebola virus proteins, and mutant proteins thereof, make to the assembly and function of viral nucleocapsids. Moreover, this protocol should be adaptable to work with other filoviruses, including Marburg virus.

2 Materials

Prepare all solutions/master mixes in Subheadings 2.2 and 2.3 using DEPC-treated water and analytical grade reagents dedicated for "RNA use only." Prepare and store all reagents at room temperature, unless otherwise indicated.

2.1 Transfection

1. HEK293 cells (ATCC CRL-1573).
2. DMEM10: 10% heat-inactivated fetal bovine serum, 50 U/mL penicillin, 50 μg/mL streptomycin, 2 mM L-glutamine in Dulbecco's modified Eagle's medium (DMEM).
3. DMEM5: 5% heat-inactivated fetal bovine serum, 50 U/mL penicillin, 50 μg/mL streptomycin, 2 mM L-glutamine in Dulbecco's modified Eagle's medium (DMEM).
4. Six-well polystyrene tissue culture treated plate (non-pyrogenic).
5. Opti-MEM (Thermo Fisher Scientific).
6. Sterile 1.5-mL snap-cap microcentrifuge tubes.
7. The following plasmids, purified for transfection: p4cis-vRNA-RLuc-VP24-3×-stop, pCAGGS-T7, pCAGGS-NP-FH, pCAGGS-VP35, pCAGGS-VP24, and pCAGGS-empty vector (*see* **Note 1**).
8. TransIT LT-1 Transfection Reagent (Mirus).

2.2 RNA Immunoprecipitation and Preparation

1. Phosphate-buffered saline (PBS): 137 mM NaCl, 2.7 mM KCl, 8.1 mM Na_2HPO_4, 1.47 mM KH_2PO_4, 0.9 mM $CaCl_2$, and 0.5 mM $MgCl_2$. Prepare using ultrapure water.
2. Tube rotator.
3. Vacuum line and flask.
4. DEPC-treated water.
5. RNA IP (RIP) buffer: 150 mM NaCl, 25 mM Tris, pH 7.4, 5 mM EDTA, 0.5 mM DTT, and 1% IGEPAL CA-360. Add 1 cOmplete, EDTA-free Protease Inhibitor Cocktail Tablet per 50 mL buffer. Store at 4 °C. Prepare in DEPC-treated water. Add 100 U/mL Recombinant Ribonuclease Inhibitor immediately prior to use for RIP/RO Buffer (*see* **Note 2**).
6. 15-mL conical centrifuge tubes.
7. Ice bucket and ice.
8. 1.5-mL snap-cap microcentrifuge tubes.
9. 1-mL syringes and 22G1 needles.
10. Monoclonal mouse anti-FLAG M2 antibody (Sigma).
11. Mouse isotype control antibody MOPC21 (Sigma).
12. Ice-cold acetone.
13. TRIzol LS Reagent.
14. TRIzol Reagent.
15. 2× Laemmli Sample Buffer (Bio-Rad) with 715 mM 2-mercaptoethanol (*see* **Note 3**).
16. Dynabeads Protein G (Thermo Fisher Scientific).
17. DynaMag-2 Magnet (Thermo Fisher Scientific).
18. Direct-zol RNA MiniPrep Kit (Zymo Research).

2.3 Quantitative Real-Time PCR

1. MEGAscript T7 Transcription Kit (Thermo Fisher Scientific).
2. MEGAclear Kit (Thermo Fisher Scientific).
3. p4cis-vRNA-RLuc-VP24-3×-stop.
4. BlpI restriction endonuclease and associated buffers.
5. 2-mL screw-cap tubes.
6. QuantiFast Probe RT-PCR + Rox Vial Kit (Qiagen).
7. TaqMan primers and probe (*see* Table 1).
8. Rotor-Gene Q thermal cycler (Qiagen).
9. Rotor-Gene 0.1-mL strip tubes and caps.
10. Aluminum loading block for Rotor-Gene strip tubes, refrigerated to 4 °C.

Table 1
qRT-PCR primers

Name	Sequence
Forward primer	TTTCCAATCCAAGTCACACTG
Reverse primer	ATTGCACATACTTTTTGCCC
Probe[a]	6FAM—TCCTCAGATGTAAGCATGCAGGCAA—BBQ

[a]Probe has 5′ 6FAM fluorophore and 3′ BBQ quencher

3 Methods

The following protocol outlines the RNA IP procedure for three conditions: a “minigenome-only” control, an “isotype antibody” control, and the experimental sample. A timeline of the experiment is provided in Fig. 2. All procedures described under Subheadings 3.2 and 3.3 should be carried out in a manner that limits RNA and RNase contamination of samples (*see* **Note 4**). In Subheading 3.2, the immunoprecipitation samples should be kept on ice or at 4 °C as much as possible.

3.1 Transfection

1. On day 1, 24 h prior to transfection, seed 1.8×10^5 HEK293 cells per well of one 6-well tissue culture plate in 2 mL DMEM10. Two wells of cells are required per IP condition (*see* **Note 5**).
2. On day 2, perform the transfection as outlined in Table 2 and described below. Three transfection conditions are described: (A) minigenome (MG)-only control, (B) isotype antibody control, and (C) the experimental sample.
3. In a clean BSC, add 500 μL sterile Opti-MEM to each of three 1.5-mL snap-cap microcentrifuge tubes.
4. To tube A (MG-only control), add 0.5 μg p4cis-vRNA-RLuc-VP24-3×-stop, 0.5 μg pCAGGS-T7, and 1 μg pCAGGS-empty.
5. To tubes B and C (isotype antibody control and experimental sample), add 0.5 μg p4cis-vRNA-RLuc-VP24-3×-stop, 0.5 μg pCAGGS-T7, 0.25 μg pCAGGS-NP-FH, 0.25 μg pCAGGS-VP35, and 0.5 μg pCAGGS-VP24.
6. Add 6 μL TransIT LT-1 to each tube and vortex lightly.
7. Incubate tubes at room temperature for 20 min.
8. During incubation, remove medium from each well of the 6-well plate and replace with 1 mL fresh DMEM10.
9. Following the 20-min incubation, spin tubes briefly in a microcentrifuge and then divide mixture from each tube evenly

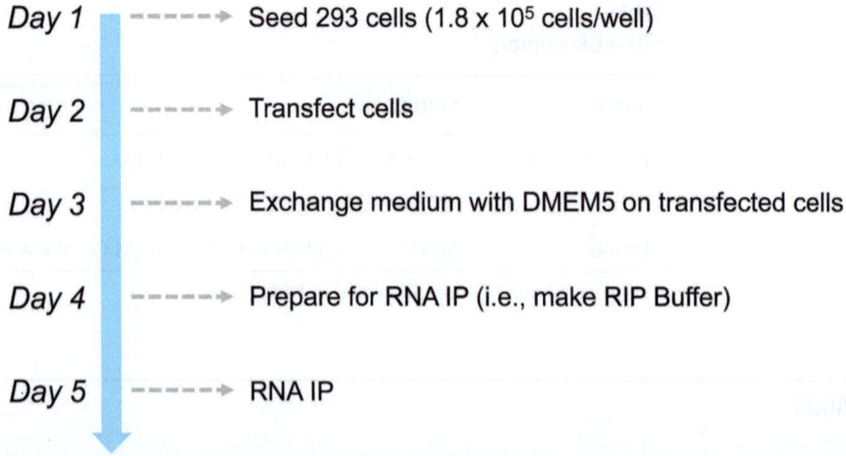

Fig. 2 A timeline for setting up an RNA IP experiment. Major tasks for each day are indicated with an arrow and dotted line. After day 5, IP RNA and whole cell lysate samples can be frozen at −80 °C until ready to process

Table 2
Transfection scheme[a]

Plasmid	Tfx A MG-only control	Tfx B isotype control	Tfx C experimental
p4cis-vRNA-RLuc-VP24-3x-stop	0.5 μg	0.5 μg	0.5 μg
pCAGGS-T7	0.5 μg	0.5 μg	0.5 μg
pCAGGS-NP-FH	–	0.25 μg	0.25 μg
pCAGGS-VP35	–	0.25 μg	0.25 μg
pCAGGS-VP24	–	0.5 μg	0.5 μg
pCAGGS-empty	1 μg	–	–
Opti-MEM volume	500 μL	500 μL	500 μL
TransIT LT-1 volume	6 μL	6 μL	6 μL

[a]For 2 wells of a 6-well plate

(~250 μL) between two wells, adding drop-wise over the surface of the well.

10. Incubate the cells in a humidified tissue culture incubator at 37 °C, 5% CO_2 for 24 h.
11. On day 3, 24 h post-transfection, remove medium from each well and replace with 2 mL fresh DMEM5 (*see* **Note 6**). Incubate the cells in a humidified tissue culture incubator at 37 °C, 5% CO_2 for an additional 48 h prior to RNAimmunoprecipitation.

3.2 RNA Immunoprecipitation and RNA Preparation

1. On day 4, the day prior to the RNA IP, prepare RIP buffer and store at 4 °C (*see* **Note 2**). It is recommended to prepare labeled tubes ahead of time, as well (*see* **Note 7**).
2. On day 5, proceed with the RNA IP as outlined below.
3. For each of the three transfection conditions, harvest the cells from the 6-well plate, according to **steps 4–8** (*see* **Note 8**).
4. Remove medium from one well, add it to a 15-mL conical centrifuge tube, and place the tube on ice.
5. Add 1000 μL PBS to the well, and pipette up and down across the cell monolayer to dislodge cells from the plate. Add cells to the same 15-mL conical centrifuge tube.
6. Add an additional 1000 μL fresh PBS to the well, and pipette up and down to collect any remaining cells. Add cells to the same 15-mL conical centrifuge tube.
7. Repeat **steps 4–6** for the second well, adding everything to the same 15-mL conical centrifuge tube. The total volume when finished should be ~8 mL.
8. Repeat **steps 4–7** for the remaining transfection conditions.
9. Pellet cells by centrifuging the 15-mL conical centrifuge tubes at 1000 × *g* for 5 min at 4 °C.
10. Aspirate supernatant from the cell pellet, and resuspend cells in 1000 μL fresh PBS (*see* **Note 9**).
11. Transfer cells to 1.5-mL snap-cap microcentrifuge tubes and place tubes on ice.
12. Pellet cells by centrifuging tubes at 1000 × *g* for 5 min at 4 °C.
13. Aspirate supernatant from cell pellet and resuspend cells in 1000 mL RIP/RO buffer (*see* **Note 2**). Pipette up and down to completely dissolve cell pellet. Place tubes back on ice.
14. Dounce each sample 4–6 times with a 1-mL syringe and a 22G1 needle (*see* **Note 10**). Place tubes back on ice.
15. Rotate tubes on tube rotator for 2 h at 4 °C (*see* **Note 11**).
16. Following the 2-h rotation, centrifuge tubes at 9000 × *g* for 10 min at 4 °C (*see* **Note 12**).
17. While avoiding the pellet, remove as much of the lysate as possible from each tube to a new 1.5-mL microcentrifuge tube (labeled R-IP). Place R-IP tubes on ice, and discard the tubes containing the pellets (*see* **Note 7**).
18. Remove 150 μL of each lysate from the R-IP tubes to new 1.5-mL microcentrifuge tubes (labeled P-WCL). Place P-WCL tubes on ice (*see* **Note 7**).
19. Remove an additional 50 μL of each lysate from the R-IP tubes to new 1.5-mL microcentrifuge tubes (labeled R-WCL). Place R-WCL tubes on ice (*see* **Note 7**).

20. To the R-IP tubes (**step 17**) for transfection condition A and C, add 2.4 μg of mouse anti-FLAG antibody.
21. To the R-IP tubes (**step 17**) for transfection condition B (the isotype antibody control), add 2.4 μg MOPC21 mouse isotype control antibody.
22. Rotate tubes on tube rotator for 2 h at 4 °C (*see* **Note 11**).
23. While tubes are rotating, acetone precipitate the whole cell lysates in the P-WCL tubes (**step 18**) according to **steps 24–29** (*see* **Note 13**).
24. Add 750 μL (five volumes) of ice-cold acetone to lysates. Invert tubes 6–8 times and incubate at −20 °C for at least 30 min up to overnight.
25. After incubation, centrifuge tubes at 18,000 × *g* for 15 min at 4 °C.
26. Aspirate supernatant and briefly allow residual acetone to evaporate (1–2 min).
27. Add 75 μL Laemmli sample buffer to pellets. Vortex tubes well and spin down briefly.
28. Boil tubes at 99 °C for 10 min; afterwards, vortex tubes well and spin down briefly.
29. Store samples at −20 °C until ready to use for SDS-PAGE.
30. To the lysates in the R-WCL tubes (**step 19**), add 150 μL TRIzol LS (three volumes). Vortex tubes well and store at −80 °C until ready to proceed with **step 61**.
31. Just prior to the end of the 2-h rotation (**step 22**), prepare Dynabeads Protein G according to **steps 32–39**.
32. Add 500 μL RIP/RO buffer to a new 1.5-mL microcentrifuge tube.
33. Mix bottle of Dynabeads until the beads are in a homogenous suspension.
34. Add 70 μL of beads directly to RIP/RO buffer from **step 32** (*see* **Note 14**).
35. Place tube on DynaMag-2 magnet and wait ~30 s for beads to pellet before aspirating supernatants.
36. Remove the tube from the magnet and add 500 μL fresh RIP/RO buffer. Gently pipette beads up and down to completely resuspend.
37. Repeat **steps 35** and **36** two more times.
38. Place tube on magnet and wait ~30 s for beads to pellet before aspirating the supernatant.
39. Remove the tube from the magnet, and add a volume of RIP/RO buffer equal to the volume of beads originally removed from the stock, 70 μL in this case.

40. Following the 2-h rotation (**step 22**), add 20 μL of prepared beads to each IP tube.
41. Rotate tubes on tube rotator for 1 h at 4 °C (*see* **Note 11**).
42. Following the 1-h rotation (**step 41**), wash and prepare the beads according to **steps 43–60**.
43. Place tubes on DynaMag-2 magnet and wait ~30 s for beads to pellet (*see* **Note 15**).
44. Open tube caps carefully and aspirate the supernatant. Ensure that vacuum line pipette tips are exchanged between aspirating each tube.
45. Remove the tubes from the magnet and add 500 μL fresh RIP/RO buffer to each tube (*see* **Note 16**).
46. Rotate tubes on tube rotator for 5 min at 4 °C for (*see* **Note 11**).
47. Place tubes on magnet and wait ~30 s for beads to pellet (*see* **Note 15**).
48. Open tube caps carefully and aspirate the supernatant. Ensure that vacuum line pipette tips are exchanged between aspirating each tube.
49. Remove the tubes from the magnet and add 500 μL fresh RIP/RO buffer to each tube.
50. Pipette the beads up and down 5–6 times until evenly dispersed.
51. Repeat **steps 47–50** one more time.
52. Ensure that the beads are evenly resuspended in RIP/RO buffer (vortex briefly, if necessary). Transfer a 75-μL aliquot from each tube into new 1.5-mL microcentrifuge tubes (labeled P-IP).
53. Place R-IP tubes back on magnet and wait ~30 s for beads to pellet (*see* **Note 15**).
54. Open tube caps carefully and aspirate the supernatant. Ensure that vacuum line pipette tips are exchanged between aspirating each tube.
55. Remove the tubes from the magnet and add 300 μL TRIzol to each tube. Vortex tubes well to completely resuspend the beads, and store at −80 °C until ready to proceed with **step 61**.
56. Place the P-IP tubes (**step 52**) on magnet and wait ~30 s for beads to pellet.
57. Open tube caps carefully and aspirate the supernatant. Ensure that vacuum line pipette tips are exchanged between aspirating each tube.
58. Remove the tubes from the magnet and add 20 μL Laemmli sample buffer. Vortex tubes well and spin down briefly.

59. Boil tubes at 99 °C for 10 min; afterwards, vortex tubes well and spin down briefly.
60. Store P-IP samples at −20 °C until ready to use for SDS-PAGE.
61. Prepare RNA from the TRIzol samples (R-WCL and R-IP) using the Direct-zol RNA MiniPrep kit according to manufacturer's instructions. Be sure to include the DNase treatment step. Elute samples once in 50 μL RNase-free water, and store at −80 °C until ready to proceed with qRT-PCR (Subheading 3.3).

3.3 Quantitative Real-Time PCR

1. Prior to beginning the qRT-PCR, ensure that you have made RNA standards according to **steps 2–6**.
2. Linearize 5 μg p4cis-vRNA-RLuc-VP24-3×-stop by digesting with BlpI restriction endonuclease. Purify DNA with a commercially available PCR purification kit.
3. Subject linearized p4cis-vRNA-RLuc-VP24-3×-stop to in vitro transcription using the MEGAscript T7 Transcription Kit according to manufacturer's directions.
4. Purify in vitro transcribed RNA using the MEGAclear Kit according to manufacturer's directions. Elute RNA twice with 50 μL RNase-free water preheated to 95 °C for a total volume of 100 μL.
5. Determine RNA concentration using a spectrophotometer (260 nm).
6. By tenfold serial dilution, make a set of five RNA standard samples ranging in concentration from 10^{-2} ng/μL to 10^{-6} ng/μL (*see* **Note 17**).
7. Setup qRT-PCR reactions as outlined in Table 3 and described in **steps 8–14**.
8. Thaw 2× QuantiFast Master Mix, primer/probe master mix (*see* **Note 18**), and RNA samples on ice. Vortex each and spin down prior to use. Keep everything on ice.
9. Make a qRT-PCR reaction master mix for four reactions (*n* + 1): In a 2-mL screw-cap tube, combine 50 μL 2× QuantiFast Master Mix, 2 μL primer/probe master mix, and 27 μL RNase-free Water. Vortex and place on ice.
10. Aliquot 5 μL of each IP RNA sample and each WCL RNA sample in duplicate into Rotor-Gene strip tubes on an aluminum loading block (*see* **Note 19**).
11. Aliquot 5 μL of RNase-free water in duplicate to two Rotor-Gene tubes.

Table 3
qRT-PCR reaction setup[a]

Reagent	[Stock]	[Final]	One reaction volume	Master mix × 4 reactions
QuantiFast master mix	2×	1×	12.5 μL	50 μL
Forward primer Reverse primer Probe	20 μM 20 μM 10 μM	0.4 μM 0.4 uM 0.2 μM	0.5 μL	2 μL
RNA sample			5 μL	–
Water			6.75 μL	27 μL
Rotor-Gene RT mix			0.25 μL	1 μL
Total reaction volume			25 μL	

[a]Adapted from manufacturer's instructions

Table 4
qRT-PCR reaction conditions[a]

Step	Temp (°C)	Duration	
Initial RT	50	10 min	
Initial denaturation/activation	95	5 min	
Denaturation	95	5 s	40 cycles
Annealing/extension	60	10 s	

[a]Adapted from manufacturer's instructions

12. Aliquot 5 μL of each of the five RNA standards in duplicate into Rotor-Gene tubes.
13. Add 1 μL of the Rotor-Gene RT Mix to the qRT-PCR reaction master mix created in **step 9** and vortex well.
14. Add 20 μL of the qRT-PCR reaction master mix to each RNA sample in the strip tubes, and place the caps on the tubes.
15. Run the samples on a Rotor-Gene thermal cycler using the conditions outlined in Table 4 (*see* **Note 20**).
16. Following the quantification of RNA in each sample by qRT-PCR, analyze each sample according to **steps 17** and **18.**
17. Calculate the average in picograms for each of the three the duplicate IP RNA and WCL RNA samples.
18. Divide the values for the IP RNA samples by the corresponding values for the WCL RNA samples to normalize the data.

4 Notes

1. The minimum proteins required to form the Ebola virus nucleocapsid are NP, VP35, and VP24. RNA, around which the nucleocapsid forms, is provided by the minigenome plasmid, p4cis-vRNA-RLuc-VP24-3x-stop, and T7, which drives transcription of the minigenome. In this case, the minigenome is composed of the Ebola virus leader and trailer, flanking the coding sequences for Renilla luciferase, VP40, $GP_{1,2}$, and VP24stp [7]. The VP24 start codon is immediately preceded by three stop codons to prevent any possible expression of this protein. Notably, in our assay, NP possesses a FLAG-HA (FH) tandem tag, allowing efficient and consistent IP and detection. T7 also possesses the K179C/M750C mutations that help reduce early termination [8]. Empty-vector pCAGGS is used to ensure that each transfection reaction receives the same total amount of DNA. These plasmids work well in this assay, although they could presumably be substituted with other plasmids more convenient for any given user. Note, however, that the qRT-PCR primers described here are specific for the 5′ noncoding region of VP40, so substitution with a minigenome plasmid that does not possess this target sequences must be accompanied by substitution of primers/probes.
2. RIP buffer should be prepared in advance to ensure that it has been thoroughly chilled to 4 °C prior to beginning the IP. Note, however, that DTT should always be added fresh immediately before beginning IP. For RIP/RO buffer, it is best to add the RNase inhibitor to smaller aliquots of RIP buffer immediately before use as needed throughout the protocol (e.g., add 25 μL of RNase inhibitor to a 10-mL aliquot of RIP buffer). This will help conserve RNase inhibitor, which is costly. RIP buffer not containing DTT or RNase inhibitor can be stored for up to 2 weeks at 4 °C.
3. Freshly prepare only enough 2× Laemmli buffer with 2-mercaptoethanol to complete your experiment.
4. Before beginning work, clean work surface, pipettes, tube blocks, and pipette tip boxes with 70% ethanol followed by a suitable RNase decontaminating solution, such as RNase AWAY (Thermo Fisher Scientific). Use dedicated aerosol-barrier pipette tips and microcentrifuge tubes that have been certified RNase-free. Use disposable gloves and change them often. Avoid leaving sample tubes open longer than necessary, and do not talk over the work surface.
5. Two wells of HEK293 cells are used per transfection condition to ensure that there is plenty of RNA to work with in downstream

analyses. We have not attempted to scale down this protocol to 12- or 24-well plates.

6. DMEM5, which contains only 5% heat-inactivated fetal bovine serum, is used to prevent the cells from over-growing.
7. We suggest labeling five sets of three tubes each. All sets should be labeled 1–3. One set of tubes should each be labeled with “R-IP” to indicate that these tubes will ultimately contain the immunoprecipitated RNA sample in TRIzol. One set of tubes should each be labeled with “R-WCL” to indicate that these tubes will hold the RNA sample from the whole cell lysates in TRIzol LS. One set of tubes should each be labeled with “P-IP” to indicate that these tubes will hold the protein sample from the IP. One set should be labeled with “P-WCL” to indicate that these tubes will hold the protein sample from the whole cell lysates. The fifth set of tubes will be used for initial lysis of the cells and will eventually be discarded.
8. The quickest way to complete this step is to use a P1000 micropipettor set to 1000 μL. Using a single tip, transfer 1000 μL medium from the well to the 15-mL conical centrifuge tube twice. Eject the tip and obtain a new one to transfer 1000 μL PBS to the well. Dislodge the cells by pipetting the PBS up and down, and then transfer to the 15-mL conical centrifuge tube. Eject the tip and obtain a new one to transfer 1000 μL fresh PBS to the well. Collect any remaining cells, and transfer to the 15-mL conical centrifuge tube. Repeat this process for each well individually.
9. This step ensures that excess medium is washed away from cell pellet. Use a vacuum line attached to a vacuum flask to aspirate the supernatant and speed up this process. Ensure that separate tips are used for each sample.
10. Avoid douncing the entire volume of buffer to prevent the formation of bubbles. Also take care not to press the needle through the bottom of the tube, and do not place your fingers, thumbs, or hands at the bottom of the tube.
11. Ensure that tube lids are tightly sealed to avoid leakage.
12. This step will separate the lysate from the insoluble material, which will be pelleted and discarded.
13. This step will generate a protein sample suitable for SDS-PAGE and subsequent Western blotting. Western blotting of protein samples may be useful in assessing protein expression in your RNA IP experiment; however, since these procedures are relatively routine, they will not be explained here.
14. A good formula for determining the quantity of beads required is as follows: 20 μL beads per IP sample + 10 μL beads.

Therefore, for three samples: (20 μL beads × 3 samples) + 10 μL beads = 70 μL beads.

15. At this point, there may be lysis buffer and beads in the caps of the 1.5-mL microcentrifuge tubes. To collect these beads, place all the tubes on the magnet, and rotate the magnet 4–6 times in both directions to allow the beads to be washed toward the magnet.

16. After the supernatant has been aspirated (**step 44**), leave the lids on all of the tubes open, and use a single pipette tip to add the fresh RIP/RO buffer. Ensure that the pipette tip does not touch the samples or the tubes to avoid cross-contamination.

17. You should have five RNA standards, 10^{-2}, 10^{-3}, 10^{-4}, 10^{-5}, and 10^{-6} ng/μL, which should provide sufficient range to interpolate and extrapolate the concentration of RNA in your samples. The exact concentrations of your standards will depend on the RNA yield from the in vitro transcription reaction. To avoid multiple freeze-thaw cycles of your RNA standards, it is best to make several small aliquots of each standard and store them at −80 °C. Since we typically use 10 μL of each standard per qRT-PCR run (i.e., 5 μL aliquots in duplicate), we find that 12-μL aliquots of each standard are sufficient.

18. To avoid multiple freeze-thaw cycles of the primers and probes, it is best to make several small aliquots of a primer/probe master mix and store them at −20 °C. We suggest making a primer/probe master mix with the forward and reverse primers at 20 μM and the probe at 10 μM.

19. The same pipette tip can be used for identical RNA samples. Be sure to add the RNA sample near the bottom of the tubes, since that will make it easier to add the qRT-PCR reaction master mix later on.

20. The user may choose to perform the qRT-PCR on a different thermal cycling platform, although we have verified this protocol for Rotor-Gene thermal cyclers only.

Acknowledgments

The authors are grateful to Brandi Williamson for her critical review of this manuscript. The authors are supported by the Division of Intramural Research, NIAID, NIH. Opinions, interpretations, conclusions, and recommendations are those of the authors and are not necessarily endorsed by the NIH.

References

1. Hendrickson DG, Hogan DJ, Herschlag D, Ferrell JE, Brown PO (2008) Systematic identification of mRNAs recruited to argonaute 2 by specific microRNAs and corresponding changes in transcript abundance. PLoS One 3(5):e2126. doi:10.1371/journal.pone.0002126
2. Rinn JL, Kertesz M, Wang JK, Squazzo SL, Xu X, Brugmann SA, Goodnough LH, Helms JA, Farnham PJ, Segal E, Chang HY (2007) Functional demarcation of active and silent chromatin domains in human HOX loci by noncoding RNAs. Cell 129(7):1311–1323. doi:10.1016/j.cell.2007.05.022
3. Khalil AM, Guttman M, Huarte M, Garber M, Raj A, Rivea Morales D, Thomas K, Presser A, Bernstein BE, van Oudenaarden A, Regev A, Lander ES, Rinn JL (2009) Many human large intergenic noncoding RNAs associate with chromatin-modifying complexes and affect gene expression. Proc Natl Acad Sci U S A 106(28):11667–11672. doi:10.1073/pnas.0904715106
4. Hendrickson DG, Hogan DJ, McCullough HL, Myers JW, Herschlag D, Ferrell JE, Brown PO (2009) Concordant regulation of translation and mRNA abundance for hundreds of targets of a human microRNA. PLoS Biol 7(11):e1000238. doi:10.1371/journal.pbio.1000238
5. Huang Y, Xu L, Sun Y, Nabel GJ (2002) The assembly of Ebola virus nucleocapsid requires virion-associated proteins 35 and 24 and posttranslational modification of nucleoprotein. Mol Cell 10(2):307–316. doi:10.1016/S1097-2765(02)00588-9
6. Watanabe S, Noda T, Kawaoka Y (2006) Functional mapping of the nucleoprotein of Ebola virus. J Virol 80(8):3743–3751. doi:10.1128/JVI.80.8.3743-3751.2006
7. Watt A, Moukambi F, Banadyga L, Groseth A, Callison J, Herwig A, Ebihara H, Feldmann H, Hoenen T (2014) A novel life cycle modeling system for Ebola virus shows a genome length-dependent role of VP24 in virus infectivity. J Virol 88(18):10511–10524. doi:10.1128/JVI.01272-14
8. Ma K, Temiakov D, Anikin M, McAllister WT (2005) Probing conformational changes in T7 RNA polymerase during initiation and termination by using engineered disulfide linkages. Proc Natl Acad Sci U S A 102(49):17612–17617. doi:10.1073/pnas.0508865102

Chapter 8

Modeling Ebolavirus Budding with Virus Like Particles

Olivier Reynard and Mathieu Mateo

Abstract

About 15 years ago, several groups initially described the release of virus like particles (VLPs) upon expression of Ebola virus VP40 in mammalian cells. Further development of the protocol later allowed for the dissection of the Ebola virus budding mechanism and for the identification of critical VP40 residues involved in this process. VLPs are now produced routinely in several laboratories as a tool to study virus entry or egress and have even been proposed as vaccine candidates against Ebola virus disease. Here we described protocols for the production and the analysis of Ebola virus VLP release.

Key words Ebola, Ebolavirus, VP40, Viruslike particles

1 Introduction

Ebola virus (EBOV) and Marburg virus are members of the family *Filoviridae*, a group of enveloped negative-strand RNA viruses responsible for severe disease in humans [1]. The EBOV matrix protein VP40 is essential for assembly and budding by supporting the incorporation of viral ribonucleocapsids into budding virus particles [2]. In the early 2000s, the groups of F. Hayes, W. Weissenhorn, and Y. Kawaoka have described the vesicular release of VP40 in the form of VLPs upon expression of this protein in HEK293T cells [2–4]. Since then, VLPs have been widely used as a surrogate of infectious EBOV particles [5] and have even been proposed as vaccine candidates [6]. When expressed alone in mammalian cells, VP40 promotes the formation of viruslike particles (VLPs) resembling filamentous virions. However, VLPs have a narrower diameter than EBOV particles (50–70 nm vs. 80 nm) and a variable length depending on their viral protein composition (VP40 alone or in combination with VP35, NP, GP), ranging from 0.5 to 2 μm, while EBOV particles of 9 μm have been documented [7, 8]. Several crystal structures of VP40 have been solved. The monomeric VP40 is formed by two functionally interrelated domains: an N-terminal oligomerization domain and a C-terminal

Thomas Hoenen and Allison Groseth (eds.), *Ebolaviruses: Methods and Protocols*, Methods in Molecular Biology, vol. 1628, DOI 10.1007/978-1-4939-7116-9_8, © Springer Science+Business Media LLC 2017

membrane-binding domain [9]. Several oligomerization states of VP40 have also been described that may be associated with different functions [10]. Dimers of VP40 are formed in the cytosol and transported to the plasma membrane. Initial steps of VP40 intracellular trafficking have been proposed to be associated with components of the cytoskeleton, notably tubulin [11–13], as well as the COPII vesicular transport system and through an interaction of VP40 with Sec24C [14, 15]. The formation of the filamentous EBOV particle depends on VP40 hexamerization at the plasma membrane [10, 16]. These different oligomeric states of VP40 can be found in VLPs (Fig. 1a). Budding and release of VLPs are triggered by the recruitment of the components of the endosomal sorting complex required for transport (ESCRT) machinery TSG101 and Nedd4 through two overlapping late domains **PTA** *PPEY*[2, 17, 18]. In addition, VP40 can assemble octamers that bind RNA in a sequence-specific manner [19, 20]. While VP40 alone is able to initiate budding of VLPs, co-expressed NP and GP, the nucleoprotein and the attachment glycoprotein of EBOV, are incorporated into VLPs and significantly enhance their release (Fig. 1b) [21]. Only the mature glycoproteins are incorporated into VLPs (Fig. 1c), thus conferring to the particles an ability to infect target cells through specific EBOV receptors [5].

VP40 can be easily engineered with fluorescent tags for imaging or with small epitope tags to facilitate its immunoprecipitation. N-terminal tagging (epitope tags, β-lactamase tag, or fluorescent proteins) of VP40 does not modify its budding capacity, and co-expression of wild-type and tagged VP40 leads to equivalent incorporation of each protein into VLPs [14, 22, 23]. However large tags require the addition of a flexible linker (*see* **Note 1**). Overexpression of tagged VP40 in EBOV-infected cells can also lead to its incorporation into viral particles. For imaging, the use of an RFP-tagged VP40 allows perfect visualization of single VLPs at high magnification without the need for time-consuming staining and washing procedures (Fig. 2a). Such fluorescent VLPs can be further used to study viral particles internalization (Fig. 2b) (for MarburgVP40, *see* [24]). Alternatively, VLPs can also be stained using lipophilic dyes as octadecyl rhodamine B chloride (R18) or DiD.

Here we describe protocols to produce and concentrate VLPs from transfected cells and two alternative methods to purify VLPs. The first method is a flotation assay that relies on the separation of particles depending on the presence of a lipid envelope, while the second is a sedimentation assay that separates VLPs from contaminants through their respective sedimentation velocity (Fig. 3). The second method, while being more time-consuming and complex, offers a better degree of purity and even can lead to the separation of different vesicular forms depending on their size.

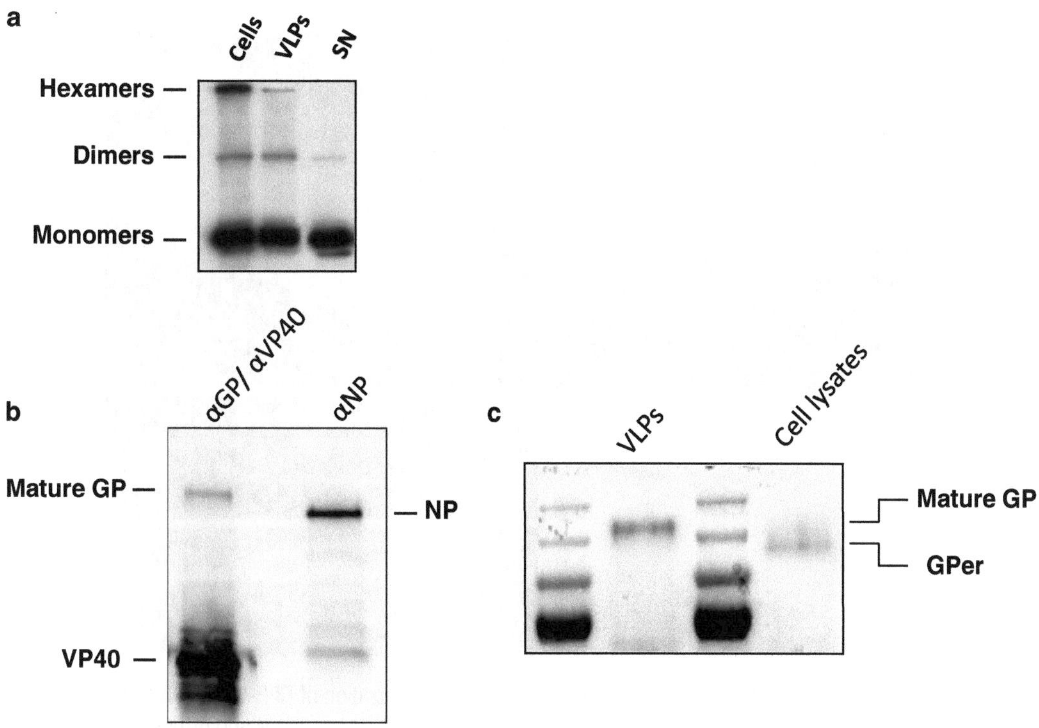

Fig. 1 Analysis of VLP production. (**a**) Immunoblot analysis of cell lysate (cells), virus-like particles (VLPs), and supernatant (SN) produced from VP40-transfected 293T cells. VLPs were concentrated by ultracentrifugation over a 20% sucrose cushion. Reproduced from [25] by permission from Oxford University Press/The Infectious Diseases Society of America (license 3931270354493). (**b**) Immunoblots showing the specific protein content of the VLPs analyzed after sucrose cushion purification (*left lane*, detection with a mix of mouse anti-VP40, clone 9B2, and anti-GP, clone 3373; right lane, detection with a mouse anti NP, clone ZDD4). (**c**) Immunoblot detected with an anti-mouse GP (clone 3373) showing the size of the glycoprotein incorporated into VLPs (GP mature only) compared to the GP present in the cell lysate, which includes both mature and incompletely processed forms found in the endoplasmic reticulum (GP mature + GPer) (*MWM* molecular weight marker).

2 Materials

2.1 Cells and Transfections

1. HEK 293T cells (ATCC #11268).
2. DMEM supplemented with 1% (v/v) 200 mM L-glutamine.
3. Fetal calf serum.
4. Transfection reagent (e.g., TurboFect, Thermo Fisher).
5. Eukaryotic expression vectors for VP40, NP, and GP.

2.2 VLP Release and Ultracentrifugation

1. PBS–CM: 0.144 g/L KH_2PO_4, 9 g/L NaCl, 0.795 g/L Na_2HPO_4-$7H_2O$, 0.1 mM $CaCl_2$, 1 mM $MgCl_2$.
2. Sucrose: 20% solution in PBS–CM (w/v) or 60% in PBS–CM (w/v) to adjust at 40%.
3. Table top low-speed centrifuge.

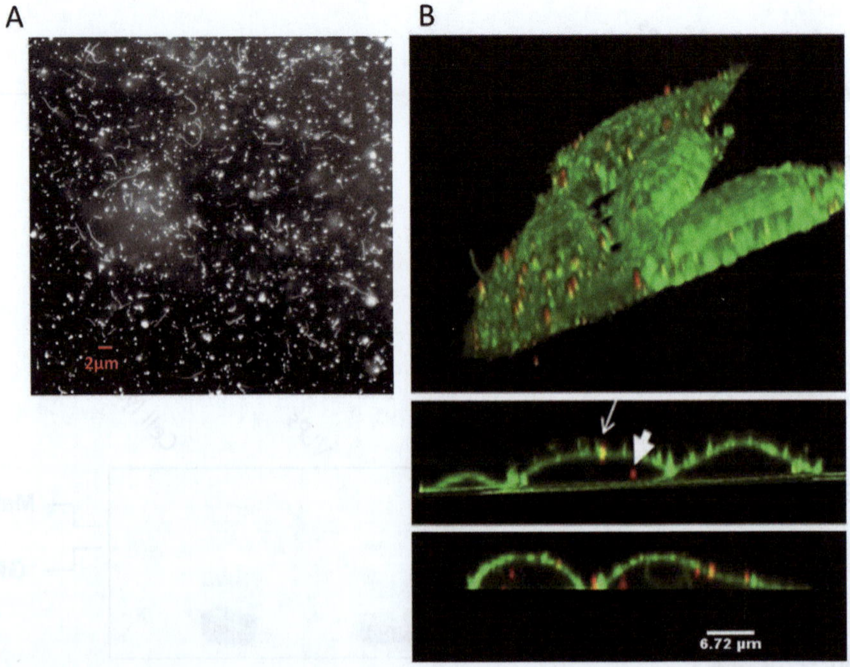

Fig. 2 Direct visualization of RFP-VLPs by optical microscopy. (**a**) Visualization of RFP-VLPs at 63× magnification using an UV inverted microscope (Zeiss Axio200). (**b**) 3D reconstruction (*upper panel*) and orthogonal slice (*lower panel*) of VeroE6 cells (cultured in an Ibidi cell culture slide) incubated in the presence of RFP-VLPs for 5 h and subsequently stained with wheat germ agglutinin coupled to Alexa 488. Pictures were taken using a Zeiss LSM510 confocal microscope and the stack was rebuilt using ImageJ software [26]. As imaging of VLPs requires the use of an oil-immersed objective, it is highly recommended to use ultrathin glass slides or culture glass slides (thickness ≈ 170 μm)

4. Ultracentrifuge with swinging rotor.
5. Gradient maker.
6. Peristaltic pump.

2.3 CO-IP Buffer

1. CO-IP buffer: 20 mM Tris–HCl, pH 7.0, 100 mM NaCl, 5 mM EDTA, 0.4% sodium deoxycholate (v/v), 1% Nonidet P40 (v/v).

2.4 Western Blot

1. Precast 4–20% gel or in-house prepared with 4% (w/v) stacking and 8% (w/v) resolving gel, prepared with acrylamide: bisacrylamide mixture (37.5:1).
2. Running buffer: Tris–glycine–SDS buffer (25 mM Tris, 192 mM glycine, 0.1% SDS).
3. Precast transfer sandwich containing 0.2 μm PVDF membrane (Bio-Rad Trans-Blot® Turbo™ Midi PVDF Transfer Packs).
4. 4× Laemmli buffer: 4% SDS (w/v), 20% glycerol (w/v), 10% 2-mercaptoethanol (w/v), 0.004% bromophenol blue (w/v), and 0.125 M Tris–HCl, pH approx. 6.8.

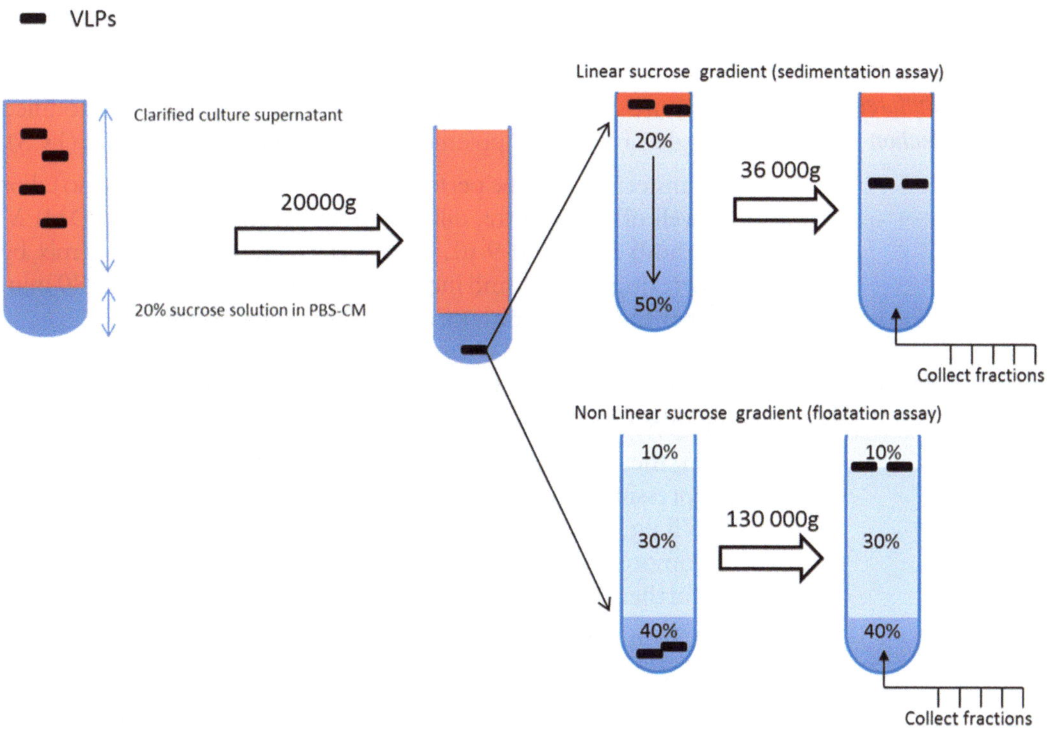

Fig. 3 Possible workflows of VLPs analysis. For VLP production, cells can be transfected in 12-well plates or up to 150cm^2 flasks (using 40 μg of total plasmid DNA). At 24 h post-transfection, the supernatant is harvested and then filtered or clarified by low-speed centrifugation before being subjected to ultracentrifugation on a 20% sucrose cushion. Then, the VLP pellet can either be analyzed by Western blot or further processed for particle purification through either a linear sucrose gradient for a sedimentation assay (migration of the VLPs from top to bottom) or using a nonlinear sucrose gradient for a flotation assay (migration of the VLPs from *bottom* to *top*)

5. Mini sodium dodecyl sulfate–polyacrylamide gel electrophoresis (SDS-PAGE) gel system (Bio-Rad Mini-Protean).
6. SDS-PAGE molecular weight standards.
7. 0.1% (v/v) Tween 20 in PBS (PBST).
8. 10% (w/v) skimmed milk in PBST.
9. Anti-VP40 monoclonal antibody (in-house production, commercial antibodies are available at IBT Bioservices).
10. Anti-mouse IgG horseradish peroxidase-conjugated secondary antibody (DAKO). Working solution is 1:25,000 in 1% milk PBST.
11. Western chemiluminescent substrate system.
12. Digital imager for chemiluminescence.

3 Methods

3.1 Cell Cultures and Transfection

1. Culture HEK 293T cells at 37 °C in Dulbecco's Modified Eagle's medium supplemented with 10% fetal calf serum (FCS).
2. Transfections can be performed using TurboFect (Thermo). For a well of 6-well plate, dilute 3 μg of plasmid in 400 μL DMEM without FCS, add 9 μL TurboFect (1 μg/3 μL ratio), mix by vortexing for 30 s, and incubate at room temperature for 30 min. Pipet the transfection mix onto the cell monolayer.

3.2 VLP Assay Budding

1. Plate 5 × 10^5 HEK293T cells per well in 6-well plates, and incubate overnight at 37 °C in the presence of 5% CO_2.
2. On the following day, transfect cells with 3 μg of plasmid expressing VP40, NP, and GP (ratio 7:3:2, in this case 1.75 μg, 0.75 μg, and 0.5 μg, respectively) under the control of a CMV promoter, and incubate at 37 °C in the presence of 5% CO_2 for a further 24 h (*see* **Note 2**).
3. At 24 h post-transfection, clarify the culture supernatant (SN) by centrifugation at 5000 × *g* for 3 min.
4. Load the supernatant on a 20% sucrose cushion in 4.5 mL ultracentrifuge tubes.
5. Centrifuge at 250,000 × *g* for 2 h at 4 °C in an ultracentrifuge.
6. Resuspend the VLP pellet in 150 μL of PBS–CM (*see* **Note 3**).
7. Add 50 μL 4× Laemmli buffer and boil samples at 95 °C for 5 min.
8. Separate samples by SDS-PAGE on a 10% polyacrylamide gel for Western blot analysis (fixed voltage at 110 V, 1 h).
9. Transfer the protein to a PVDF membrane using the Trans-Blot Turbo transfer pack (Bio-Rad), using 1.3A, 25 V, 7 min.
10. Block the membrane for at least 1 h in PBS/10% nonfat milk, then stain the membrane using anti-VP40 antibodies (diluted in PBS/ 0.1% nonfat milk) for 1 h, and then wash three times in PBS with 0.1% Tween 20 for 5 min. Afterward, stain with horseradish peroxidase (HRP)-conjugated secondary antibodies diluted in PBS/0.1% nonfat milk for 45 min. Wash three times in PBS with 0.1% Tween 20 for 5 min, followed by a final wash in PBS for 10 min.
11. Detect proteins using an HRP substrate such as chemiluminescence (ECL) HRP substrate with stable light output for mid-femtogram-level detection (as in Fig. 1).

3.3 Trypsin Protection Assay

1. Prepare and purify VLPs as in Subheading 3.2, **steps 1–5**.
2. Resuspend VLPs in PBS and split the sample in three equal portions, and add 2 mg/mL trypsin, 2 mg/mL trypsin with 1% Triton X100 (positive control), or 2 mg/mL trypsin with 1% Triton X100 + trypsin inhibitor (negative control), respectively (*see* **Note 4**).
3. Incubate samples for 15 min at 37 °C.
4. Add preheated (95 °C) 4× Laemmli buffer for Western blot analysis as in Subheading 3.2, and immediately boil samples, proceeding with Subheading 3.2, **steps 7–10**.

3.4 VLP Purification by Nonlinear Sucrose Gradient (Flotation Assay)

1. Clarify the SN of VP40-transfected cells as in Subheading 3.2, **steps 1–3** (*see* **Note 5**).
2. Adjust the clarified SN to 40% sucrose.
3. Load 12 mL of SN adjusted to 40% sucrose at the bottom of a 38 mL ultracentrifuge tube and layer first 21 mL of 30% sucrose and then 5 mL of 10% sucrose solutions in PBS–CM on top of it (*see* **Note 6**).
4. Centrifuge the samples at 130,000 × *g* for 20 h in an appropriate swinging rotor.
5. Collect gradient fractions from the bottom to the top (*see* **Note 6**) and analyze by Western blot using anti-VP40 antibodies. Bottom fractions (40% sucrose fractions) contain soluble proteins, while top fractions (10% sucrose fractions) contain membranes and membrane-associated proteins.
6. Analyze the fractions by SDS-PAGE followed by Western blot using anti-VP40 antibodies as in Subheading 3.2, **steps 7–10**.

3.5 VLPs Purification by Linear Sucrose Gradient (Sedimentation Assay)

1. Produce and pellet VLPs as indicated in Subheading 3.1, **steps 1–6**.
2. Load VLPs onto a 20–50% linear sucrose gradient generated by mixing a 20% and a 50% sucrose solution with a gradient maker and a peristaltic pump (*see* **Note** 7). Centrifuge at 36,000 × *g* for 14 h in an SW41 rotor.
3. Collect 0.5 mL fractions using a fraction collector plugged to a peristaltic pump (*see* **Note 8**).
4. Analyze the fractions by SDS-PAGE followed by Western blot using anti-VP40 antibodies as in Subheading 3.2, **steps 7–10**.

4 Notes

1. Sample construct with a flexible linker: Start codon–RFP–(glycine–serine)$_6$–VP40 ORF–stop codon.

2. After 24 h, protein expression will cause cell rounding but should not cause detachment.
3. Add 150 μL PBS–CM and incubate on ice for 30 min, and then resuspend the pellet thoroughly by pipetting.
4. The enveloped nature can be verified by performing a trypsin protection assay in which the proteins embedded in a lipid envelope will then be fully resistant to trypsin proteolysis, but susceptible to digestion following treatment with a detergent such as Triton X100 (Subheading 3.3).
5. Supernatants can also be filtered through 0.45 μm filters, but large VLPs may be retained by the filter.
6. The 40% sucrose can be loaded first, as described, or alternatively the 30% and 10% sucrose solutions could be added first and the 40% sucrose sample solution added through the first two layers with a fine pipet (e.g., a glass Pasteur pipet). Proceeding this way will generate very clean solution interfaces that will result in cleaner results at analysis.
7. In the absence of gradient maker, linear gradients can alternatively be generated using the diffusion method. Typically, divide the gradient into six equal portions (prepare 20, 26, 32, 38, 44, and 50% sucrose solutions). Load first the 20% solution into the tube, then add the 26% solution at the bottom of the 20% layer, and so on up to the final 50% solution. Close the tube tightly with parafilm, and carefully lay it horizontally for 3 h at 4 °C; then, carefully move it back to a vertical position, and load the VLP sample on top.
8. If a fraction collector is not available, the bottom of the tube can be perforated perpendicular to the tube with a red-hot 21-gauge needle, in such a way that the needle will finally be placed inside the tube at 1 or 2 mm above the bottom of the tube with the needle bevel orientated toward the top. This should allow for a drip collection of the sample.

References

1. Feldmann H, Sanchez A, Geisbert T (2013) Filoviridae: Marburg and Ebola viruses. In: Fields BN, Knipe DM, Howley PM (eds) Fields virology, 6th edn. Lippincott Williams & Wilkins, Philadelphia, pp 923–956
2. Harty RN, Brown ME, Wang G, Huibregtse J, Hayes FP (2000) A PPxY motif within the VP40 protein of Ebola virus interacts physically and functionally with a ubiquitin ligase: implications for filovirus budding. Proc Natl Acad Sci U S A 97(25):13871–13876. doi:10.1073/pnas.250277297
3. Jasenosky LD, Neumann G, Lukashevich I, Kawaoka Y (2001) Ebola virus VP40-induced particle formation and association with the lipid bilayer. J Virol 75(11):5205–5214. doi:10.1128/JVI.75.11.5205-5214.2001
4. Timmins J, Scianimanico S, Schoehn G, Weissenhorn W (2001) Vesicular release of ebola virus matrix protein VP40. Virology 283(1):1–6. doi:10.1006/viro.2001.0860
5. Hoenen T, Groseth A, Kolesnikova L, Theriault S, Ebihara H, Hartlieb B, Bamberg S, Feldmann H, Stroher U, Becker S (2006)

Infection of naive target cells with virus-like particles: implications for the function of ebola virus VP24. J Virol 80(14):7260–7264. doi:10.1128/JVI.00051-06

6. Warfield KL, Bosio CM, Welcher BC, Deal EM, Mohamadzadeh M, Schmaljohn A, Aman MJ, Bavari S (2003) Ebola virus-like particles protect from lethal Ebola virus infection. Proc Natl Acad Sci U S A 100(26):15889–15894. doi:10.1073/pnas.2237038100
7. Noda T, Sagara H, Suzuki E, Takada A, Kida H, Kawaoka Y (2002) Ebola virus VP40 drives the formation of virus-like filamentous particles along with GP. J Virol 76(10):4855–4865
8. Beniac DR, Melito PL, Devarennes SL, Hiebert SL, Rabb MJ, Lamboo LL, Jones SM, Booth TF (2012) The organisation of ebola virus reveals a capacity for extensive, modular polyploidy. PLoS One 7(1):e29608. doi:10.1371/journal.pone.0029608
9. Dessen A, Volchkov V, Dolnik O, Klenk HD, Weissenhorn W (2000) Crystal structure of the matrix protein VP40 from Ebola virus. EMBO J 19(16):4228–4236
10. Timmins J, Schoehn G, Kohlhaas C, Klenk HD, Ruigrok RW, Weissenhorn W (2003a) Oligomerization and polymerization of the filovirus matrix protein VP40. Virology 312(2):359–368
11. Han Z, Harty RN (2005) Packaging of actin into Ebola virus VLPs. Virol J 2:92
12. Noda T, Ebihara H, Muramoto Y, Fujii K, Takada A, Sagara H, Kim JH, Kida H, Feldmann H, Kawaoka Y (2006) Assembly and budding of Ebolavirus. PLoS Pathog 2(9):e99. doi:10.1371/journal.ppat.0020099
13. Ruthel G, Demmin GL, Kallstrom G, Javid MP, Badie SS, Will AB, Nelle T, Schokman R, Nguyen TL, Carra JH, Bavari S, Aman MJ (2005) Association of ebola virus matrix protein VP40 with microtubules. J Virol 79(8):4709–4719. doi:10.1128/JVI.79.8.4709-4719.2005
14. Reynard O, Nemirov K, Page A, Mateo M, Raoul H, Weissenhorn W, Volchkov VE (2011a) Conserved proline-rich region of Ebola virus matrix protein VP40 is essential for plasma membrane targeting and virus-like particle release. J Infect Dis 204(Suppl 3):S884–S891. doi:10.1093/infdis/jir359
15. Yamayoshi S, Noda T, Ebihara H, Goto H, Morikawa Y, Lukashevich IS, Neumann G, Feldmann H, Kawaoka Y (2008) Ebola virus matrix protein VP40 uses the COPII transport system for its intracellular transport. Cell Host Microbe 3(3):168–177
16. Ruigrok RW, Schoehn G, Dessen A, Forest E, Volchkov V, Dolnik O, Klenk HD, Weissenhorn W (2000) Structural characterization and membrane binding properties of the matrix protein VP40 of Ebola virus. J Mol Biol 300(1):103–112
17. Licata JM, Simpson-Holley M, Wright NT, Han Z, Paragas J, Harty RN (2003) Overlapping motifs (PTAP and PPEY) within the Ebola virus VP40 protein function independently as late budding domains: involvement of host proteins TSG101 and VPS-4. J Virol 77(3):1812–1819
18. Timmins J, Schoehn G, Ricard-Blum S, Scianimanico S, Vernet T, Ruigrok RW, Weissenhorn W (2003b) Ebola virus matrix protein VP40 interaction with human cellular factors Tsg101 and Nedd4. J Mol Biol 326(2):493–502
19. Gomis-Ruth FX, Dessen A, Timmins J, Bracher A, Kolesnikowa L, Becker S, Klenk HD, Weissenhorn W (2003) The matrix protein VP40 from Ebola virus octamerizes into pore-like structures with specific RNA binding properties. Structure 11(4):423–433
20. Hoenen T, Volchkov V, Kolesnikova L, Mittler E, Timmins J, Ottmann M, Reynard O, Becker S, Weissenhorn W (2005) VP40 octamers are essential for Ebola virus replication. J Virol 79(3):1898–1905. doi:10.1128/JVI.79.3.1898-1905.2005
21. Licata JM, Johnson RF, Han Z, Harty RN (2004) Contribution of ebola virus glycoprotein, nucleoprotein, and VP24 to budding of VP40 virus-like particles. J Virol 78(14):7344–7351
22. Martinez O, Johnson J, Manicassamy B, Rong L, Olinger GG, Hensley LE, Basler CF (2010) Zaire Ebola virus entry into human dendritic cells is insensitive to cathepsin L inhibition. Cell Microbiol 12(2):148–157. doi:10.1111/j.1462-5822.2009.01385.x
23. Tscherne DM, Manicassamy B, Garcia-Sastre A (2010) An enzymatic virus-like particle assay for sensitive detection of virus entry. J Virol Methods 163(2):336–343. doi:10.1016/j.jviromet.2009.10.020
24. Schudt G, Kolesnikova L, Dolnik O, Sodeik B, Becker S (2013) Live-cell imaging of Marburg virus-infected cells uncovers actin-dependent transport of nucleocapsids over long distances. Proc Natl Acad Sci U S A 110(35):14402–14407. doi:10.1073/pnas.1307681110
25. Reynard O, Reid SP, Page A, Mateo M, Alazard-Dany N, Raoul H, Basler CF, Volchkov VE (2011b) Unconventional secretion of Ebola virus matrix protein VP40. J Infect Dis 204(Suppl 3):S833–S839. doi:10.1093/infdis/jir305
26. Schneider CA, Rasband WS, Eliceiri KW (2012) NIH image to ImageJ: 25 years of image analysis. Nat Methods 9(7):671–675

Chapter 9

Modeling the Ebolavirus Life Cycle with Transcription and Replication-Competent Viruslike Particle Assays

Nadine Biedenkopf and Thomas Hoenen

Abstract

Ebolaviruses are the causative agent of a severe hemorrhagic fever with high case fatality rates, for which no approved specific therapy is available. As biosafety level 4 (BSL4) agents, work with live ebolaviruses is restricted to maximum containment laboratories. Transcription and replication-competent viruslike particle (trVLP) systems are reverse genetics-based life cycle modeling systems that allow researchers to model virtually the entire ebolavirus life cycle outside of a maximum containment laboratory. These systems can be used to dissect the virus life cycle, and thus increase our understanding of virus biology, as well as for more applied uses such as the screening and development of novel antivirals, and thus represent powerful tools for work on ebolaviruses.

Key words Ebolaviruses, Filoviruses, Reverse genetics, Life cycle modeling system, Transcription and replication-competent viruslike particle system, trVLPs

1 Introduction

Most ebolaviruses cause severe hemorrhagic fevers in humans and nonhuman primates, with case fatality rates of up 90% [1]. While there has been significant progress in the development of countermeasures against ebolaviruses over the last years, there still is no approved specific therapy or vaccine available [2], and ebolaviruses are classified as biosafety level 4 (BSL4) agents. Therefore, all work with live virus is restricted to only a few maximum containment facilities worldwide, posing a significant obstacle for research on these viruses.

Reverse genetics systems, which we define as systems that generate viral genomes or genome analogues from cDNA plasmids, can be divided into full-length clone systems that allow the generation of recombinant viruses, and life cycle modeling systems, which model either individual aspects or the whole virus life cycle under BSL1 or BSL2 conditions (*see* **Note 1**) [3]. These life cycle modeling systems employ miniature versions of the viral genome

Thomas Hoenen and Allison Groseth (eds.), *Ebolaviruses: Methods and Protocols*, Methods in Molecular Biology, vol. 1628, DOI 10.1007/978-1-4939-7116-9_9, © Springer Science+Business Media LLC 2017

called minigenomes (Fig. 1a), in which some or all viral open reading frames (ORFs) have been removed and replaced by reporter ORFs. However, these minigenomes still contain the noncoding genome-termini, called the leader and trailer, which harbor the signals for recognition of the minigenomes by ebolavirus proteins as authentic templates for viral genome replication, transcription, and packaging into nascent particles [4]. Basic minigenome systems allow the study of ebolavirus genome replication and transcription. In order to model additional aspects of the virus life cycle, these systems have been expanded into transcription and replication-competent viruslike particle (trVLP) systems [5–7], with different versions reflecting virtually every aspect of the virus life cycle (Table 1). However, all of these systems can be safely used under BSL1 or BSL2 conditions (*see* **Note 1**).

In a trVLP system (Fig. 1b, c), minigenome RNAs are initially transcribed in mammalian cells (called producer or p0 cells) either by the exogenous T7 polymerase or by cellular polymerases, with ribozyme sequences ensuring authentic transcript termini (for details, the interested reader is referred to [8]). These initially "naked" minigenomes are illegitimately encapsidated by the ebolavirus nucleoprotein NP and then recognized by the ribonucleoprotein complex (RNP) proteins NP, L (viral polymerase), VP35 (polymerase cofactor), and VP30 (transcriptional activator). This leads to viral genome replication and transcription, and thus the generation of reporter mRNAs and ultimately reporter activity, reflecting these processes. In addition, the other ebolavirus structural proteins VP40 (matrix protein, responsible for morphogenesis and budding), $GP_{1,2}$ (surface glycoprotein, responsible for attachment and fusion), and VP24 (nucleocapsid-associated protein, involved in nucleocapsid morphogenesis) are also expressed in a trVLP system, leading to the formation of trVLPs that can incorporate minigenome-containing nucleocapsids and mediate entry into target cells. In the case of trVLP systems featuring a monocistronic minigenome, these proteins are overexpressed from expression plasmids (Fig. 1b), which can be somewhat problematic because VP40 and VP24 are inhibitors of (mini-)genome replication and transcription [9, 10] and because this results in a large number of noninfectious trVLPs being produced [11]. In contrast, in tetracistronic trVLP systems, these proteins are expressed as product of viral genome replication and transcription from the minigenome itself, which encodes the ORFs for the reporter

Fig. 1 (continued) **(c)** Tetracistronic trVLP assay. Similar to a monocistronic trVLP system, in a tetracistronic trVLP system minigenomes are replicated and transcribed, and trVLPs containing these minigenomes are produced; however, the source of the viral proteins VP40, $GP_{1,2}$, and VP24 is the minigenome instead of expression plasmids. In contrast to a monocistronic minigenome, multiple infectious cycles can be modeled without transfecting plasmids encoding for VP40, $GP_{1,2}$, and VP24 in target cells. Copyright (for panel C) © 2014, American Society for Microbiology. Reprinted from [7] with permission of the American Society for Microbiology

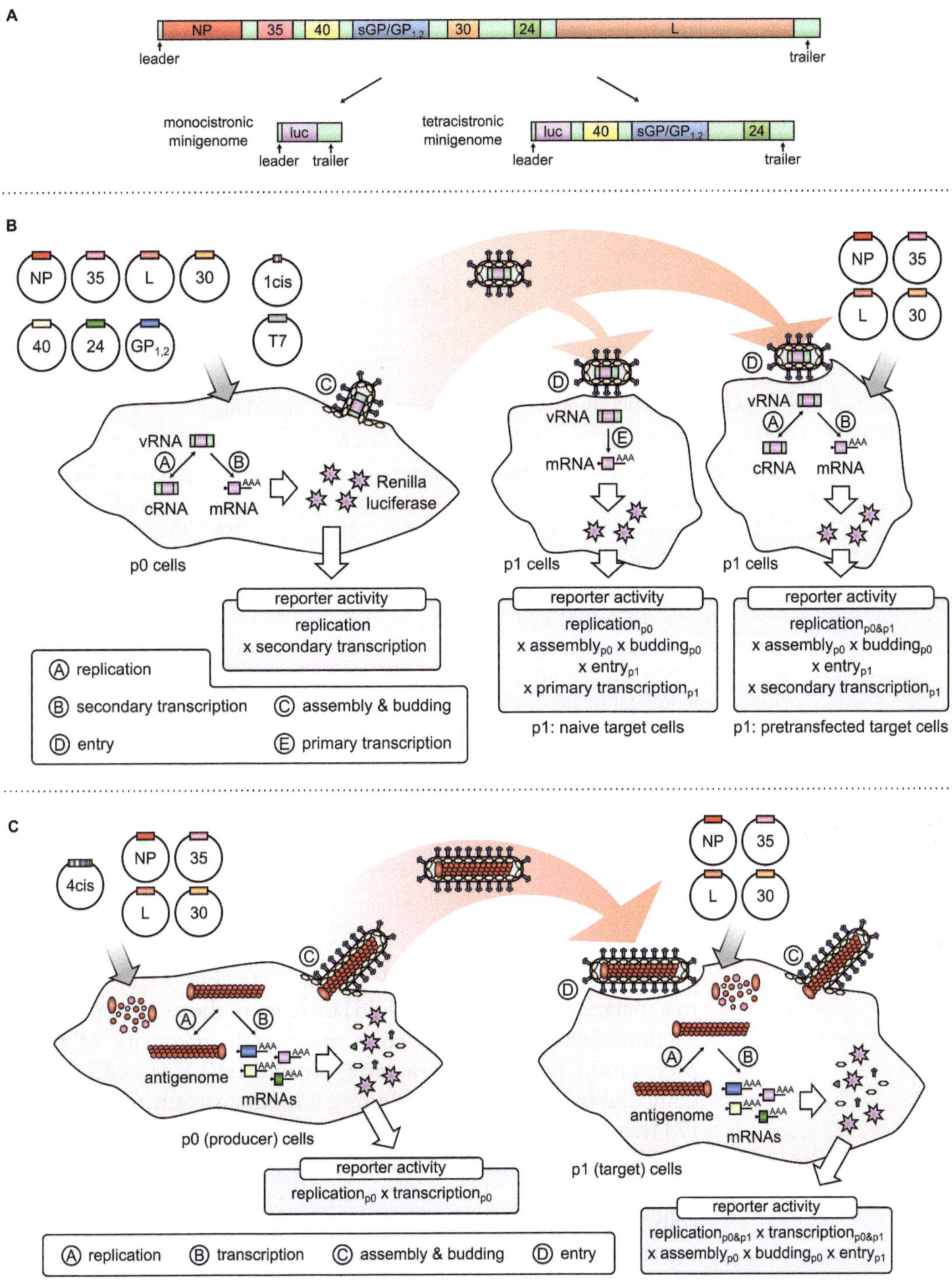

Fig. 1 trVLP system overview. (**a**) Minigenomes. An ebolavirus full-length genome and derived mono- and tetracistronic minigenomes are shown. (**b**) Monocistronic trVLP assay. A monocistronic minigenome is expressed in mammalian cells together with the viral proteins NP, VP35, VP30, and L. After initial transcription by a coexpressed T7 polymerase the minigenome is replicated and transcribed by these proteins. Additional expression of VP40, $GP_{1,2}$, and VP24 leads to the formation of trVLPs, which can infect target cells.

Table 1
Overview of different trVLP systems. Basic properties for each system, including the steps modeled, but also the artificial aspects, are listed

	Classical monocistronic trVLP system	Monocistronic trVLP system with naïve target cells	Tetracistronic trVLP system
Minigenome	Monocistronic	Monocistronic	Tetracistronic
Target cells	Pretransfected	Naïve	Pretransfected
Multiple infectious cycles	No	No	Yes
Modeled aspects in producer cells	Genome replication and secondary transcription, budding		
Modeled aspects in target cells	Entry, genome replication and secondary transcription	Entry, primary transcription	Entry, genome replication and secondary transcription, budding
Artificial aspects in producer cells	Initial minigenome transcription, illegitimate encapsidation, plasmid-driven expression of RNP proteins, plasmid-driven kinetically unregulated overexpression of VP40, $GP_{1,2}$, and VP24		Initial minigenome transcription, illegitimate encapsidation, plasmid-driven expression of RNP proteins
Artificial aspects in target cells	Plasmid-driven gene expression of RNP proteins		Plasmid-driven gene expression of RNP proteins

protein as well as for VP40, $GP_{1,2}$, and VP24 (Fig. 1a, c). This leads to a marked (more than 100-fold) increase in the ratio of infectious to noninfectious trVLPs, while the overall infectivity of trVLP preparation produced in a tetracistronic trVLP system remains comparable to that produced using a monocistronic trVLP system [7] (*see* **Note 2**).

trVLPs can be used to infect target cells (called target or p1 cells) and deliver the minigenome they carry into these target cells (Fig. 1b). In the case of naïve target cells, the minigenome is subject to primary transcription mediated by the RNP components brought into target cells within trVLPs, but no further genome replication or secondary transcription (using RNP proteins produced in target cells) takes place. In contrast, if target cells are pretransfected with expression plasmids encoding the RNP proteins, minigenomes are replicated and undergo secondary transcription (however, primary transcription cannot be assessed independently in this case). Further, if tetracistronic minigenomes are used (Fig. 1c) to infect RNP protein pretransfected target cells, VP40, $GP_{1,2}$, and VP24 are also produced once genome replication and transcription takes place. Thus, new trV-LPs are generated, which can be transferred to another set of target cells (p2 cells), thereby allowing us to model multiple infectious cycles under BSL1 or BSL2 conditions [7, 12]. However, as with the monocistronic trVLP system, infection is restricted to cell pretransfected to express the RNP proteins, and thus they are also safe to use under BSL1/2 conditions.

While ebolavirus trVLP systems have many advantages, namely, the fact that they can be safely used outside of maximum containment laboratories, and the fact that they allow researchers to dissect the virus life cycle and, depending on the system used, to analyze its individual aspects in isolation, they also have some shortcomings that need to be considered. Particularly, they take advantage of cellular processes in order to function that are not necessarily requirements for live ebolaviruses. Particularly, in all these systems, at least some of the ebolavirus proteins have to be expressed from plasmids, whereas the ebolavirus life cycle is, of course, independent of plasmid-driven gene expression. Since plasmid-driven gene expression relies on the cellular Pol-II machinery for transcription of mRNAs, whereas a role of this machinery for viral genome replication and transcription during infection is not known to exist, influences on the Pol-II machinery as a result of experimental perturbations could lead to results falsely suggesting a direct influence on the ebolavirus life cycle. Further, in contrast to expression from viral genomes in context of an infection, plasmid-driven expression cannot be as well-regulated in terms of expression levels and kinetics, potentially skewing results. Use of the tetracistronic trVLP system partially alleviates this issue, at least for VP40, $GP_{1,2}$, and VP24. In producer cells, additional artificial aspects without an equivalent in the virus life cycle include the initial minigenome transcription either by T7 polymerase or the cellular Pol-I, which leads to the production of naked minigenomes with the potential to form dsRNA structures, and the illegitimate encapsidation of these minigenomes by NP. However, an influence of these steps can be excluded by introducing experimental perturbations in target cells rather than in producer cells (*see* **Note 3**). Finally, minigenomes can be significantly shorter than full-length genomes, and recent work has demonstrated that there can be dramatic differences in experimental results depending on the length of the minigenomes being used [7].

Nevertheless, trVLP systems represent extremely powerful tools to further our understanding of ebolavirus biology [7, 13] and also have the potential to be used for high-throughput screening approaches for the discovery and the development of novel countermeasures [14, 15]. Here, we will provide details on the use of ebolavirus monocistronic and tetracistronic trVLP systems with a Renilla luciferase reporter encoded by the minigenome and a

Firefly luciferase as a control reporter to assess cell viability and effects on plasmid-driven gene expression in producer cells and to exclude cellular carry-over from producer cells to target cells.

2 Materials

1. HEK 293 cells and HuH7 cells (*see* **Note 4**).
2. Dulbecco's Modified Eagle's Medium (DMEM) with 10% (v/v), 5% (v/v) or 0% fetal bovine serum (FBS, heat-inactivated 30 min at 56 °C) and 1% L-glutamine (Q, 2 mM) and 1% penicillin/streptomycin (PS, 100 U/mL/100 μg/mL). These three formulations will be referred to as $DMEM_{10\%}$, $DMEM_{5\%}$, and $DMEM_{0\%}$, respectively.
3. Opti-MEM (Thermo Fisher Scientific).
4. 6-well plates.
5. DNA plasmids pCAGGS-NP, pCAGGS-VP35, pCAGGS-VP30, pCAGGS-L, pCAGGS-VP24, pCAGGS-VP40, pCAGGS-$GP_{1,2}$, pCAGGS-T7, pCAGGS-luc2 (Firefly luciferase), pCAGGS-GFP (or another fluorescent protein), p1cis-vRNA-Rluc, or p4cis-vRNA-Rluc.
6. Transit-LT1 transfection reagent (Mirus Bio).
7. High-speed ultracentrifuge (e.g., OptimaTM L-100K) with swing-out ultracentrifugation rotor (e.g., SW41 Ti or SW60 Ti) and corresponding clear, thin-walled centrifugation tubes (*see* **Note 5**).
8. 20% Sucrose in 1× TNE:1 mM EDTA (pH 8), 150 mM NaCl, 10 mM Tris–HCl (pH 7.5), sterile filtered.
9. Phosphate-buffered saline without $CaCl_2$ and $MgCl_2$ (PBS_{def}):140 mM NaCl, 10 mM Na_2HPO_4, 2.7 mM KCl, 1.8 mM KH_2PO_4.
10. GloLysis buffer, BrightGlo and RenillaGlo reagents (Promega), or equivalent reagents (*see* **Note 6**).
11. Opaque white or black 96-well plates, flat bottom (*see* **Note 7**).
12. Luminometer capable of reading 96-well plates.

3 Methods

3.1 Transfection of trVLP Components in p0 Cells

1. Split 293 cells (*see* **Note 4**) into 6-well plates in a volume of 3 mL $DMEM_{10\%}$ per well for a confluency of ~50% to 60% on the next day (*see* **Note 8**). Incubate cells at 37 °C and 5% CO_2 in a humidified incubator.

Table 2
DNA amounts for transfection of producer (p0) cells

Plasmid	Amount per well (6-well format) Monocistronic trVLP assay	Tetracistronic trVLP assay
pCAGGS-NP	125 ng	125 ng
pCAGGS-L (or pCAGGS-GFP)	1000 ng	1000 ng
pCAGGS-VP35	125 ng	125 ng
pCAGGS-VP30	75 ng	75 ng
pCAGGS-VP24	30 ng	–
pCAGGS-VP40	250 ng	–
pCAGGS-GP1,2	250 ng	–
pCAGGS-luc2	25 ng	25 ng
p1cis-vRNA-RLuc	250 ng	–
p4cis-vRNA-RLuc	–	250 ng

2. Pipette DNA following the amounts in Table 2 into a 1.5 mL reaction tube (*see* **Note 9**). Prepare a negative control by replacing the pCAGGS-L plasmid with an expression plasmid for GFP or another fluorescent reporter or with empty pCAGGS vector.
3. In a biosafety cabinet, add 100 μL/well Opti-MEM to the tube. Vortex, and briefly spin down. Add 7.5 μL/well Transit-LT1 to the tube (vortex gently prior to use). Vortex gently, and incubate for 15–30 min.
4. In the meantime, change the medium on the cells to 2 mL $DMEM_{5\%}$.
5. After 15–30 min, mix the formed transfection complexes gently by pipetting, and add 100 μL/well dropwise to the cells. Try to cover the whole well with the droplets.
6. Rock the plates from side to side and back and forth—do not rotate or swirl the plates, to avoid uneven distribution of the transfection complexes. Return the cells to the incubator.
7. After 24 h, change the medium to 3 mL $DMEM_{5\%}$. If you have substituted L against a fluorescent reporter in the negative control, check for fluorescence to obtain an estimate for the efficiency of transfection (this should be >50%). Return the cells to the incubator.

8. Incubate the cells for additional 48 h (72 h post transfection).
9. At 72 h post transfection, collect the supernatant containing the trVLPs in a sterile conical 15 mL tube (*see* **Note 10**). For the trVLPs, proceed with Subheading 3.2.
10. Continue with lysis of the cells and reporter assay for luciferase expression as described in Subheading 3.5.

3.2 Purification of trVLPs by Ultracentrifugation

1. Spin down cell debris from supernatants for 10 min at 3000 × *g* and 4 °C.
2. Add 1 mL sterile 20% sucrose/TNE to the bottom of the ultracentrifugation tube without touching the wall.
3. Carefully pipette 3 mL of precleared supernatant onto the sucrose cushion without disturbing the interface (*see* **Notes 10** and **5**).
4. Adjust precisely the weight of each tube to balance them. Adjust, if necessary, with sterile PBS_{def}.
5. Spin trVLPs for 2 h in a high-speed ultracentrifuge at 160,000 × *g* at 4 °C (e.g., in a SW60 Ti at 40,000 rpm).
6. After centrifugation, carefully decant the supernatant. Dry the tube with a tissue without touching the pellet at the bottom.
7. Resuspend the trVLP pellet in 500 μL $DMEM_{0\%}$ while minimizing production of air bubbles (*see* **Note 11**).
8. Continue with Subheading 3.3 (for infection of naïve target cells) or Subheading 3.4 (for infection of pretransfected target cells).

3.3 Infection of Naïve Target Cells (p1) by trVLPs

1. Split HuH7 cells (*see* **Note 4**) into 6-well plates in a volume of 3 mL $DMEM_{10\%}$ per well for a confluency of ~50% on the next day (*see* **Note 12**). Incubate cells at 37 °C and 5% CO_2 in a humidified incubator.
2. Remove supernatant from the HuH7 cells.
3. Wash the cells once with 1 mL $DMEM_{0\%}$.
4. Add 500 μL of trVLPs dropwise to the cells. Rock the plates from side to side and back and forth to evenly distribute the trVLPs.
5. Incubate cells at 37 °C and 5% CO_2 in a humidified incubator for 1 h.
6. Carefully rock the plates from side to side and back and forth every 15 min to avoid drying out of the cells.
7. After incubation, add 3 mL $DMEM_{5\%}$ to the cells.
8. Incubate cells at 37 °C and 5% CO_2 in a humidified incubator for an additional 60 h.
9. Continue with Subheading 3.5.

Table 3
DNA amounts for pretransfection of target cells (p1) by RNP plasmids

Plasmid	Amount per well (6-well format)
pCAGGS-NP	125 ng
pCAGGS-L	1000 ng
pCAGGS-VP35	125 ng
pCAGGS-VP30	75 ng

3.4 Infection of Pretransfected Target Cells (p1) by trVLPs

1. Split HuH7 cells (*see* **Note 4**) into 6-well plates in a volume of 3 mL $DMEM_{10\%}$ per well for a confluency of ~50% on the next day (*see* **Note 12**). Incubate cells at 37 °C and 5% CO_2 in a humidified incubator.
2. Carry out transfection of the HuH7 cells four to 6 h prior to infection.
3. Pipette DNA following the amounts in Table 3 into a 1.5 mL reaction tube (*see* **Note 9**). Prepare a negative control by replacing the pCAGGS-L plasmid with an expression plasmid for GFP or another fluorescent reporter or with empty pCAGGS vector.
4. In a biosafety cabinet, add 100 μL/well Opti-MEM to the tube. Vortex, and briefly spin down. Add 2.8 μL/well Transit-LT1 to the tube (vortex gently prior to use). Vortex gently, and incubate for 15–30 min.
5. In the meantime, change the medium on the cells to 2 mL $DMEM_{5\%}$.
6. After 15–30 min, mix the formed transfection complexes gently by pipetting, and add 100 μL/well dropwise to the cells. Try to cover the whole well with the droplets.
7. Rock the plates from side to side and back and forth—do not rotate or swirl the plates, to avoid uneven distribution of the transfection complexes. Return cells to the incubator.
8. After 4–6 h, remove the supernatant from cells.
9. Wash the cells once with 1 mL $DMEM_{0\%}$.
10. Add 500 μL trVLPs dropwise to the cells. Rock the plates from side to side and back and forth to evenly distribute the trVLPs.
11. Incubate cells at 37 °C and 5% CO_2 in a humidified incubator for 1 h.
12. Carefully rock the plates from side to side and back and forth every 15 min to avoid drying out of the cells.

13. After incubation, add 3 mL $DMEM_{5\%}$ to the cells.
14. Incubate cells at 37 °C and 5% CO_2 in a humidified incubator for additional 60 h.
15. Continue with Subheading 3.5.

3.5 Measuring Reporter Activity in Producer (p0) and Target (p1) Cells

1. For p0 cells, add 200 μL/well of 1× GloLysis buffer (diluted in H_2O) to each well after supernatants have been collected. For p1 cells, take off the supernatant and add 200 μL/well of 1× GloLysis buffer to each well. Incubate for 10 min at room temperature.
2. Set pipette to 100 μL, and wash the cells into the GloLysis buffer (*see* **Note 13**). Then transfer the lysates into a 1.5 mL tube.
3. Spin down the lysates for 3 min at 10,000 × *g* and 4 °C, and transfer the supernatants to a fresh tube. Avoid disturbing the pellet of cell debris.
4. Thaw 40 μL/well BrightGlo reagent and prepare 40 μL/well RenillaGlo reagent (thaw RenillaGlo buffer, and add 1% RenillaGlo substrate).
5. For each sample, pipet 40 μL BrightGlo reagent into the wells of an opaque white or black (*see* **Note** 7) 96-well plate, and pipet 40 μL RenillaGlo reagent into separate wells (*see* **Note 14**).
6. Add 40 μL of each sample to corresponding wells with BrightGlo reagent or RenillaGlo reagent. Be careful not to carry over anything between the two different wells (i.e., do not reuse the same tip for pipetting the same sample into the different reagents).
7. After 5–10 min, measure the reporter activity in a luminometer. Generally, an integration time of 0.5–1 s should provide sufficiently strong signals (*see* **Note 15**). While the absolute values are dependent on the plate color and on the specific luminometer used, positive (+L) controls should be 100- to 1000-fold stronger than the negative (−L) controls.
8. Normalize Renilla values to Firefly values to compensate for slight differences in plasmid-driven gene expression, or report both values independently (*see* **Note 16**).

4 Notes

1. Regulations for the biosafety classification of minigenome and trVLP systems vary from country to country. One major factor contributing to these differences in assessment is the classification of the mammalian cells in which these systems are used. For example, in the United States, 293 cells, which are the most commonly used cell type in minigenome and trVLP

systems, are considered risk group 2, so that such systems have to be used under BSL2 conditions, whereas in Germany, the same cells are considered risk group 1, and thus these systems in Germany are considered safe for use under biosafety level 1 conditions.

2. In addition, in our experience, the tetracistronic trVLP system seems to be somewhat easier to handle/establish than the monocistronic trVLP system and to be more robust with respect to experimental variations.
3. For the same reason, it can be wise to study virus genome replication and transcription in target cells infected with trVLPs rather than by using a classical minigenome system, albeit at the cost of being no longer able to clearly distinguish effects on genome replication and transcription from effects on particle entry.
4. While 293 cells are most often used as p0 cells due to their good transfectability, infection of p1 cells by trVLPs is mainly performed in HuH7 cells.
5. For harvesting the supernatant from a 6-well (3 mL volume), centrifugation at 40,000 rpm using a Beckmann Coulter SW60 Ti rotor is suitable (max. volume 4 mL). For upscaling the amount of trVLPs, supernatants can be pooled and centrifuged using a SW40 Ti (max. volume 12 mL, in which case 2 mL of 20% sucrose/TNE cushion should be used) or SW32 (max. volume 32 mL, in which case 5 mL of 20% sucrose/TNE cushion should be used). For large volumes of supernatants, it may be easier to fill the ultracentrifugation tube with the supernatant and underlay it carefully with the 20% sucrose/TNE cushion.
6. Luciferase Reagent Note Luciferase assay reagents include reagents for Renilla luciferase measurement (EBOV-specific minigenome) and Firefly luciferase measurement (transfection control), which can be purchased from various companies. Additionally, a luciferase compatible lysis buffer is required that ensures measurement of both luciferases. Here we describe the procedure using Promega reagents; if reagents from other manufacturers are used, procedures might have to be slightly adjusted.
7. While white plates result in about 100× stronger signals than black plates, they also tend to be more susceptible to problems with crosstalk. For this reason, and because signals produced by the Ebola minigenome/trVLP systems tend to be rather high, in most cases the use of black plates is preferable.
8. Usually, 8×10^5 HEK 293 cells per well will result in ~60% confluency the next day.
9. In our experience, this step can be done outside a biosafety cabinet without jeopardizing the experiment (particularly since

the preparation of plasmid DNA is also performed outside a biosafety cabinet). All further steps should be performed in a biosafety cabinet (with the exception of cell harvest and measuring of luciferase activity).

10. It is important to use filtered tips and sterile solutions (20% sucrose in 1× TNE, PBS) once the transfection is completed (e.g., for harvesting the supernatants, preparing the ultracentrifugation tubes, resuspension of trVLPs) to minimize the risk of contamination with detergents which would result in defective trVLPs.

11. Optionally, an aliquot of the purified trVLPs can be taken for Western blot analysis. Resuspension of purified trVLPs should then be performed in 50 μL $DMEM_{0\%}$ to increase their concentration. Take 10 μL of trVLPs for Western blot analysis. Take the remaining 40 μL trVLPs, add 460 μL $DMEM_{0\%}$ (final volume 500 μL), and continue with infection of p1 cells as described in Subheadings 3.3 and 3.4.

12. Usually, 4×10^5 HuH7 cells per well will result in ~50% confluency the next day.

13. Make sure that this is done as uniformly as possible for different wells, in order to ensure an even degree of lysis among the wells.

14. Crosstalk between wells can be an issue, particularly since the Firefly luciferase values tend to be much higher than the Renilla luciferase values. Therefore, it makes sense to cluster all the Renilla wells and all the Firefly wells, but to keep at least one empty row between them.

15. Ensure that values are in the linear range of your luminometer; otherwise dilute the samples in PBS_{def}.

16. Normalization to a control reporter can be somewhat problematic: If differences in the control reporter activity between wells are due to slight experimental variations during sample harvest, one can assume that the relationship between the effect on the control reporter activity and the minigenome reporter activity is linear, making normalization a valid procedure to compensate for this. Similarly, if differences in control reporter activity between wells are due to differences in cell number (e.g., because of slight experimental variations during cell seeding or toxicity of tested compounds), normalization is valid. However, if such differences are due to different efficacies of plasmid-driven gene expression between different wells, the effect on control reporter activity is not in a linear relationship to the effect on minigenome reporter activity [10], in which case it is better to report both Firefly and Renilla luciferase activity independently rather than to normalize results.

References

1. Feldmann H, Geisbert TW (2011) Ebola haemorrhagic fever. Lancet 377(9768):849–862. doi:10.1016/S0140-6736(10)60667-8
2. Mendoza EJ, Qiu X, Kobinger GP (2016) Progression of Ebola therapeutics during the 2014-2015 outbreak. Trends Mol Med 22(2):164–173. doi:10.1016/j.molmed.2015.12.005
3. Hoenen T, Groseth A, de Kok-Mercado F, Kuhn JH, Wahl-Jensen V (2011) Minigenomes, transcription and replication competent virus-like particles and beyond: reverse genetics systems for filoviruses and other negative stranded hemorrhagic fever viruses. Antivir Res 91(2):195–208. doi:10.1016/j.antiviral.2011.06.003
4. Muhlberger E, Weik M, Volchkov VE, Klenk HD, Becker S (1999) Comparison of the transcription and replication strategies of marburg virus and Ebola virus by using artificial replication systems. J Virol 73(3):2333–2342
5. Watanabe S, Watanabe T, Noda T, Takada A, Feldmann H, Jasenosky LD, Kawaoka Y (2004) Production of novel ebola virus-like particles from cDNAs: an alternative to ebola virus generation by reverse genetics. J Virol 78(2):999–1005
6. Hoenen T, Groseth A, Kolesnikova L, Theriault S, Ebihara H, Hartlieb B, Bamberg S, Feldmann H, Stroher U, Becker S (2006) Infection of naive target cells with virus-like particles: implications for the function of ebola virus VP24. J Virol 80(14):7260–7264. doi:10.1128/JVI.00051-06
7. Watt A, Moukambi F, Banadyga L, Groseth A, Callison J, Herwig A, Ebihara H, Feldmann H, Hoenen T (2014) A novel life cycle Modeling system for Ebola virus shows a genome length-dependent role of VP24 in virus infectivity. J Virol 88(18):10511–10524. doi:10.1128/JVI.01272-14
8. Hoenen T (2017) Minigenome systems for filoviruses. In: Salvato MS (ed) Hemorrhagic fever viruses: methods and protocols. Springer (New York)
9. Watanabe S, Noda T, Halfmann P, Jasenosky L, Kawaoka Y (2007) Ebola virus (EBOV) VP24 inhibits transcription and replication of the EBOV genome. J Infect Dis 196(Suppl 2):S284–S290. doi:10.1086/520582
10. Hoenen T, Jung S, Herwig A, Groseth A, Becker S (2010) Both matrix proteins of Ebola virus contribute to the regulation of viral genome replication and transcription. Virology 403(1):56–66. doi:10.1016/j.virol.2010.04.002
11. Spiegelberg L, Wahl-Jensen V, Kolesnikova L, Feldmann H, Becker S, Hoenen T (2011) Genus-specific recruitment of filovirus ribonucleoprotein complexes into budding particles. J Gen Virol 92(Pt 12):2900–2905. doi:10.1099/vir.0.036863-0
12. Hoenen T, Watt A, Mora A, Feldmann H (2014) Modeling the lifecycle of Ebola virus under biosafety level 2 conditions with virus-like particles containing tetracistronic minigenomes. J Vis Exp 91:52381. doi:10.3791/52381
13. Biedenkopf N, Lier C, Becker S (2016) Dynamic phosphorylation of VP30 is essential for Ebola virus life cycle. J Virol 90(10):4914–4925. doi:10.1128/JVI.03257-15
14. Hoenen T, Feldmann H (2014) Reverse genetics systems as tools for the development of novel therapies against filoviruses. Expert Rev Anti-Infect Ther 12(10):1253–1263. doi:10.1586/14787210.2014.948848
15. Wenigenrath J, Kolesnikova L, Hoenen T, Mittler E, Becker S (2010) Establishment and application of an infectious virus-like particle system for Marburg virus. J Gen Virol 91(Pt 5):1325–1334. doi:10.1099/vir.0.018226-0

Chapter 10

Assays to Measure Suppression of Type I Interferon Responses by Filovirus VP35 Proteins

Priya Luthra and Christopher F. Basler

Abstract

Innate immunity is the first line of defense against virus infections and is marked by production of type I interferons (IFN), a family of cytokines that includes IFN-β and several IFN-αs. For the filoviruses and many other RNA viruses that replicate in the cytoplasm, the RIG-I-like pattern recognition receptors (RLRs) are potential triggers of IFN production. To counteract such innate antiviral responses, many viruses encode proteins that antagonize RLR signaling. Ebola virus (EBOV) and other filoviruses produce VP35 proteins that block IFN induction via RLR signaling. We describe here cell-based reporter gene assays that quantify the IFN-antagonist function of filovirus VP35 proteins by assessing activation of the IFN-β promoter.

Key words VP35 protein, Luciferase assay, Interferon antagonism, IRF-3 phosphorylation

1 Introduction

Ebolavirus and *Marburgvirus* are genera within the family *Filoviridae*. Filovirus family members can cause severe hemorrhagic fever characterized by uncontrolled virus replication and excessive inflammation [1]. The severity of these infections is likely aided by the potent suppression of innate antiviral immunity by filoviral gene products including the Ebola virus (EBOV) VP24 protein, the Marburg virus (MARV) VP40 protein, and the filoviral VP35 proteins (as reviewed in [2]).

VP35 is a multifunctional viral protein, serving as an antagonist of the innate immune response and also functioning as a non-enzymatic cofactor for the viral RNA-dependent RNA polymerase (as reviewed in [3]). Suppression of innate immune responses by VP35 is mediated through inhibition of the retinoic acid-inducible gene I (RIG-I) signaling pathway that triggers the production of antiviral cytokines, IFN-α/β [4]. The importance of such suppression for EBOV is underlined by the fact that preactivation of RIG-I signaling prior to infection reduces EBOV titers by approximately

Thomas Hoenen and Allison Groseth (eds.), *Ebolaviruses: Methods and Protocols*, Methods in Molecular Biology, vol. 1628, DOI 10.1007/978-1-4939-7116-9_10, © Springer Science+Business Media LLC 2017

1000-fold [5]. Furthermore, recombinant EBOVs possessing mutations that disrupt VP35 IFN-antagonist function are highly attenuated in cells capable of mounting an IFN-α/β response and are avirulent in rodent models [6, 7].

The VP35 proteins interfere with RLR signaling cascades at multiple levels. The carboxy-terminal IFN inhibitory domain (IID) of VP35 plays a crucial role in inhibiting RIG-I in a manner that largely correlates with its dsRNA-binding capacity [4, 8]. The dsRNA-binding domain can sequester immune-stimulatory dsRNAs that would otherwise activate RLR. The same residues required for dsRNA-binding activity contribute to an interaction with host protein PACT which has RIG-I-activating activity. Interaction with PACT prevents PACT-mediated interaction with and activation of RIG-I [9]. The disruption of the VP35 double-stranded RNA (dsRNA)-binding activity by mutation substantially impairs its suppression of IFN-α/β responses by RIG-I activators [4, 7, 8]. In addition to these dsRNA-binding-dependent mechanisms, VP35 has been demonstrated to act as a decoy substrate for the kinases IKKε and TBK1 such that it prevents their interaction with and phosphorylation of interferon regulatory factor 3 (IRF-3), a transcription factor critical for induction of IFN-β gene expression [10]. VP35 also interacts with the transcription factor IRF-7 and the machinery that SUMOylates IRF-7 (Ubc9 and PIAS1), thereby disabling IRF-7-dependent type I IFN transcription [11, 12].

The IFN inhibition functions of VP35 protein can be assessed by several methods. A straightforward method is outlined in this chapter. Luciferase reporter assays, in which promoter and enhancer elements are placed upstream of luciferase coding sequences, provide highly sensitive and semiquantitative methods to assess gene expression. A commonly used luciferase gene is from the firefly (*Photinus pyralis*). This gene encodes a 61-kDa enzyme that oxidizes D-luciferin in the presence of ATP, oxygen, and Mg(++), yielding light when enzyme and luciferase substrate are combined [13]. The reporter activity within a transfected cell population (the light signal) is then proportional to the steady-state mRNA level. When placed downstream of the IFN-β promoter, the luciferase gene responds to RLR activation. In the assay outlined here, cells are simultaneously transfected with either an empty vector or a protein expression plasmid for VP35, a plasmid encoding the inducible IFN-β-firefly luciferase reporter gene, and a constitutively expressed *Renilla* luciferase expression plasmid, which serves as a control for transfection efficiency and for normalization of the firefly luciferase values. Transfected cells are either mock-treated or treated with an inducer of RLR signaling, which will activate the IFN-β promoter and thereby turn on firefly luciferase expression.

We describe alternate means to activate the RLR pathway. Sendai virus (SeV) strain Cantell is a potent RIG-I activator [14];

therefore, SeV infection 1 day post-transfection is a straightforward means to activate the pathway. Alternatively, we describe stimulation via a second transfection with low molecular weight (LMW) polyI:C, which preferentially activates RIG-I signaling. If one wishes to preferentially activate MDA5, the second major RLR, infection with encephalomyocarditis virus or transfection of high molecular weight polyI:C can be used [15]. Finally, we describe activation induced by the overexpression of full-length RIG-I, which typically results in some activation of the pathway which is further enhanced by SeV infection or polyI:C transfection, or transfection of an amino-terminal domain of RIG-I encoding its two caspase recruitment domains (CARDs), which acts as a constitutive activator of the pathway [16]. When activating with the CARD domains alone, the initial activation steps of the pathway are not assayed. VP35-mediated inhibition is typically less potent under these conditions relative to when SeV or polyI:C transfection is used [8]. This presumably reflects the fact that a major component of VP35 inhibition is to block the initial activation of RLRs. This permutation of the assay allows one to evaluate the inhibitory effects of VP35 that occur post-RLR activation which, for example, would reflect inhibition at the level of IKKε or TBK1. The protocol can be adapted to test wild-type or mutant versions of any filovirus VP35 or any other protein that may modulate RIG-I signaling.

2 Materials

2.1 IFN-β Luciferase Reporter Assay

1. HEK293T cells.
2. Lipofectamine 2000 (Invitrogen).
3. Opti-MEM serum-free medium.
4. Dulbecco's modified Eagle's medium (DMEM) supplemented with 10% fetal bovine serum (FBS).
5. pCAGGS expression plasmids for VP35 protein [7] and empty pCAGGS vector plasmid.
6. Reporter plasmid consisting of the IFN-β promoter followed by the firefly luciferase gene (IFNβ-FF) [17] and also the control *Renilla* luciferase reporter plasmid phRL-TK (Promega). The phRL-TK vector contains the herpes simplex virus thymidine kinase (HSV-TK) promoter and will constitutively express *Renilla* luciferase expression.
7. Activators of RIG-I-mediated IFN-β promoter activation: (a) Sendai virus, strain Cantell (stocks of the virus were generated by growing the virus in 10-day-old embryonated eggs for 2 days at 37 °C) [18]; (b) low molecular weight polyinosinic-polycytidylic

acid (polyI:C) with an average size of 0.2–1 kb (InvivoGen); and (c) expression plasmids that produce either full-length RIG-I or the N-terminal CARD domains of RIG-I (previously described in [9, 19]).

8. Dual-Luciferase kit (Promega).
9. General tissue culture supplies including tissue culture flasks and 24-well cell culture and 96-well tissue culture plates.
10. Luminometer for 96-well plates (i.e., tissue culture-treated white opaque polystyrene plates).

3 Methods

3.1 Transfection of IFN-β Luciferase Reporter Assay Components

1. Prepare the DNA mixtures for transfection (Solution A). If using SeV or polyI:C to activate RLR signaling, prepare the mixture as shown in Table 1. If using expression of RIG-I or the N-terminal CARD domains of RIG-I to activate RLR signaling, prepare the DNA mixtures as shown in Table 2. These DNA mixtures contain increasing amounts (5 ng up to 500 ng) of empty expression plasmid pCAGGS and pCAGGS-based plasmids expressing wild-type or mutant VP35 proteins (*see* **Note 1**) such that the total mass of these plasmids is 500 ng (or equal to the maximum amount of VP35 plasmid one intends to use) per transfected well. In addition, add 100 ng of IFN-β-firefly luciferase reporter plasmid (IFN-β-FF) and 10 ng of the *Renilla* luciferase expression plasmid pRL-TK per well. If using RIG-I plasmid-based RLR activation, 10 ng of either the RIG-I expression plasmid or the RIG-I N-terminal CARD domain plasmid is also added (*see* **Note 2**). In both cases bring the DNA mixture to a final volume of 25 μL using Opti-MEM (*see* **Notes 3** and **4**).
2. Prepare the Lipofectamine 2000 mix (Solution B) as shown in Table 3. Combine 1.2 μL of Lipofectamine 2000 and 25 μL Opti-MEM per well (*see* **Notes 3** and **4**). Incubate Solution B for 5 min at room temperature.
3. Form the DNA-Lipofectamine 2000 (DNA-LP2K) complexes. Mix the Solutions A and B and incubate the mixture at room temperature for 20 min.
4. Trypsinize and resuspend the HEK293T cells in 10% FBS-DMEM media at a concentration of 1×10^6 cells/mL.
5. Add 50 μL of the DNA-LP2K complexes per well of a 96-well plate. The transfection can be performed directly in the 96-well luminometer plates if desired.
6. Add 100 μL of the cell suspension (1×10^5 cells per well) to the DNA-LP2K complexes in each well.
7. Place the plates in a 37 °C, 5% CO_2 incubator.

Table 1

Solution A for transfection of one well of a 96-well plate (1×10^5 cells/well) using SeV or polyI:C as IFN-β activators

	Plasmid	pCAGGS (ng)	pCAGGS-VP35 (ng)	IFN-β-FF (ng)	phRL-TK (ng)	Total DNA (ng)	Opti-MEM (μL) (total)
1	pCAGGS	500	0	100	10	610	25
2	pCAGGS	500	0	100	10	610	25
3	pCAGGS-VP35	450	50	100	10	610	25
4	pCAGGS-VP35	0	500	100	10	610	25

Table 2

Solution A for transfection of one well of a 96-well plate (1×10^5 cells/well) using RIG-I or RIG-I CARD as IFN-β activators

	Plasmid	pCAGGS (ng)	pCAGGS-VP35 (ng)	IFN-β-FF-(ng)	RIG-I or RIG-I-CARD[a] (R) (ng)	phRL-TK (ng)	Total DNA (ng)	Opti-MEM (μL) (total)
1	pCAGGS	500	0	100	0	10	620	25
2	pCAGGS + R	500	0	100	10	10	620	25
3	pCAGGS-VP35 + R	450	50	100	10	10	620	25
4	pCAGGS-VP35 + R	0	500	100	10	10	620	25

[a]Expression plasmid producing the N-terminus of RIG-I encoding the CARD domains

8. If the samples have been transfected with the N-terminal CARD domains of RIG-I or full-length RIG-I to induce the IFN-β promoter, a separate induction step is not needed, and the luciferase assay can be read 20 h after transfection (*see* Subheading 3.3). If using other RLR stimuli (i.e., SeV infection or polyI:C transfection), proceed with Subheading 3.2 on the following day.

3.2 Stimulation with RLR Activators

The following day, cells are mock-treated or treated with an inducer of RLR signaling. As noted above, different inducers including SeV and polyI:C can be used to induce RIG-I. Potent induction of IFN-β gene expression by SeV reflects replication of defective interfering viral RNAs [18], while LMW polyI:C is a synthetic double-stranded RNA (dsRNA) analog that can be transfected to stimulate IFN-β gene expression.

Table 3

Solution B for transfection of one well of 96-well plate (1 × 10^5 cells/well)

	Per well (μL)
Opti-MEM	25
LP2K	1.2

The DNA-Lipofectamine 2000 (DNA-LP2K) ratio used can vary between 1:1 and 1:2)

1. If using SeV infection to stimulate RLR signaling, remove 50 μL of media from the transfected wells without disturbing the cells. Add 100 μL of a 1:100 dilution of the SeV stock in DMEM supplemented with 10% FBS (*see* **Note 5**). Leave the virus on the cells and place the plates in a 37 °C, 5% CO_2 incubator.
2. If using LMW polyI:C for activation, a second transfection, must be performed. For this transfection, carefully remove the media from the wells without disturbing the monolayer and add 100 μL of fresh 10% FBS-DMEM media (*see* **Note 6**). Prepare the transfection mix according to the instructions in Subheading 3.1 and Tables 1 and 3 using 200 ng of LMW polyI:C per well in Solution A and 0.4 μL Lipofectamine 2000 in Solution B. Add 50 μL of the DNA-LP2K complexes to each well and place the plates in a 37 °C, 5% CO_2 incubator.
3. 20 h post SeV infection or polyI:C transfection, the cells can be lysed and analyzed for firefly and *Renilla* luciferase expression using the Dual-Luciferase assay kit (Promega) (*see* **Note 7**) described in Subheading 3.3.

3.3 Measurement of Luciferase Reporter Activity

1. Dilute 5× Passive Lysis Buffer (supplied with Dual-Luciferase kit) to 1× with water and equilibrate to room temperature before use.
2. Remove the media from the cells carefully without disturbing the attached monolayer of cells.
3. Add 30 μL of 1× Passive Lysis Buffer to the cells and rock the plates on an orbital shaker for 15 min at room temperature.
4. Reconstitute the lyophilized Luciferase Assay Substrate in Luciferase Assay Buffer II (LARII) as per the manufacturer's instructions.
5. Prepare the desired amount of Stop & Glo Reagent by adding the 50× Stop & Glo substrate to final 1× concentration. For example, add 20 μL of Stop & Glo substrate to 1 mL of Stop & Glo Buffer to make 1× solution.
6. Make sure the assay reagents are at ambient temperature (20–25 °C) before use. Maintaining the optimum temperature is critical for the luciferase activity.

7. If using a plate reader with injectors, set the injectors of the luminometer to dispense 30 μL of LARII and Stop & Glo Reagent. For measurement, use a 1 s delay with a 2 s read time. The light intensity of the reaction is constant for about 1 min with a half-life of 10 min.
8. Place the luminometer plate with the lysates in the reader and read the plate.
9. The firefly luciferase activities are normalized to the *Renilla* luciferase activities. The results are typically presented as relative to the signal obtained in the mock-induced, empty vector transfection to allow calculation of fold induction (Fig. 1). For SeV "mock" represents treatment of cells with just media

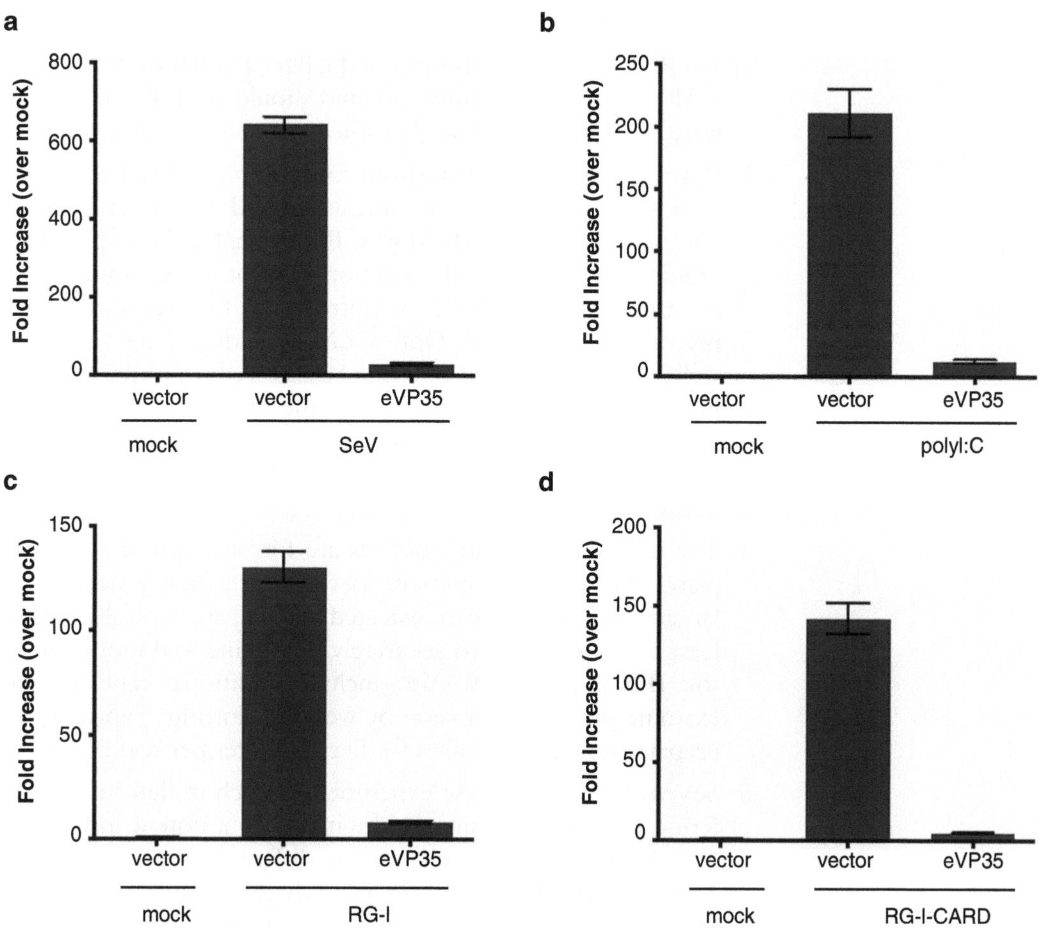

Fig. 1 Inhibition of IFN-β activation by the Ebola virus VP35 protein. IFN-β promoter reporter gene activity in the absence or presence of EBOV VP35 (eVP35, 250 ng) was assessed following infection with SeV (**a**), transfection with low molecular weight (LMW) polyI:C (**b**), expression of RIG-I (**c**), or the N-terminal RIG-I CARD (**d**) domains, as indicated. IFN-β-firefly luciferase activities were normalized to *Renilla* luciferase activities. Fold induction was determined by setting the mock-treated, empty vector (mock) transfection to a value of 1. The error bars indicate standard deviation of three replicates

without virus, and for polyI:C or RIG-I, "mock" treatment represents treatment of cells with only transfection media without any polyI:C or RIG-I.

4 Notes

1. Luciferase assays can be very useful tools to compare IFN inhibitory activities of different viral proteins or viral protein variants (e.g., VP35 or VP35 mutants). For this purpose, it is highly recommended to perform a dose response curve by transfecting different amounts of expression plasmid. Additionally, in such cases it is important to also evaluate protein expression to provide a correlation between luciferase activity and viral protein expression. This can be assessed by western blotting.
2. For RLR activation, amounts of the RIG-I or RIG-I N-terminal CARD domain expression plasmid should be in the 10–50 ng range, but 10 ng is typically sufficient to elicit a robust signal.
3. When performing transfections for multiple wells, it is recommended to prepare a master mix with the IFN-β-luc and phRL-TK plasmids in Opti-MEM first. For example, if an experiment requires five different transfection reactions (e.g., four different amounts of VP35 plus a control) with four replicates each, prepare a DNA mix in Opti-MEM by adding 2 μg (100 ng/well) of IFN-β-luc and 200 ng (10 ng/well) of phRL-TK into Opti-MEM to achieve a final volume of 500 μL. Divide 100 μL of this DNA mix into four separate transfection tubes and then add the additional plasmids (VP35 or mutants) to the desired amounts.
4. The indicated plasmid amounts are for one well of a 96-well plate. Scale up all components of the transfection as needed for larger wells. For each transfected sample, we typically run at least three replicates to accurately determine fold induction of the IFN-β promoter. Also include additional replicates to examine protein expression by western blotting. Typically, we prepare transfection mixes for five replicates per condition.
5. SeV to be used in these experiments is rich in defective interfering particles, explaining why it is such a potent inducer of RIG-I signaling. Typically, we test different quantities (1:100, 1:1000, 1:10,000 dilutions) of newly grown stocks of SeV to identify an amount that yields potent IFN-β promoter activation [18].
6. For the use of LMW polyI:C as a RIG-I inducer, one can also optimize the amounts, but we typically use 100–1000 ng/well. It is recommended to generate a master mix to simplify the transfection. For example, if LMW polyI:C is to be trans-

fected into 10 wells, then 2 μg (200 ng/well) is diluted into 250 μL of Opti-MEM (25 μL/well). For Solution B, 4 μL of LP2K is added to 250 μL of Opti-MEM. Transfections are performed as described in **steps 2** and **3** in Subheading 3.1.

7. Other luciferase reagents can also be used such as Dual-Glo Luciferase Assay Systems (Promega) which avoids the need for separate lysis step and generates a stable luminescence signal that can be read over a period of 2 h.

References

1. Sanchez A, Wagoner KE, Rollin PE (2007) Sequence-based human leukocyte antigen-B typing of patients infected with Ebola virus in Uganda in 2000: identification of alleles associated with fatal and nonfatal disease outcomes. J Infect Dis 196(Suppl 2):S329–S336. doi:10.1086/520588
2. Messaoudi I, Amarasinghe GK, Basler CF (2015) Filovirus pathogenesis and immune evasion: insights from Ebola virus and Marburg virus. Nat Rev Microbiol 13(11):663–676. doi:10.1038/nrmicro3524
3. Basler CF, Amarasinghe GK (2009) Evasion of interferon responses by Ebola and Marburg viruses. J Interferon Cytokine Res 29(9):511–520. doi:10.1089/jir.2009.0076
4. Cardenas WB, Loo YM, Gale M Jr, Hartman AL, Kimberlin CR, Martinez-Sobrido L, Saphire EO, Basler CF (2006) Ebola virus VP35 protein binds double-stranded RNA and inhibits alpha/beta interferon production induced by RIG-I signaling. J Virol 80(11):5168–5178. doi:10.1128/JVI.02199-05
5. Spiropoulou CF, Ranjan P, Pearce MB, Sealy TK, Albarino CG, Gangappa S, Fujita T, Rollin PE, Nichol ST, Ksiazek TG, Sambhara S (2009) RIG-I activation inhibits ebolavirus replication. Virology 392(1):11–15. doi:10.1016/j.virol.2009.06.032
6. Hartman AL, Ling L, Nichol ST, Hibberd ML (2008) Whole-genome expression profiling reveals that inhibition of host innate immune response pathways by Ebola virus can be reversed by a single amino acid change in the VP35 protein. J Virol 82(11):5348–5358. doi:10.1128/JVI.00215-08
7. Prins KC, Binning JM, Shabman RS, Leung DW, Amarasinghe GK, Basler CF (2010) Basic residues within the ebolavirus VP35 protein are required for its viral polymerase cofactor function. J Virol 84(20):10581–10591. doi:10.1128/JVI.00925-10. JVI.00925-10 [pii]
8. Leung DW, Prins KC, Borek DM, Farahbakhsh M, Tufariello JM, Ramanan P, Nix JC, Helgeson LA, Otwinowski Z, Honzatko RB, Basler CF, Amarasinghe GK (2010) Structural basis for dsRNA recognition and interferon antagonism by Ebola VP35. Nat Struct Mol Biol 17(2):165–172. doi:10.1038/nsmb.1765
9. Luthra P, Ramanan P, Mire CE, Weisend C, Tsuda Y, Yen B, Liu G, Leung DW, Geisbert TW, Ebihara H, Amarasinghe GK, Basler CF (2013) Mutual antagonism between the Ebola virus VP35 protein and the RIG-I activator PACT determines infection outcome. Cell Host Microbe 14(1):74–84. doi:10.1016/j.chom.2013.06.010
10. Prins KC, Cardenas WB, Basler CF (2009) Ebola virus protein VP35 impairs the function of interferon regulatory factor-activating kinases IKKepsilon and TBK-1. J Virol 83(7):3069–3077. doi:10.1128/JVI.01875-08
11. Chang TH, Kubota T, Matsuoka M, Jones S, Bradfute SB, Bray M, Ozato K (2009) Ebola Zaire virus blocks type I interferon production by exploiting the host SUMO modification machinery. PLoS Pathog 5(6):e1000493. doi:10.1371/journal.ppat.1000493
12. Kubota T, Matsuoka M, Chang TH, Tailor P, Sasaki T, Tashiro M, Kato A, Ozato K (2008) Virus infection triggers SUMOylation of IRF3 and IRF7, leading to the negative regulation of type I interferon gene expression. J Biol Chem 283(37):25660–25670. doi:10.1074/jbc.M804479200
13. Greer LF 3rd, Szalay AA (2002) Imaging of light emission from the expression of luciferases in living cells and organisms: a review. Luminescence 17(1):43–74. doi:10.1002/bio.676. [pii]
14. Kato H, Sato S, Yoneyama M, Yamamoto M, Uematsu S, Matsui K, Tsujimura T, Takeda K, Fujita T, Takeuchi O, Akira S (2005) Cell type-specific involvement of RIG-I in antiviral

response. Immunity 23(1):19–28. doi:10.1016/j.immuni.2005.04.010

15. Kato H, Takeuchi O, Sato S, Yoneyama M, Yamamoto M, Matsui K, Uematsu S, Jung A, Kawai T, Ishii KJ, Yamaguchi O, Otsu K, Tsujimura T, Koh CS, Reis e Sousa C, Matsuura Y, Fujita T, Akira S (2006) Differential roles of MDA5 and RIG-I helicases in the recognition of RNA viruses. Nature 441(7089):101–105
16. Yoneyama M, Kikuchi M, Natsukawa T, Shinobu N, Imaizumi T, Miyagishi M, Taira K, Akira S, Fujita T (2004) The RNA helicase RIG-I has an essential function in double-stranded RNA-induced innate antiviral responses. Nat Immunol 5(7):730–737. doi:10.1038/ni1087
17. Basler CF, Wang X, Muhlberger E, Volchkov V, Paragas J, Klenk HD, Garcia-Sastre A, Palese P (2000) The Ebola virus VP35 protein functions as a type I IFN antagonist. Proc Natl Acad Sci U S A 97(22):12289–12294. doi:10.1073/pnas.220398297
18. Baum A, Sachidanandam R, Garcia-Sastre A (2010) Preference of RIG-I for short viral RNA molecules in infected cells revealed by next-generation sequencing. Proc Natl Acad Sci U S A 107(37):16303–16308. doi:10.1073/pnas.1005077107
19. Rajsbaum R, Versteeg GA, Schmid S, Maestre AM, Belicha-Villanueva A, Martinez-Romero C, Patel JR, Morrison J, Pisanelli G, Miorin L, Laurent-Rolle M, Moulton HM, Stein DA, Fernandez-Sesma A, tenOever BR, Garcia-Sastre A (2014) Unanchored K48-linked polyubiquitin synthesized by the E3-ubiquitin ligase TRIM6 stimulates the interferon-IKKepsilon kinase-mediated antiviral response. Immunity 40(6):880–895. doi:10.1016/j.immuni.2014.04.018

Chapter 11

Nonradioactive Northern Blot Analysis to Detect Ebola Virus Minigenomic mRNA

Kristina Brauburger, Tessa Cressey, and Elke Mühlberger

Abstract

In this chapter, we describe the detection of Ebola virus minigenomic mRNA using a nonradioactive Northern hybridization. This protocol comprises all steps beginning with the synthesis of a digoxigenin-labeled riboprobe, harvest of transcribed mRNA from cells transfected with the Ebola virus minigenome system, separation of mRNA species by denaturing RNA gel electrophoresis, transfer of the mRNA to nylon membranes by vacuum blotting, and finally the detection of minigenome-specific mRNA through hybridization with a labeled riboprobe directed against the reporter gene.

This method allows the direct study of *cis*-acting regulatory regions as well as *trans*-acting factors involved in Ebola virus minigenome transcription compared to the indirect measurement of reporter protein activity that additionally reflects translational effects (*see* Chapter 6 in this book for details).

Key words Nonradioactive Northern hybridization, Northern blot, Vacuum blot, Ebola virus, Filoviruses, Minigenome, mRNA detection, mRNA purification, Denaturing RNA gel electrophoresis, Digoxigenin-labeled RNA probe

1 Introduction

Northern blotting refers to a technique widely used to detect RNA which was developed in 1977 [1, 2]. In the protocol we describe here, RNA molecules are isolated from cells, separated by size using denaturing agarose gel electrophoresis [3] and subsequently transferred to a positively charged nylon membrane by fast downward alkaline blotting [4] using a vacuum blotter [5]. Selected RNAs are then detected on the membrane by hybridization with a riboprobe that is complementary to part of or the entire RNA sequence of interest [6]. Non-segmented negative-sense RNA viruses such as Ebola virus (EBOV) produce three types of RNA: full-length negative-sense genomic RNA, full-length positive-sense antigenomic RNA, and mono- as well as bicistronic mRNAs. This set of RNAs can be recapitulated in minigenome systems as described in Chapter 6 in this book. Northern blot analysis is a

Thomas Hoenen and Allison Groseth (eds.), *Ebolaviruses: Methods and Protocols*, Methods in Molecular Biology, vol. 1628,
DOI 10.1007/978-1-4939-7116-9_11, © Springer Science+Business Media LLC 2017

powerful technique to discriminate between these different RNA species and helps to identify minigenome replication and transcription products. Here we describe an application of this technique for the nonradioactive detection of EBOV minigenome mRNA based on a protocol used in several previous studies [6–9]. Single-stranded negative-sense RNA probes that are directed against the sequence of the reporter gene contained within the minigenome are synthesized by in vitro transcription with digoxigenin-UTP, thus incorporating digoxigenin (DIG) into the riboprobe. Hybridized blots are treated with an anti-DIG antibody conjugated to alkaline phosphatase. The antibody will bind to areas of the blot where the DIG-labeled riboprobe has hybridized with the minigenome mRNA. The hybridized probe on the membrane can be detected with a chemiluminescent substrate for alkaline phosphatase and visualized on x-ray film or using a chemiluminescence imaging system.

2 Materials

2.1 Preparation of Template DNAs

1. For probe transcription: plasmid containing the reporter gene of the EBOV minigenome in negative-sense orientation under the control of the T7 RNA polymerase promoter (*see* **Note 1** and Fig. 1a).
2. For positive control RNA: plasmid containing the reporter gene in positive-sense orientation under the control of the T7 RNA polymerase promoter (*see* **Notes 1** and **2**; Fig.1b).
3. Restriction enzyme: any appropriate enzyme that linearizes the template DNA by cutting either within or shortly after the reporter gene sequence. The restriction enzyme does not need to be a single cutter. However, it is important that it does not separate the T7 RNA polymerase promoter from the reporter gene sequence (Fig. 1).
4. Restriction enzyme buffer.
5. PCR Purification Kit.
6. RNase-free water.
7. Agarose and gel electrophoresis equipment.

2.2 In Vitro Transcription of Digoxigenin-Labeled Riboprobe

For this step, we routinely use the DIG RNA Labeling Kit (Roche).

1. 10× NTP labeling mixture containing digoxigenin-labeled UTP: 10 mM ATP, 10 mM CTP, 10 mM GTP, 6.5 mM UTP, 3.5 mM DIG-11-UTP (Roche).
2. T7 RNA polymerase and transcription buffer (Roche).
3. RNase-free DNase I (10 U/μL).

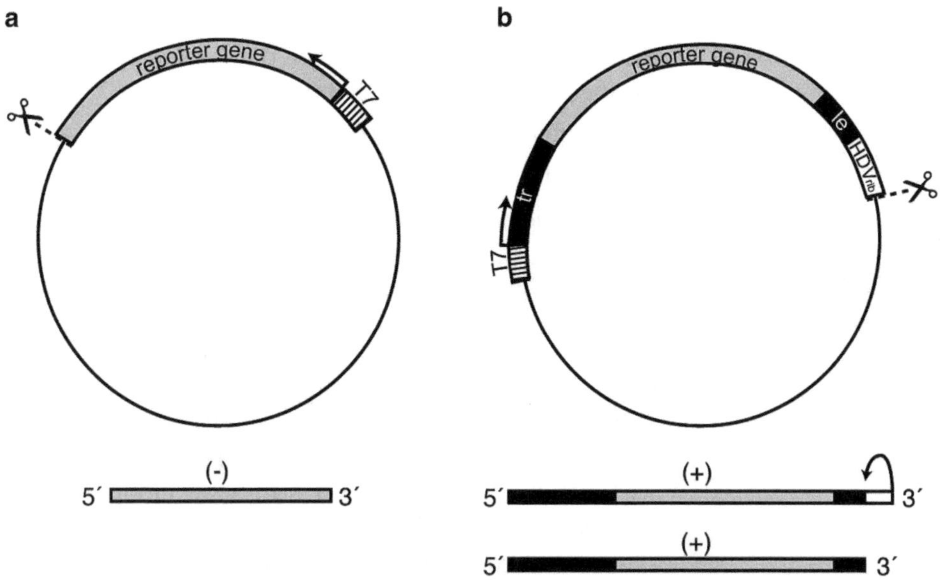

Fig. 1 Schematic representation of templates used for transcription of the riboprobe, the positive control RNA, and the minigenome mRNA. (**a**) *Top*, plasmid used for in vitro transcription of the negative-sense riboprobe; *bottom*, in vitro transcribed negative-sense riboprobe. *Scissors mark* the location of the restriction enzyme site used for linearization of the plasmid DNA prior to in vitro transcription. *Striped box*, T7 RNA polymerase promoter. The *arrow* marks the start and direction of T7 RNA polymerase transcription. *Gray box*, reporter gene. (−), negative-sense orientation. (**b**) *Top*, minigenome plasmid, which is also used for in vitro transcription of the positive-sense control RNA; *bottom*, positive-sense RNA used as a control for the riboprobe. le (*black box*), leader; tr (*black box*), trailer; HDV_{rib} (*white box*), hepatitis delta virus ribozyme. Other labeling as described in (**a**). The transcribed positive-sense RNA contains the sequence of the HDV_{rib}, which will autocatalytically cleave itself off (cleavage site marked by the *bent arrow*), resulting in the two transcripts shown. (+), positive-sense orientation

4. Plasmid or PCR fragment containing the probe template under the control of the T7 RNA polymerase promoter (*see* Subheading 3.1).
5. RNase inhibitor (20 U/μL).
6. Nuclease-free water.
7. RNA Purification Kit (e.g., RNeasy Mini Kit (Qiagen)).
8. For storage of riboprobe: RNase inhibitor (40 U/μL), 100 mM DTT.

2.3 In Vitro Transcription of Positive-Sense Control RNA

For this step, we use the AmpliScribe T7-Flash Transcription Kit (Epicentre).

1. 100 mM ATP, 100 mM CTP, 100 mM GTP, and 100 mM UTP (Epicentre).
2. 100 mM DTT.
3. T7 RNA polymerase and transcription buffer (Epicentre).
4. RNase-free DNase I (Epicentre).

5. Plasmid or PCR fragment containing the reporter gene in positive-sense orientation under the control of the T7 RNA polymerase promoter (*see* Subheading 3.1).
6. RNase inhibitor (40 U/μL) (Epicentre).
7. Nuclease-free water.
8. RNA Purification Kit (e.g., RNeasy Mini Kit (Qiagen)).

2.4 Quality Control of In Vitro Transcription Reactions

2.4.1 Control of RNA Quality

1. RNase-free ultrapure agarose.
2. Freshly made 1× TAE buffer: 40 mM Tris, 20 mM acetic acid, 1 mM EDTA. For a stock solution of 50× TAE buffer, dissolve 242 g Tris base in ultrapure H_2O, and add 57.1 mL glacial acetic acid and 100 mL of 500 mM EDTA (pH 8.0). Adjust the final volume to 1 L with ultrapure H_2O and autoclave. The stock solution can be diluted 50:1 with ultrapure H_2O and used as a 1× working solution. Both the stock and working solution are stored at room temperature.
3. RNase-free agarose gel tray, agarose gel electrophoresis chamber (*see* **Note 3**), power supply.
4. RiboRuler High Range RNA ladder (Fermentas).
5. Gel loading buffer II: 95% formamide, 18 mM EDTA, 0.025% SDS, 0.025% xylene cyanol, 0.025% bromophenol blue (*see* **Note 4**).
6. Thermoblock or PCR cycler.
7. Ice.
8. Nucleic acid stain (e.g., ethidium bromide, GelRed (Biotium)).
9. Gel documentation system."

2.4.2 Quality Control of Labeled Riboprobe

1. Nylon membrane, positively charged (e.g., Roche), 6 × 2 cm.
2. Whatman paper.
3. UV transilluminator (302 nm).
4. Lab rocker, refrigerated centrifuge.
5. Buffer I: 100 mM maleic acid, 150 mM NaCl, pH 7.5. Dissolve 11.7 g maleic acid and 8.8 g sodium chloride in 800 mL ultrapure H_2O. Adjust the pH with NaOH solution to pH 7.5 and fill up with ultrapure H_2O to 1 L. Autoclave buffer and store at room temperature.
6. Blocking buffer: buffer I containing 2% Blocking Reagent (Roche). For a 10% stock solution, dissolve 10 g Blocking Reagent (Roche) in 100 mL buffer I. Autoclave, store 50 mL frozen at −20 °C and store 50 mL at 4 °C for immediate use. Dilute to 2% in buffer I before use.
7. Anti-digoxigenin-AP Fab fragments (Roche).
8. Buffer I containing 0.3% Tween 20 (*see* **Note 5**).

9. Buffer III: 100 mM Tris, 100 mM NaCl, 50 mM $MgCl_2$, pH 9.5. Dissolve 12.1 g Tris and 5.8 g NaCl in 800 mL ultrapure H_2O. Adjust the pH to 9.5 with NaOH solution and fill up to 900 mL before autoclaving. Before use, add 1/10 Volume of 500 mM $MgCl_2$ stock solution.
10. CDP-Star solution (Roche).
11. Shrink-wrap plastic film.
12. X-ray film.

2.5 mRNA Isolation from Transfected Cells

1. Cells transfected with the EBOV minigenome system components (*see* Chapter 6 for a complete description of the properties of these constructs).
2. Phosphate-buffered saline, containing 154 mM NaCl, 5 mM Na_2HPO_4, 1 mM KH_2PO_4, deficient in $CaCl_2$ and $MgCl_2$ ($PBS_{deficient}$ (PBS_{def})) (sterile and cold (4 °C)).
3. Frozen plastic cool pack, sterile cell scrapers.
4. Sterile 1 mL syringes with needles (0.55 × 25 mm (24G × 1)).
5. RNeasy Mini Kit (Qiagen).
6. MicroPoly(A)Purist Kit (Ambion).

2.6 Denaturing RNA Gel Electrophoresis

1. 5× MOPS (3-(*N*-morpholino)propanesulfonic acid) running buffer: 100 mM MOPS (pH 7.0), 40 mM sodium acetate, 5 mM EDTA in RNase-free water. For 500 mL of 5× MOPS running buffer, dissolve 10.5 g MOPS in RNase-free water and adjust pH with NaOH to 7.0. Add 6.7 mL of 3 M sodium acetate (pH 7.0) and 10 mL of 250 mM EDTA stock solution (pH 8.0). All used stock solutions should be RNase-free. Use sterile plastic pipettes for the addition of sodium acetate and EDTA to the MOPS buffer. Store protected from light in a dark bottle or wrapped with aluminum foil (*see* **Note 6**).
2. RNase-free agarose gel tray, agarose gel electrophoresis chamber (*see* **Note 3**), power supply.
3. 1.8% denaturing RNA gel.
4. NanoDrop.
5. RNA samples.
6. RiboRuler High Range RNA ladder (Fermentas) (*see* **Note 7**).
7. Positive-sense control RNA (*see* Subheading 3.3).
8. 6× RNA sample loading buffer: 50% glycerol, 1 mM EDTA, 0.25% bromophenol blue, 0.25% xylene cyanol FF (*see* **Note 8**).
9. RNA gel loading buffer: for one sample, combine 2 μL RNase-free water, 1 μL 5× MOPS running buffer, 1.75 μL 37% formaldehyde, 5 μL deionized formamide (*see* **Notes 9** and **10**).
10. Thermoblock.
11. Ice.

2.7 Transfer of RNA to Nylon Membrane Using a Vacuum Blotter

1. Vacuum blotter (*see* **Note 11**).
2. Nylon membrane, positively charged (e.g., Roche), cut to the full size of the agarose gel.
3. Whatman paper, cut to the size of the nylon membrane (*see* **Note 12**).
4. Transfer buffer: 3 M NaCl, 8 mM NaOH. Transfer buffer has to be freshly prepared. For 200 mL, add 120 mL of 5 M NaCl stock solution (RNase-free) in an RNase-free graduated glass cylinder. Add 1.6 mL of 1 M NaOH with a sterile plastic pipette and fill up to 200 mL with RNase-free water. Store in RNase-free glass bottle.
5. Methanol (*see* **Note 13**).
6. 200 mM phosphate buffer, pH 6.8. 1 M phosphate buffer stock: prepare 1 M Na_2HPO_4 with RNase-free water and adjust pH to 6.8 with RNase-free 1 M NaH_2PO_4. Dilute to 200 mM with RNase-free water before use.
7. UV transilluminator (302 nm).

2.8 Prehybridization and Hybridization with DIG-Labeled Riboprobes

1. Hybridization oven with matching hybridization glass tubes (*see* **Notes 3** and **14**).
2. Hybridization solution [6]: 6× SSC, 5× Denhardt's solution, 0.1% SDS, 0.5 mg/mL sonicated herring sperm DNA. Add 50 mL RNase-free water to an RNase-free graduated glass cylinder (*see* **Note 3**), add 30 mL 20× SSC (3 M NaCl, 300 mM sodium citrate, pH 7.0; RNase-free) and 10 mL 50× Denhardt's solution (1% Ficoll (type 400), 1% polyvinylpyrrolidone, and 1% bovine serum albumin; sterile filtered; RNase-free). Make solution up to 94 mL with RNase-free water and finally add 1 mL 10% SDS (RNase-free). Mix well. Hybridization buffer can be aliquoted and stored in RNase-free 50 mL tubes at −20 °C. Note that the denatured herring sperm DNA (10 mg/mL) must be added shortly before the prehybridization step to bring the hybridization solution to the final volume (*see* Subheading 3.8.1, **step 6**).
3. DIG-labeled riboprobe (*see* Subheading 3.2).
4. Thermoblock.
5. Ice.

2.9 Detection of DIG-Labeled Riboprobes via Chemiluminescence

1. Lab rocker, refrigerated centrifuge.
2. Washing buffer: 0.1× SSC, 0.1% SDS. Add 5 mL of 20× SSC stock solution to a 1 L graduated cylinder and fill up to 990 mL with ultrapure water. Add 10 mL of 10% SDS stock solution and mix. Store at room temperature.

3. Buffer I: 100 mM maleic acid, 150 mM NaCl, pH 7.5 (*see* Subheading 2.4.2).
4. Blocking buffer: buffer I containing 2% Blocking Reagent (*see* Subheading 2.4.2).
5. Anti-digoxigenin-AP Fab fragments (Roche).
6. Buffer I containing 0.3% Tween 20 (*see* Subheading 2.4.2, **Note 5**).
7. Buffer III: 100 mM Tris, 100 mM NaCl, 50 mM $MgCl_2$, pH 9.5 (*see* Subheading 2.4.2).
8. CDP-Star solution (Roche).
9. Vacuum sealer and shrink-wrap plastic film.
10. Film development cassette.
11. X-ray film.

3 Methods

General remarks: for preparation of all buffer solutions used in Subheading 3.1–3.8, we use commercially available RNase-free water and RNase-free glassware and perform work in an RNase-free lab space (*see* **Note 3**). Carry out all steps at room temperature unless otherwise specified.

3.1 Preparation of Template DNAs

1. Linearize 5 μg of template DNA with 20 units of the appropriate restriction enzyme in a total volume of 50 μL following the manufacturer's recommendation for complete digest of the given DNA amount.
2. Purify the digested DNA using a PCR Purification Kit following the manufacturer's instructions and elute with RNase-free water.
3. Load 1/10 volume of the purified, linearized template DNA on a 1% agarose gel to check if the plasmid is completely digested. Complete DNA digestion is indicated by the appearance of a single DNA band in the gel.

3.2 In Vitro Transcription of Digoxigenin-Labeled Riboprobe

1. Follow the manufacturer's instructions for the DIG RNA labeling kit used.
2. For Roche's DIG RNA Labeling Kit, combine the following components in an RNase-free 0.5 mL tube on ice in the order specified:

   x μL RNase-free water to reach a total reaction volume of 20 μL.

   1 μg of purified template DNA (linearized plasmid or PCR fragment; *see* Subheading 3.1).

   2 μL 10× dNTP labeling mixture.

2 μL 10× transcription buffer.
1 μL RNase-inhibitor (20 U/μL).
2 μL T7 RNA polymerase (10 U/μL).

3. Mix gently and spin down briefly.
4. Incubate 2 h at 37 °C in a PCR cycler.
5. Add 2 μL of DNase I (10 U/μL) recombinant, RNase-free.
6. Incubate for 15 min at 37 °C.
7. Purify RNA following the RNA cleanup protocol of the RNeasy Mini Kit.
8. Elute twice with 50 μL of RNase-free water (total volume of 100 μL). Add 1 μL RNase inhibitor (40 U/μL) and 3.2 μL DTT (100 mM; final concentration 8 mM) to the riboprobe and prepare 6 μL aliquots that are stored at −80 °C or −20 °C.

3.3 In Vitro Transcription of Positive-Sense Control RNA

1. Follow the manufacturer's instructions for the transcription kit used.
2. For the AmpliScribe T7-Flash Transcription Kit, combine the following components at room temperature in an RNase-free 0.5 mL tube, in the order given:

x μL RNase-free water to reach a total reaction volume of 20 μL.
1 μg of purified template DNA (linearized plasmid or PCR fragment; *see* Subheading 3.1).
2 μL 10× AmpliScribe T7-Flash reaction buffer.
1.8 μL ATP, CTP, GTP, UTP (100 mM).
0.5 μL RNase inhibitor (40 U/μL).
2 μL AmpliScribe T7-Flash Enzyme Solution.

3. Incubate at 37 °C for 60 min in a PCR cycler.
4. Add 1 μL (1 MBU) of RNase-free DNase I.
5. Flick tube to mix and spin down briefly.
6. Incubate 15 min at 37 °C in a PCR cycler.
7. Purify RNA following the RNA cleanup protocol of the RNeasy Mini Kit.
8. Elute 1× with 50 μL of RNase-free water. Aliquot RNA to 5 μL aliquots and store at −80 °C or −20 °C.

3.4 Quality Control of In Vitro Transcription Reactions

3.4.1 Control of RNA Quality

1. To assess the quality of the in vitro transcribed riboprobe and control RNA, check 0.5 μL and 1 μL of each RNA on a non-denaturing agarose gel.
2. Prepare a 1% agarose gel in fresh 1x TAE buffer.
3. Mix 0.5 μL and 1 μL of RNA with 10 μL of gel loading buffer II.
4. Mix 1 μL of RNA ladder with 10 μL of gel loading buffer II.

5. Heat all samples for 3 min at 95 °C and place on ice for 3 min before loading on the agarose gel.
6. The labeled riboprobe should show as a single clear band after staining with a nucleic acid stain (e.g., ethidium bromide, GelRed (Biotium)).
7. If a minigenome containing a reporter gene was used as a template for the transcription of the positive-sense control RNA, two RNA bands are expected: the higher molecular weight band represents the in vitro transcript containing the ribozyme sequence, while the lower band represents the transcript from which the ribozyme has been cleaved off (Figs. 1 and 2).

3.4.2 Control of Riboprobe Labeling Efficiency

1. Perform a serial dilution of the riboprobe from 10^{-1} to 10^{-4} in RNase-free water in volumes of 10 μL. Keep the RNA on ice.
2. On a piece of positively charged nylon membrane, draw four circles (ø 0.5 cm) with a pencil and label each with the respective dilution.
3. Drop 1 μL of each riboprobe dilution in the respective circle.
4. Air-dry the membrane on Whatman paper.
5. To crosslink the RNA to the membrane, place the membrane on an UV-transilluminator (cleaned with RNase AWAY) with the RNA-side facing the UV lamps and crosslink for 3 min at 302 nm.
6. Proceed with all steps described in Subheading 3.9 starting from **step 2** in Subheading 3.9.1.
7. On the x-ray film, the dots should show a clearly visible signal at least up to the 10^{-3} dilution after a 30 s exposure time.

3.5 mRNA Isolation from Transfected Cells

1. Transfect ca. 70% confluent BSR-T7/5 cells with the minigenome system components (1.0 μg pT/L_{EBO}, 0.5 μg pT/NP_{EBO}, 0.5 μg pT/$VP35_{EBO}$, and 0.1 μg pT/$VP30_{EBO}$ [7], along with 1.5 μg of a minigenome construct that contains the minigenome under the control of the T7 RNA polymerase promoter) as described in detail in Chapter 6.
2. 48 h post transfection, place the transfected cells on a pre-chilled plastic cool pack (−20 °C).
3. Carefully wash the transfected cells twice with 1 mL per well of cold (4 °C) PBS_{def} using sterile plastic pipettes.
4. Remove remaining PBS_{def} completely with 1000 μL filter tips.
5. Add 600 μL of RLT buffer containing 1% beta-mercaptoethanol (RNeasy Mini Kit) to each well and scrape cells using sterile cell scrapers.
6. Transfer cell lysates to 2 mL tubes placed on ice.

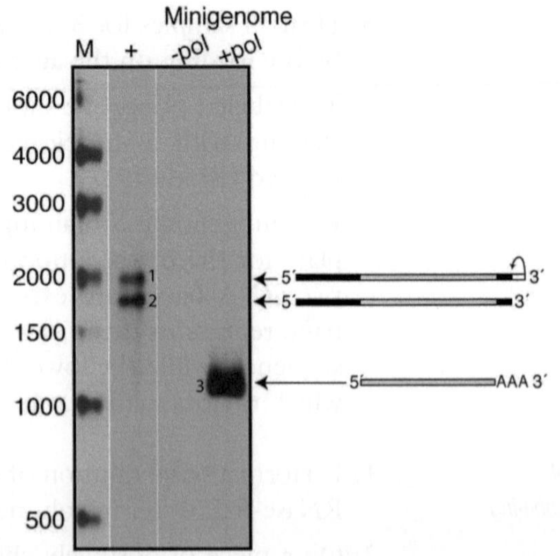

Fig. 2 Northern blot result showing the detection of EBOV-specific minigenomic RNA. Sample result of a Northern hybridization. M, RNA size ladder; +, positive control RNA. As explained in the text and in Fig. 1, transcription of the positive control RNA results in two bands due to the autocatalytic cleavage of the ribozyme sequence (band 1, RNA containing the ribozyme sequence; band 2, RNA without ribozyme sequence). In the lanes marked with minigenome, transcribed EBOV-specific minigenome mRNA was detected with a negative-sense riboprobe directed against the CAT gene (DIG-BS/CAT, [8]). Transcription activity of the EBOV minigenome is shown in the presence (+pol) and absence (−pol) of the EBOV polymerase. Band 3, minigenomic mRNA. The structure and size of the different RNA species are illustrated on the right as described in Fig. 1. All lanes are from the same blot, but the marker lane was exposed for a longer period of time. Other lanes not relevant to this chapter were excised

7. Shear the cell lysates on ice by passing through sterile needles with a 1 mL syringe. Change needles and syringes for each sample (*see* **Note 15**).
8. Add 600 μL of ice-cold 70% ethanol to each cell lysate, vortex (*see* **Note 16**).
9. Purify total RNA of samples by following the "Animal-Cell/RNA-Isolation-protocol" of the RNeasy Mini Kit up to the elution step.
10. Elute total RNA from the columns with 250 μL RNase-free water.
11. Add 250 μL of 2× binding solution (MicroPoly(A)Purist Kit).
12. Follow the protocol for "Poly(A) RNA Isolation from Total RNA" starting with step A.2.b of the MicroPoly(A)Purist Kit up to the elution step (E.1.d).
13. To precipitate the mRNA, add 20 μL RNase-free 5 M ammonium acetate, 1 μL glycogen (*see* **Note 17**), and 550 μL 100% ethanol to the eluate (200 μL).
14. Incubate tubes at −20 °C overnight.

3.6 Denaturing RNA Gel Electrophoresis

1. Prepare a 1.8% denaturing agarose gel. For casting a gel that fits in a tray of 6 x 10 cm, weigh 0.7 g nuclease-free ultrapure agarose in a 100 mL RNase-free glass Erlenmeyer flask. Add 28 mL RNase-free water and weigh before melting agarose by heating in a microwave oven. Weigh melted agarose solution again and add back evaporated water. Cool down below 60°C and add 7.2 mL of 5x MOPS running buffer. Place Erlenmeyer flask in fume hood and add 1.2 mL of 37% formaldehyde. Use sterile plastic pipettes for addition of MOPS running buffer and formaldehyde (*see* **Note 9**).
2. Transfer the gel to an RNase-free gel electrophoresis chamber and fill the chamber with 1× MOPS running buffer.
3. Spin the RNA samples for 20 min at 4 °C at maximum speed in a table top centrifuge and carefully remove the supernatant.
4. Add 1 mL of 70% ethanol (−20 °C) to each pellet and spin again for 10 min at 4 °C, maximum speed.
5. Carefully remove the supernatant with 1000 μL filter tips and leave approximately 50 μL on the pellet. Then, use extra long 100 μL pipette tips to remove the remaining supernatant. Be careful not to remove the pellet.
6. Air-dry RNA pellets at room temperature for 5 min.
7. Resuspend pellets in 10 μL RNase-free water (*see* **Note 18**).
8. Measure RNA concentration of 2 μL of each sample on a NanoDrop.
9. Mix 100 ng of each sample with 10 μL of RNA gel loading buffer.
10. For the RNA ladder, mix 1 μL of RNA ladder with 10 μL of RNA gel loading buffer.
11. For the positive control RNA: mix 3 μL of a 10^{-4} dilution with 10 μL RNA gel loading buffer.
12. Denature the RNA by incubating for 15 min at 65 °C in a heat block.
13. Place tubes on ice for 3 min.
14. Briefly centrifuge all samples to collect any condensate.
15. Add 1.5 μL of 6× RNA sample loading buffer to each sample.
16. Load the samples on the gel; leave the first and last wells empty.
17. Run the gel at 3 V/cm until the dye line has entered the gel. Then increase the voltage to 5–6 V/cm.
18. Run the gel until the dark blue bromophenol blue band has just run out.

3.7 Transfer of RNA to Nylon Membrane Using a Vacuum Blotter

1. Label the nylon membrane with experiment number and date at the bottom.
2. On a shaker placed inside the fume hood, shake the membrane for 5 min in methanol in an RNase-free box.
3. Transfer the wet membrane to a fresh RNase-free box containing transfer buffer and shake for at least 10 min.
4. Assemble Whatman paper, membrane, and the RNA gel on the vacuum blotter as follows (bottom to top) (Fig. 3; *see* **Note 19**):
   (a) Porous white plastic plate from vacuum blotter (smooth side facing up).
   (b) One piece of Whatman paper, cut to a size bigger than the cutout on the blue plastic sheet ("mask" in picture) used for blotting; mark location of the cutout in the mask on the Whatman paper with a pencil, and soak Whatman paper with transfer buffer.
   (c) Nylon membrane soaked in transfer buffer; fit membrane over pencil markings on Whatman paper, so that it overlaps the markings on each side.
   (d) Blue plastic sheet from vacuum blotter ("mask"); fit cutout on the membrane.
   (e) Soak the membrane with transfer buffer before adding the gel on top. Avoid any air bubbles between the membrane and the gel by slowly sliding the gel from the gel tray onto the membrane. It is important that the gel overlaps the cutout on the mask on all sides.
5. Switch on the vacuum blotter and blot at 50 mbar for 1 h 45 min.
6. Throughout the blotting, constantly add transfer buffer on top of the gel (the gel must not dry out) (*see* **Note 20**).
7. Following blotting: take off the gel, label the location of the cutout on the membrane with a pencil, and disassemble the vacuum transfer unit (*see* **Note 21**).
8. Neutralize the membrane by shaking it for 15 min in phosphate buffer.
9. Dry the membrane on Whatman paper.
10. To crosslink the RNA to the membrane, transfer the membrane to an UV-transilluminator (cleaned with RNase AWAY) with the RNA-side facing the UV-lamps and crosslink for 3 min at 302 nm (*see* **Note 22**).

3.8 Hybridization with DIG-Labeled Riboprobes

3.8.1 Prehybridization

1. Preheat hybridization oven to 65 °C.
2. Preheat 19 mL of the hybridization solution without herring sperm DNA to 65 °C in the hybridization oven (the hybridization temperature depends on the length and nucleotide composition of the riboprobe and might require adjustments when using other probes).

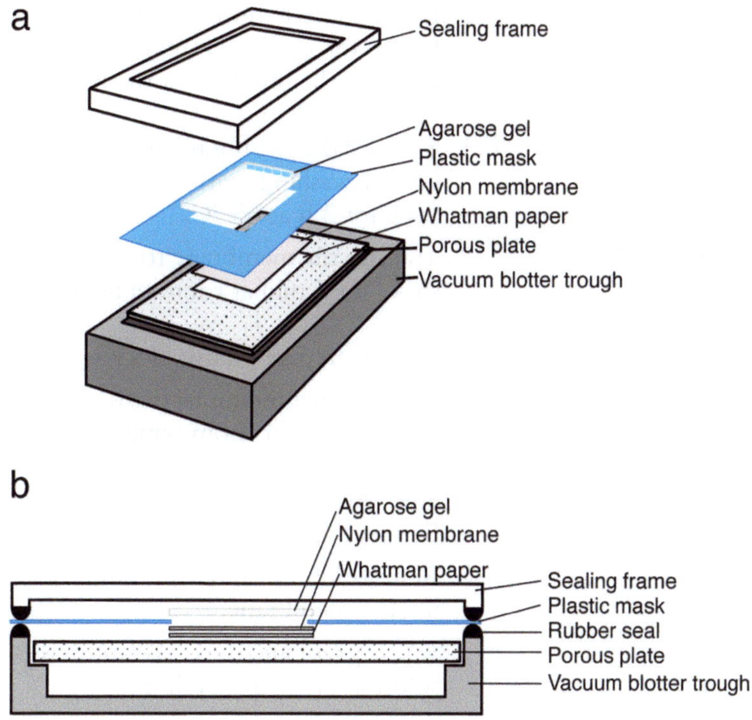

Fig. 3 Schematic drawing of vacuum blotter assembly for Northern blot. (**a**) Frontal view. (**b**) Side view. For a detailed description of the assembly, refer to the Subheading 3.7, **step 4**

3. Denature 1 mL of sonicated herring sperm DNA at 95 °C for 15 min and place quickly on ice for at least 5 min, or until use.
4. In a baked and RNase-free glass hybridization tube, carefully add the nylon membrane with the RNA blotted onto it facing toward the inside of the tube. Ensure that the blot slides down all the way to the bottom of the tube.
5. Add 19 mL preheated hybridization solution.
6. Add 1 mL denatured herring sperm DNA to the hybridization solution.
7. Rotate tubes in the hybridization oven for at least 6 h at 65 °C (*see* **Note 23**).

3.8.2 Hybridization

1. Denature 5 µL of the riboprobe for 3 min at 95 °C and place on ice for at least 3 min.
2. Remove the hybridization solution from the prehybridization tube with an RNase-free plastic pipette and add back 3 mL.
3. Add 3 µL of denatured riboprobe directly to the hybridization solution without touching the membrane.
4. Rotate tubes in the hybridization oven overnight at 65 °C.

5. Prepare one hybridization tube filled to the top with washing buffer for each membrane and place in it the hybridization oven for the duration of the hybridization to preheat.

3.9 Detection of DIG-Labeled Riboprobes via Chemiluminescence

3.9.1 Washing

1. Transfer the membrane to a hybridization tube filled with preheated washing buffer and rotate tubes in the hybridization oven for 2 h at 65 °C.
2. Transfer the membrane to a plastic box containing buffer I and shake for 5 min at room temperature.

3.9.2 Blocking

1. Freshly prepare 75 mL blocking buffer (*see* Subheading 2.4.2).
2. Transfer membrane to plastic box containing blocking buffer and incubate at room temperature on the shaker for 1 h.

3.9.3 Detection

1. Spin down the anti-DIG AP Fab fragments at 3400 x g for 5 min at 4 °C in a table top centrifuge.
2. Remove 35 mL blocking buffer from the plastic box containing the membrane.
3. Add 2 μL of anti-DIG-AP antibody to the remaining blocking solution (1: 20,000 dilution).
4. Shake for 30 min at room temperature.
5. Wash three times for 10 min/each with buffer I containing 0.3% Tween 20.
6. Equilibrate the membrane for 5 min with buffer III at room temperature on a shaker.
7. Prepare CDP-Star solution (Roche): dilute CDP-Star 1:200 with buffer III (per blot, prepare 995 μL buffer III + 5 μL CDP-Star).
8. Dry membrane briefly on Whatman paper.
9. Place the membrane on the inverted lid of a 6-well plate and add 1 mL CDP-Star solution.
10. Incubate for 5 min at room temperature.
11. Dry membrane quickly on Whatman paper and place between sheets of shrinkable plastic foil.
12. Shrink-wrap in plastic foil.
13. Tape the wrapped membrane in a film cassette (*see* **Note 24**).
14. Expose x-ray film.
15. Start with an exposure time of 5 min and adjust exposure depending on the strength of the signal up to overnight.
16. Develop exposed film using accessible equipment.

4 Notes

1. Other bacteriophage RNA polymerase promoters that could be used include the SP6 and T3 RNA polymerase promoters, but these may require some adaptations of the in vitro transcription protocol. As a further alternative to cloning the reporter gene in an appropriate vector under the control of a bacteriophage RNA polymerase promoter (we use the pBluescript II KS vector), a PCR product can be generated using primers that contain at least 20 nucleotides of the reporter gene-specific sequence and the sequence of an RNA polymerase promoter provided as a 5′ overhang.
2. We use a plasmid encoding the EBOV minigenome containing a CAT reporter gene in positive-sense orientation under the control of the T7 RNA polymerase promoter (Fig. 1b). Transcription of the positive-sense minigenome RNA will result in two bands of defined length in the Northern blot, which can also serve as a size marker.
3. Glassware should be thoroughly cleaned with soap and water in a dish washer and subsequently baked overnight (or a minimum of 4 h) at 200 °C to remove all RNases. If possible, all work with RNA should be performed on a designated RNA lab bench with a pipette set reserved exclusively for RNA work. Plasticware should be cleaned with an RNase-destroying substance, e.g., RNase AWAY. In addition, we autoclave plastic bottle lids. RNase-free filter tips should be used for pipetting. We also use autoclaved, nuclease-free 1.5 and 2 mL tubes.
4. This buffer is also commercially available. Buffer from Ambion worked well in our hands.
5. Tween 20 is very viscous. Cut 1000 μL filter tip with scissors (cleaned before with RNase AWAY) to widen the tip opening and pipette very slowly. Wash out the pipette tip in the buffer solution and mix on magnetic stirrer.
6. MOPS buffer can be autoclaved before addition of sodium acetate and EDTA. However, it then changes color. If not autoclaved, color can be used as an indicator for freshness. If straw-colored, it can still be used; if the color gets darker, it should be made fresh.
7. For unknown reasons, this ladder seems to unspecifically bind to the negative-sense riboprobe directed against the chloramphenicol acetyltransferase (CAT) gene and thus will appear on the x-ray film (Fig. 2). However, we have not tested whether the ladder appears on the x-ray film when other probes are used.
8. This buffer is also commercially available. Buffer from Boston Biolabs worked well in our hands.

9. 37% formaldehyde and formamide are hazardous chemicals. Follow appropriate safety precautions and dispose of waste according to local hazardous waste disposal regulations.
10. Prepare enough buffer for all samples (include 10% extra to account for pipetting errors).
11. We use the Appligene vacuum blotter, model 230600 from Boekel Scientific. It comes with several plastic sheets that are used as masks for the blotting procedure (Fig. 3). Prepare plastic sheets by cutting out a rectangle that is about 1 cm smaller than the width of the agarose gel and 2 cm shorter than the length of the gel. The agarose gel needs to fit on the cutout with 0.5 cm overlap on each side and at least 1 cm at the top and the bottom of the gel. The gel wells should overlap with the mask.
12. Precut at least three pieces per gel and store in an RNase-free box.
13. Methanol is a hazardous chemical. Follow appropriate safety precautions and dispose of waste according to local hazardous waste disposal regulations.
14. Hybridization tubes should be RNase-free (*see* **Note 3**). It is important to check the rim of the tube for cracks before use. Tubes with cracks should not be used for the hybridization, as even small cracks can lead to leaking of the hybridization solution. Ensure that all tube lids contain a rubber sealing before use. Lids can be treated with RNase AWAY and autoclaved to ensure they are free of RNases.
15. This step can be substituted by extensive vortexing at full speed for at least 30 s.
16. Samples can be stored at −80 °C for several weeks following this step.
17. We recommend using GlycoBlue™ coprecipitant (Ambion), as its blue color increases the visibility of the only loosely attached pellet after precipitation.
18. Alternatively, and if RNA concentration is expected to be low, the pellets can be directly dissolved in 10 μL of RNA gel loading buffer.
19. Do not let the membrane dry out. Always keep it wetted with transfer buffer.
20. Check the pressure throughout blotting and if necessary readjust, as it might change if leaks occur. Check for hissing sounds that indicate leaks, and if necessary, seal any leaks with plastic wrap.
21. Following disassembly of the vacuum blotter, immediately and thoroughly clean all parts of the vacuum blotter of residues

from the transfer buffer. Especially the porous plate needs to be carefully cleaned and soaked with water to wash all buffer residues off the pores. We attach a small hose to the water faucet to force the tap water through the plate.

22. If not proceeding directly to prehybridization, shrink-wrap the blot in RNase-free plastic foil and store at 4 °C overnight or for a couple of days. We have no experience with longer storage periods.
23. Add a hybridization tube filled with a volume of water that equals the volume of the prehybridization/hybridization/washing solution on the opposite side of the rotating holder to keep it in balance.
24. In order to identify the orientation of the x-ray film, we use fluorescent stickers placed on one side of the film cassette. Alternatively, one edge of the x-ray film may be cut.

Acknowledgement

This work was supported by the National Institute of Allergy and Infectious Diseases of the National Institutes of Health under award numbers U01-AI082954, R03-AI114293, and UC6AI058618.

References

1. Hayes PC, Wolf CR, Hayes JD (1989) Blotting techniques for the study of DNA, RNA, and proteins. BMJ 299:965–968
2. Alwine JC, Kemp DJ, Stark GR (1977) Method for detection of specific RNAs in agarose gels by transfer to diazobenzyloxymethyl-paper and hybridization with DNA probes. Proc Natl Acad Sci U S A 74:5350–5354
3. Green MR, Sambrook J (2012) Molecular cloning: a laboratory manual, 4th edn. Cold Spring Harbor Lab, Cold Spring Harbor, NY, pp 388–393
4. Chomczynski P (1992) One-hour downward alkaline capillary transfer for blotting of DNA and RNA. Anal Biochem 201:134–139. doi:10.1016/0003-2697(92)90185-A
5. Olszewska E, Jones K (1988) Vacuum blotting enhances nucleic acid transfer. Trends Genet 4:92–94. doi:10.1016/0168-9525(88)90095-9
6. Grosfeld H, Hill MG, Collins PL (1995) RNA replication by respiratory syncytial virus (RSV) is directed by the N, P, and L proteins; transcription also occurs under these conditions but requires RSV superinfection for efficient synthesis of full-length mRNA. J Virol 69:5677–5686
7. Mühlberger E, Weik M, Volchkov VE, Klenk HD, Becker S (1999) Comparison of the transcription and replication strategies of Marburg virus and Ebola virus by using artificial replication systems. J Virol 73:2333–2342
8. Mühlberger E, Lötfering B, Klenk HD, Becker S (1998) Three of the four nucleocapsid proteins of Marburg virus, NP, VP35, and L, are sufficient to mediate replication and transcription of Marburg virus-specific monocistronic minigenomes. J Virol 72:8756–8764
9. Brauburger K, Boehmann Y, Tsuda Y, Hoenen T, Olejnik J, Schümann M, Ebihara H, Mühlberger E (2014) Analysis of the highly diverse gene borders in ebola virus reveals a distinct mechanism of transcriptional regulation. J Virol 88:12558–12571. doi:10.1128/JVI.01863-14

Part III

Studying Infectious Ebolaviruses In Vitro

Chapter 12

A Semi-automated High-Throughput Microtitration Assay for Filoviruses

Claire Marie Filone, David Miller, and Victoria Wahl-Jensen

Abstract

The 50% tissue culture infectious dose ($TCID_{50}$) endpoint dilution assay is one of the gold standard methods for measuring filovirus infectivity. We have increased virology microtitration assay throughput at biosafety level (BSL)-4 by implementing automated liquid handling and semi-automated assay endpoint readout. Utilization of automated liquid handling for cell plating and virus dilution along with optimization of the assay endpoint readout, using a luminescent-based cell viability assay and an automated plate reader, has improved workflow efficiency, reduced operator burden and assay time, decreased assay variability, and increased data return.

Key words Filovirus, Ebola, Marburg, Microtitration, Semi-automated, High-throughput screening (HTS), Tissue culture infectious dose ($TCID_{50}$), Viability, Luminescent, Luminescence, Assay, Endpoint, Biosafety level 4 (BSL-4), Liquid handling, Efficiency, Dilution

1 Introduction

Since the discovery of Marburg virus and Ebola virus (EBOV) in 1967 and 1976, respectively, the biological assays utilized to quantify infectious filoviruses have changed remarkably little over time. The two biological assays that are most frequently utilized for measuring the amount of infectious filovirus in materials are the plaque assay and the endpoint tissue culture infectious dose 50 ($TCID_{50}$) assay [1–5]. While the plaque assay is a quantitative assay based on the ability of a single infectious virus particle to generate a macroscopic area of cytopathology on otherwise healthy cells, the endpoint $TCID_{50}$ assay is typically used when distinct plaques are difficult to discern, but some level of cytopathic effect (CPE) is still evident in the cells. Following serial dilution of the virus and inoculation of replicate samples onto healthy cells, an absolute qualitative judgment is made whether infection occurred following an appropriate incubation period in the $TCID_{50}$ assay. The $TCID_{50}$ assay assesses the point at which 50% of the cells in a culture are

Thomas Hoenen and Allison Groseth (eds.), *Ebolaviruses: Methods and Protocols*, Methods in Molecular Biology, vol. 1628, DOI 10.1007/978-1-4939-7116-9_12, © Springer Science+Business Media LLC 2017

infected [6]. Both Marburg virus and Ebola virus readily induce cytopathic effects in several cell lines [7–9], although Vero and Vero E6 cells have remained the cell lines of choice for both the plaque assay and $TCID_{50}$ assay.

While traditional, manual virology titration methods have long been a gold standard for measuring infectious filovirus titers, the assays are not amenable to evaluation of large numbers of samples, the analysis is subjective, and manual data recording may lead to transcription errors. The incorporation of high-throughput screening (HTS) elements (process designs and robotics) into virology assays in the BSL-4 laboratory has led to improved workflow efficiency, reduced operator burden, reduced assay time, decreased assay variability, and increased data return.

In addition to decreasing the filovirus microtitration assay length, an automated assay readout has several advantages. The luminescence-based viability assay readout is quantitative; therefore, the subjectivity of the operator is removed. It takes significant, time-consuming operator training to ensure visual reads have the least variability possible. The use of a quantitative readout reduces this training burden and alleviates the subjective determination of CPE. Furthermore, an automated approach with electronic data collection alleviates the need for data transcription. The traditional, visual determination of CPE is recorded onto data sheets, which are then transcribed into a spreadsheet for Log $TCID_{50}$/mL calculations. This process has potential for user error when recording onto the data collection sheet and again when transcribing into the spreadsheet. The automated plate reader records quantitative data in an electronic format, avoiding the potential data entry errors present with the visual assay.

Thus, we have established a semi-automated microtitration assay for filoviruses that included a luminescence-based viability reagent which was measured as a quantitative endpoint for microtitration coupled to an automated readout. Use of this method reduced the normally recommended length of the EBOV microtitration assay, without a change in endpoint titer. This assay allowed for data collection in a higher throughput format, in less time, and removed the operator subjectivity in determining the level of CPE and is therefore recommended for filovirus microtitration assays.

2 Materials

96-well plates: Black tissue culture treated optical bottom.

Robotic dilution system: Precision Microplate Pipetting System (BioTek).

Liquid dispensing system: VIAFILL Dispensing System (INTEGRA).

Cells: Vero E6 cells, African green monkey kidney (ATCC CRL-1586).

Cell culture components (*see* **Note 1**): Complete growth medium consisting of Minimum Essential Medium (MEM) with Earle's salts and L-Glutamine supplemented with 10% (v/v) heat-inactivated fetal bovine serum (FBS), 2 mM GlutaMAX supplement (Life Technologies), 0.1 mM nonessential amino acids solution (NEAA, Invitrogen), 1 mM sodium pyruvate and 1% (v/v) Antibiotic-Antimycotic solution (Life Technologies). Cell culture splits are performed using TrypLE Select stable trypsin replacement enzyme (Invitrogen/Gibco).

Cell viability reagent: CellTiter-Glo (CTG, Promega).

Plate reader: SpectraMax M5 Multi-Mode Microplate Reader (Molecular Devices), or SpectraMax Paradigm Multi-Mode Microplate Reader with LUM 384 luminescence cartridge (Molecular Devices).

Software: SoftMax Pro (Molecular Devices), Precision Power (BioTek).

Disinfectant: A disinfectant approved for use in your BSL4-laboratory (e.g., Micro-Chem Plus 1:20 solution).

3 Methods

All operations are to be performed inside a biological safety cabinet with the following exceptions: microscopic evaluation, centrifugation or incubations.

3.1 Microtitration

Seed Vero E6 cells in 96-well plates (3×10^5 cells/mL in 100 μL/well; ~95–100% confluence at 24 h after plating) 1 day prior to date of microtitration experiment. Cell seeding must be uniform to prevent bias in cytopathic effect readout. It is recommended to use an automated system such as the VIAFILL liquid dispensing system (Fig. 1) for seeding cells to minimize cell seeding variability (*see* **Note 2**).

1. Observe the cellular morphology in the 96-well plates under a microscope before performing the assay to make sure the cells appear healthy and are at the appropriate confluency. Plates seeded at 3×10^5 cells/mL in 100 μL/well 24 h prior to use appear visually to be ~95–100% confluent, which is recommended for the assay. Wells should also be visually assessed for any signs of contamination (turbidity, cell death, etc.) and cell health (no vacuoles, nonadherent cells, etc.). Do not use if the wells have signs of contamination or if the cells appear unhealthy.
2. Prepare the negative control to be used in this assay. This should consist of all the components which comprise the virus suspension except for the virus itself (e.g., complete growth medium with additives listed above).

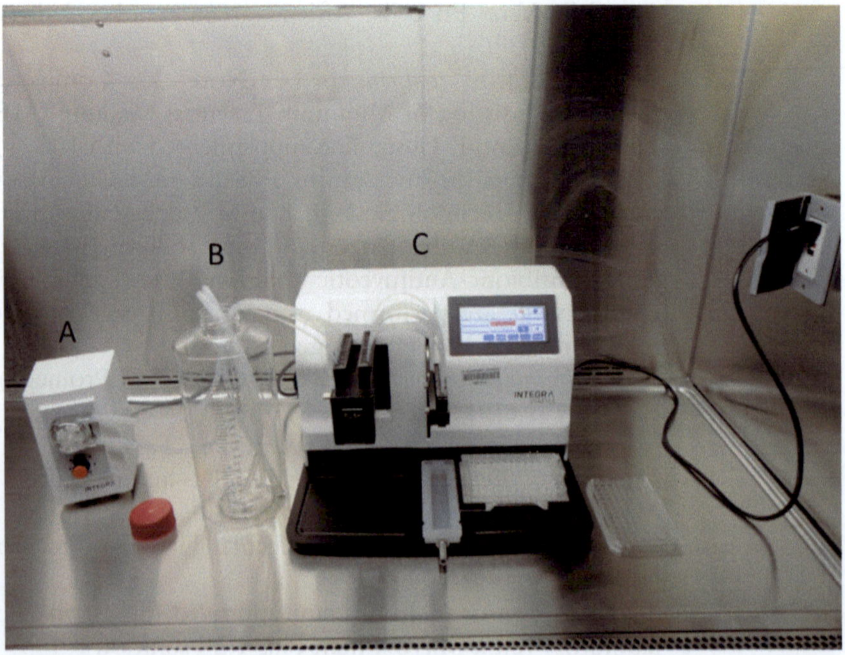

Fig. 1 INTEGRA VIAFILL reagent dispenser. A peristaltic pump (*A*) mixes cell slurry housed in a sterile, disposable bottle (*B*). The dispensing unit (*C*) pumps a measured volume of cell slurry into each well of a cell culture plate. All components are housed in a BSC to ensure sterility

3. For samples and positive/negative controls: Using a multichannel pipette, remove all of the cell culture medium from the "top" row of the dilution series (in Fig. 2, Row 2, columns B-G). Then add 111 μL of your sample or control to each of the wells in this row. The process controls are wells that are not altered from the time cells are initially seeded (i.e., outer wells) (*see* **Note 3**).
4. Serially dilute the top row by transferring 11 μL from each well into the next row followed by a mixing step (pipetting up and down five times without disturbing the monolayer or causing air bubbles) followed by a tip change to prevent carryover of virus. Disinfectant should be drawn in and out of the tips before discarding as an added safety precaution. Repeat this process down the plate for as many dilutions as are required (e.g., nine times as diagrammed in Fig. 2). Discard 11 μL from the last dilution well into disinfectant to maintain equal volumes between wells. Although the dilutions may be completed manually using a multichannel pipette, the precision, accuracy, and length of time will be improved by completing the dilutions using robotics, such as the BioTek Precision Microplate Pipetting System (Precision) (*see* **Note 4**) cited above and shown in Fig. 3. An appropriate program for the Precision will include mixing the sample in each well during dilutions, discarding the tips between each dilution, and drawing disinfectant in and out before tip disposal.

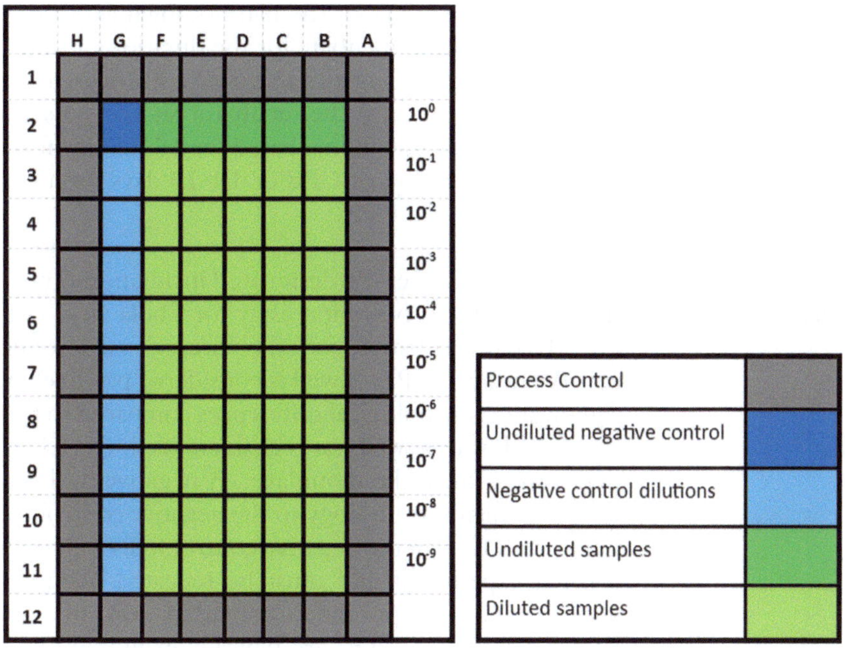

Fig. 2 Sample plate map for 96-well plate microtitration assay. Cells are seeded onto the entire plate 1 day prior to the experiment. Outer wells are utilized as process controls. Undiluted negative control (*dark blue box*) and undiluted samples (*dark green boxes*) are loaded into column 2, rows B–G. Serial dilutions are performed down the plate as described in step #5

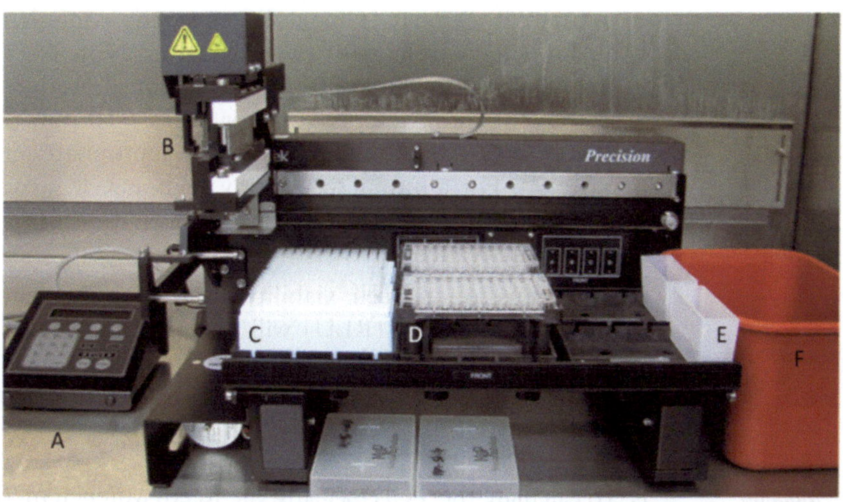

Fig. 3 Precision Microplate Pipetting System (BioTek). The machine is operated using a small control panel (*A*). Once initiated, the pipette head (*B*) picks up pipette tips (*C*) and dilutes the samples contained in the cell culture plates (*D*, black plates not shown). After each dilution, the pipette tips are rinsed in disinfectant (*E*) and disposed in a sharps container (*F*). The robot is housed in a BSC to ensure sterility and biosafety

5. Place the 96-well plate(s) in an incubator set at the appropriate temperature and % CO_2 for the cells. When available, an actively humidified incubator set at 95% humidity should be used for long-duration assays to reduce evaporation in the

wells along the edges of the plate (known as "edge effects"). Allow the plates to incubate for the duration of time appropriate for the virus being titrated. For a filovirus microtitration assay in Vero E6 cells, the incubator should be set at 37 °C, 95% RH, 5% CO_2, and incubated for 13–15 day for a manual, visual determination of CPE or 10–13 days for a cell viability reagent readout of CPE.

3.2 Manual $TCID_{50}$ Data Collection

1. Assess the CPE after the designated incubation time. The standard 96-well microtitration assay for Ebola virus (EBOV) has a 13–15 day incubation period before a visual determination of EBOV-induced CPE. A well is considered positive if it exhibits significant morphological differences compared to the negative control. Positive wells are characterized by CPE and cell detachment from the monolayer. A negative well is characterized by similar morphology to the negative control. Any experiment with CPE in the negative control wells will be considered a failed experiment and must be repeated. Assessors should also notate any potential contamination or cytotoxicity. Cytotoxic effects (CTE) are typically exhibited by changes in the medium color (red to yellow), cloudiness, and/or medium evaporation due to edge effects.

3.3 Cell Viability $TCID_{50}$ Data Collection

1. Follow manufacturer's guidelines on processing plates (*see* **Note 5**). Following the appropriate incubation time, add a volume of CellTiter-Glo (CTG) reagent equivalent to the medium in the wells (100 μL) is added (*see* **Note 6**). The plates are shaken for 2 min at 70 rpm, and then allowed to incubate at ambient room temperature for at least 10 min. Plates are then read on a luminometer using the appropriate settings (*see* **Note 7**). For a luminescent assay such as CTG, each well is read from 0.1 to 0.5 s/well, depending on the sensitivity of the microplate reader.
2. Determination of cell viability using a microplate reader: Relative light units (RLU) will be averaged for the negative control wells. Any sample wells that are more than three standard deviations from the negative control value will be considered positive (Fig. 4) (*see* **Notes 8** and **9**).

3.4 Reed–Muench $TCID_{50}$ Calculation

1. To calculate the $TCID_{50}$ using the Reed–Muench method, first count the number of positive and negative wells at each dilution. Positive wells are defined as CPE if performing manual reads and three times the SD from the negative control value when using CTG-based luminescence (*see* **Note 9**).

   Notional example:

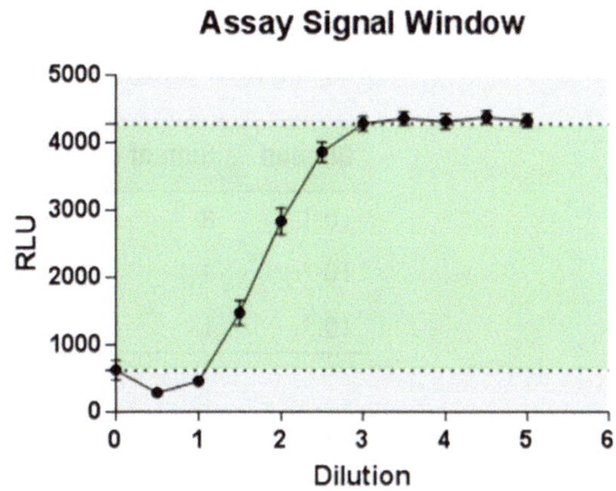

Fig. 4 Representative assay illustrating signal window definition. An example of the difference between the RLU signal in wells with maximum virus-induced CPE and wells with no virus-induced CPE (Signal Window) is shown within the *green box*. For assay qualification, this signal window (the mean high-signal divided by the mean low signal) must have a value greater than 3

Dilution	Wells	Positive	Negative
10^{-1}	(+)(+)(+)(+)	4	0
10^{-2}	(+)(+)(+)(-)	3	1
10^{-3}	(+)(-)(-)(-)	1	3

2. Add the number of positive wells from the bottom up (highest to lowest dilution), adding each number to the previous value. For this example, you would add 1 + 3 = 4, then 4 + 4 = 8. For the Negative row you would add the number of negative wells from the top down (lowest to highest dilution). In this example you would add as follows: 0, 0 + 1 = 1, 1 + 3 = 4.

   Calculations can easily be performed in an Excel spreadsheet.

Dilution		Positive	Negative		Sum of +	Sum of –
10^{-1}	↑	4	0	↓	8	0
10^{-2}	↑	3	1	↓	4	1
10^{-3}	↑	1	3	↓	1	4

3. For each dilution calculate the ratio of positive wells to total wells at each dilution.

Dilution	Sum of +	Sum of –	Ratio
10^{-1}	8	0	8/8
10^{-2}	4	1	4/5
10^{-3}	1	4	1/5

4. Calculate the percent value corresponding to this ratio.

Dilution	Ratio	Percent
10^{-1}	8/8	100%
10^{-2}	4/5	80%
10^{-3}	1/5	20%

5. The $TCID_{50}$ dilution is the dilution at which exactly 50% of the wells are calculated to have CPE. Note that this is not necessarily the same as having 2 out of the 4 wells showing actual CPE. If no dilution has a percentage exactly equal to 50% then calculate the proportionate distance (PD) between the dilution with a percentage value below 50% and the dilution above 50% according to the equation below:
   Proportionate distance =

$$\frac{(\%CPE\ at \text{ dilution next above } 50\%) - 50\%}{(\%CPE\ at \text{ dilution next above } 50\%) - (\%CPE\ at \text{ dilution next below } 50\%)}$$

In this example:
PD = (80–50)/(80–20)
(80–50)/(80–20) = (30/60)
(30/60) = 0.5

The Log of the dilution at which 50% of the samples displayed CPE can then be calculated as:

$\log_{10}$ (50% CPE titer) = ($\log_{10}$ of the next dilution above 50% CPE-PD)

In the example above:
CPE titer = ($\log_{10}10^{-2}$)–0.5
($\log_{10}10^{-2}$)–0.5 = –2.0–0.5
–2.0–0.5 = –2.5

Thus in this example the dilution at which 50% CPE was calculated is $10^{-2.5}$

6. The total $TCID_{50}$ titer of the original sample added to the first well is found by taking the inverse of the 50% CPE dilution:

$$\frac{1}{10^{-2.5}} = 10^{2.5}$$

7. To calculate the titer of the original stock in $TCID_{50}$ per mL, the $TCID_{50}$ value from the previous step must be divided by the volume (in mL) of the virus stock used to create the first dilution. For example if 200 μL (0.2 mL) of sample was used to create the first dilution, the $TCID_{50}$/mL is:

   $(10^{2.5} TCID_{50}/0.2\ mL) = (316/0.2)$
   $(316/0.2) = 1580\ TCID_{50}/mL$
   $\log_{10}(1580) = 3.20$
   Original sample titer = $10^{3.20}\ TCID_{50}/mL$

8. Note that the limit of quantification (LOQ) for this assay is $0.5n + 1$ positive wells in the lowest dilution, where n is the total number of wells in the lowest dilution. For example, if 4 wells are used, the (LOQ) is 3 positive (displaying CPE) wells (if 0.2 mL of sample was used in each well, the resulting virus titer would be 1.03 $TCID_{50}$/mL). If less than $0.5n + 1$ positive wells are observed in the lowest dilution, the titer for that sample would be considered "below (LOQ)".

4 Notes

1. All media additives are cited as final concentration.
2. To qualify uniformity of dispensing, a cell viability reagent can be used to quantitate the number of cells in each well (seeded manually or with an automated system), and the Coefficient of Variation (CV), also known as Relative Standard Deviation (RSD), which describes the frequency of assay data distribution can be determined using standard spreadsheet software [10]. A CV can be chosen to qualify the assay; a CV of <25% represents the upper threshold of acceptable variation based on previously published cell-based assay qualification [11]. Following the addition of the cell viability reagent CellTiter-Glo (CTG) to the plates, luminescent relative light units (RLUs) are quantified on a plate reader which can be used to calculate CV.
3. Avoid using the outer wells of the plate in the experimental design. The length of time necessary to achieve consistent CPE induced by filovirus infection allows for significant evaporation of cell culture medium in the outer wells. An actively humidified incubator reduces the damage to the cell monolayer due

to evaporative effects; however, there are still decreases in cell viability in the outer wells that may affect virus $TCID_{50}$ calculations.

4. The most efficient setup for diluting large numbers of plates uses two Precision units inside a large BSC with another six foot Class IIA2 BSC in the same room to manually load samples onto the plates. For efficiency, this setup requires two laboratorians: one to add samples to the plates and another loading the plates onto the robots, filling out data sheets, managing the robot consumables and waste, and placing diluted plates into the incubator.
5. After adding CTG to the assay, there is a finite amount of time in which the assay must be read. The stability of the CTG-induced luminescent signal was established to determine the maximum length of time between CTG addition and RLU quantification in which the luminescent signal decreased less than 10% from maximum RLU. This determines the length of time the data must be collected to maintain consistency between plates. CTG was added to 96-well plates containing Vero cells, per manufacturer's instructions, luminescence read at 10 min post-CTG addition (Time 0) and every hour for the subsequent 7 h (Fig. 5). A stable CTG-induced luminescent signal was seen for at least 2 h when plates remained at room temperature in the light. Therefore, plates can be read up to 2 h post CTG addition without compromising data quality.
6. Strain or variant specific growth kinetics may require the assay to be run longer than what is optimal for the high throughput method using cell viability reagent. In this case, the volume in the wells may be increased from 100 μL to 200 μL with a corresponding increase in dilution volume (11 μL to 22 μL). The plates can then be scored visually.
7. A bottom read on the luminometer is recommended due to condensation on the inside lids of the plates after incubation.
8. To qualify this assay for EBOV, a time-course was completed, processing 3 plates per day from days 7 to 14 post microtitration infection. The CV was calculated on the uninfected wells for each plate throughout the time course. The CVs fell within qualification parameters, with values less than 25%, on each day. The assay dynamic range (Signal Window) was also measured by comparing the lowest CTG-based luminescence reading (maximum virus CPE) to the highest reading (no virus-induced CPE; Fig. 4). For 96-well plates, days 10–13 qualified with Signal Window values >3 (Fig. 6a). Figure 6b shows the RLU values for each dilution across the plates on Day 8 and Day 12. Although the Log $TCID_{50}$/mL calculated for each is similar, the assay dynamic range is not sufficient on

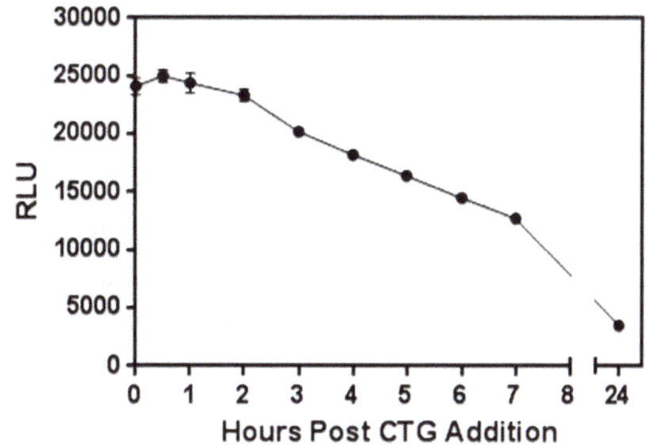

Fig. 5 CellTiter-Glo is stable for 2 h post addition. Luminescence from Vero cells lysed in CTG was read hourly over a time course to determine that a 10% decrease in signal occurs after 2 h of incubation at room temperature in the light. The experiment was completed in triplicate, data averaged from 8 wells containing undiluted cells (3×10^4 cells/well) on each plate. Error bars represent 95% confidence interval (CI) from the mean values from triplicate plates

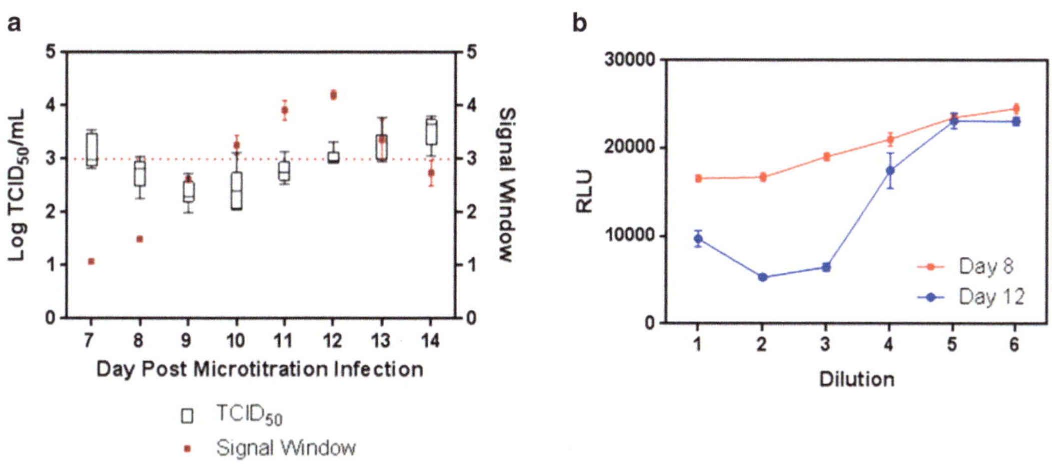

Fig. 6 EBOV/Mak-C05 microtitrations can be read using CTG-based luminescence on days 10–13 post-infection. (**a**) The Log $TCID_{50}$/mL value for each sample per day is graphed (*whisker plot, left Y axis*). Data show a line at the median, with the box hinges at 25th and 75th percentiles. The *whiskers* denote the 5th and 95th percentiles, with outliers. The signal window (*red dot plot, right Y axis*) shows the average (mean) signal window for each day, with 95% CI error bars. (**b**) The averaged raw RLU values for seven replicates read on days 8 (*red*) or 12 (*blue*) are graphed for each dilution, with 95% CI error bars

Day 8 to qualify. Therefore, an EBOV microtitration assay on Vero E6 cells can be read using CTG-based luminescence on days 10 through 13 post-infection for 96-well plates. Method optimization is required for each virus, reagent and cell type used to obtain the largest dynamic range and highest reproducibility to ensure the lowest assay variability.

9. When numerical data are collected using a cell viability reagent, the Log $TCID_{50}$/mL may be calculated several ways. If the Reed–Muench method is used, the Log $TCID_{50}$/mL value will not significantly differ between the titer calculated using a visual estimation of CPE or a calculated percent decrease from negative control wells, as described here. Alternatively, the cell viability reagent data may be graphed as the raw luminescence versus the dilution factor. A nonlinear regression curve fit analysis with a sigmoidal dose-response (variable slope) algorithm may be performed, and the EC_{50} value provided can be used for the Log $TCID_{50}$/mL. However, this number is typically 1.3 Log lower than the Reed–Muench method; therefore, a conversion factor is necessary to compare data collected using different methods.

Acknowledgment

This work was funded under Contract No. HSHQDC-07-C-00020 awarded by the Department of Homeland Security (DHS) Science and Technology Directorate (S&T) for the management and operation of the National Biodefense Analysis and Countermeasures Center (NBACC), a Federally Funded Research and Development Center. The views and conclusions contained in this document are those of the authors and should not be interpreted as necessarily representing the official policies, either expressed or implied, of the DHS or S&T. In no event shall DHS, NBACC, S&T, or Battelle National Biodefense Institute have any responsibility or liability for any use, misuse, inability to use, or reliance upon the information contained herein. DHS does not endorse any products or commercial services mentioned in this publication.

References

1. Moe JB, Lambert RD, Lupton HW (1981) Plaque assay for Ebola virus. J Clin Microbiol 13(4):791–793
2. Titenko AM, Novozhilov SS, Andaev EI, Borisova TI, Kulikova EV (1992) Ebola virus reproduction in cell cultures. Vopr Virusol 37(2):110–113
3. Ustinova EN, Shestopalov AM, Bakulina LF, Chepurnov AA (2003) Titration of Ebola and Marburg viruses by plaque formation under semi liquid agar. Vopr Virusol 48(1):43–44
4. Shurtleff AC, Biggins JE, Keeney AE, Zumbrun EE, Bloomfield HA, Kuehne A, Audet JL, Alfson KJ, Griffiths A, Olinger GG, Bavari S, Filovirus Animal Nonclinical Group Assay Working G (2012) Standardization of the filovirus plaque assay for use in preclinical studies. Virus 4(12):3511–3530. doi:10.3390/v4123511
5. Smither SJ, Lear-Rooney C, Biggins J, Pettitt J, Lever MS, Olinger GG Jr (2013) Comparison of the plaque assay and 50% tissue culture infectious dose assay as methods for measuring filovirus infectivity. J Virol Methods 193(2):565–571. doi:10.1016/j.jviromet.2013.05.015
6. Reed LJ, Muench H (1938) A simple method of estimating the fifty percent endpoints. Am J Hyg 27(3):493–497
7. Schnittler HJ, Mahner F, Drenckhahn D, Klenk HD, Feldmann H (1993) Replication of

Marburg virus in human endothelial cells. A possible mechanism for the development of viral hemorrhagic disease. J Clin Invest 91(4):1301–1309. doi:10.1172/JCI116329

8. Maass G, Muller J, Seemayer N, Haas R (1969) Production of kidney tissue cultures from African green monkeys, experimentally infected with the causative agent of Frankfurt-Marburg-syndrome. Am J Epidemiol 89(6): 681–690
9. Kuhn JH (2008) Cultivation of filoviruses. In: Calisher CH (ed) Filoviruses. A compendium of 40 years of epidemiological, clinical, and laboratory studies. Springer-Verlag, Wien, pp 171–173
10. Iversen PW, Beck B, Chen YF, Dere W, Devanarayan V, Eastwood BJ, Farmen MW, Iturria SJ, Montrose C, Moore RA, Weidner JR, Sittampalam GS (2004) HTS assay validation. In: Sittampalam GS, Coussens NP, Nelson H et al (eds) Assay guidance manual. Eli Lilly & Company and the National Center for Advancing Translational Sciences, Bethesda
11. Birmingham A, Selfors LM, Forster T, Wrobel D, Kennedy CJ, Shanks E, Santoyo-Lopez J, Dunican DJ, Long A, Kelleher D, Smith Q, Beijersbergen RL, Ghazal P, Shamu CE (2009) Statistical methods for analysis of high-throughput RNA interference screens. Nat Methods 6(8):569–575. doi:10.1038/nmeth.1351

Chapter 13

Generation of Recombinant Ebola Viruses Using Reverse Genetics

Allison Groseth

Abstract

Reverse genetics systems encompass a wide array of tools aimed at recapitulating some or all of the virus life cycle. In their most complete form, full-length clone systems allow us to use plasmid-encoded versions of the ribonucleoprotein (RNP) components to initiate the transcription and replication of a plasmid-encoded version of the complete viral genome, thereby initiating the complete virus life cycle and resulting in infectious virus. As such this approach is ideal for the generation of tailor-made recombinant filoviruses, which can be used to study virus biology. In addition, the generation of tagged and particularly fluorescent or luminescent viruses can be applied as tools for both diagnostic applications and for screening to identify novel countermeasures. Here we describe the generation and basic characterization of recombinant Ebola viruses rescued from cloned cDNA using a T7-driven system.

Key words Ebola virus, Reverse genetics, Full-length clone system, Infectious clone system, Recombinant virus, Virus rescue

1 Introduction

Reverse genetics is an approach to molecular biology that involved the investigation of resulting phenotypes based on changes made to the encoding sequences. For filoviruses, such reverse genetics systems can be broadly divided into (1) life-cycle modeling systems, which use genome analogues to model various aspects of the virus life cycle (for details, *see* Chaps. 6 and 9), and (2) full-length clone systems (also known as infectious clone systems), which allow the generation of recombinant viruses entirely from cDNA [1].

Full-length clone plasmids consist of a full-length cDNA copy of the virus (e.g., Ebola virus (EBOV)) genome under the control of a promoter that allows expression of the corresponding genomic RNA in mammalian cells (Fig. 1). In all systems reported to date for filoviruses, the transcription of the viral genome has been accomplished using T7 RNA polymerase. However, other strategies based on transcription by RNA

Thomas Hoenen and Allison Groseth (eds.), *Ebolaviruses: Methods and Protocols*, Methods in Molecular Biology, vol. 1628, DOI 10.1007/978-1-4939-7116-9_13, © Springer Science+Business Media LLC 2017

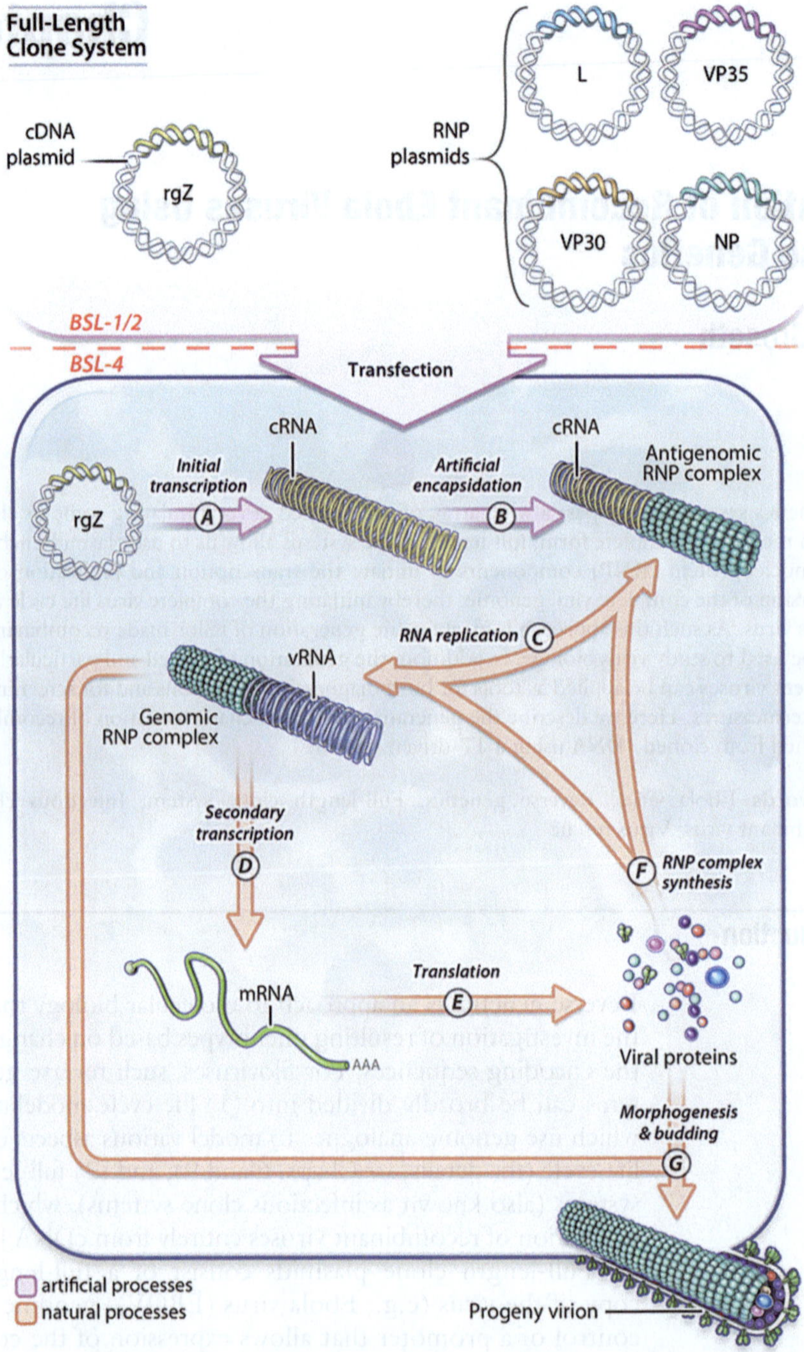

Fig. 1 Schematic of the rescue procedure. Cells are transfected with a full-length plasmid (rgZ) and the RNP plasmids encoding L, VP35, VP30, and NP. Initial transcription (*A*) either by cellular Pol I or Pol II or alternatively by T7 RNA polymerase provided from a plasmid (not shown) results in a "naked" full-length genome (cRNA), which is subsequently encapsidated (*B*) and can then serve as a template for genome replication (*C*) and transcription (*D*). This results in translation of viral proteins (*E*), which further encapsidate newly replicated genomes (*F*) and lead to the generation of recombinant virus particles (*G*). Reprinted from [1] with permission from Elsevier

polymerase I or II have been employed very successfully for other virus families [2–10] and should in principle also be feasible for members of the filovirus family. Further, the genome is generally cloned in antigenomic or cRNA orientation, such that a cRNA rather than a vRNA is expressed. While this appears to be less critical for filoviruses than for some other systems (e.g., paramyxoviruses), slightly better rescue efficiencies have been suggested using this approach [9]. Further coexpression of the filovirus ribonucleoprotein complex (RNP) proteins allows them to complex with the viral cRNA to convert it into a helical nucleocapsid [11, 12], which is the active template for viral transcription and translation. For ebolaviruses, the minimal components required for this process are (1) the nucleoprotein (NP), which encapsidates the viral genome [13]; (2) virion protein 35 (VP35), which serves as a polymerase cofactor [13]; (3) viral protein VP30 (VP30), which for ebolaviruses acts as transcriptional activator [13] and is involved in mRNA editing of the ebolavirus glycoprotein mRNA [14]; and (4) the viral polymerase (L) [13]. Once complexed in this fashion, the nucleocapsid can then serve as an authentic viral template for genome replication, and be replicated into vRNA genomes, which in turn can be transcribed into mRNAs coding for each of the viral gene products, thereby initiating the remainder of the virus life cycle (e.g., secondary transcription, RNP packaging, egress and budding of new virions). Naturally, since rescue of full-length clones results in the formation of infectious replicating viruses, the virus rescue needs to be performed in a maximum containment (BSL4) laboratory.

While the need for BSL4 containment restricts the number of facilities where this kind of work can be conducted, and thus the number of studies that have so far availed themselves of this technology remains limited, it clearly provides an extremely powerful approach for investigating virus biology. To date these systems have been used to examine VP40 function and details of the filovirus budding mechanism [15, 16], to study GP production and processing [17, 18], and to analyze the function of VP30 [19, 20], as well as details of genome replication, transcription, and translation [21–23]. Further, the generation of recombinant Ebola viruses can allow us to study the contributions of individual genes or functions to virus pathogenicity [24, 25]. In addition to these applications to the study of virus biology, recombinant virus rescue can be very useful in developing tools for applied research, including both diagnostics and the screening of antiviral countermeasures. In particular, this includes the production of reporter-expressing viruses (e.g., viruses expressing fluorescent proteins or luciferases) [26, 27].

In this chapter we focus on the protocol used to rescue infectious EBOV from a corresponding full-length clone systems, although similar protocols are also applicable for members of the other Ebola virus species [24] as well as Marburg virus [28, 29].

2 Materials

2.1 Initial Transfection (For Rescue of Recombinant Virus)

1. Cells for transfection and initial virus rescue, e.g., Vero and 293 or Huh7 cells (*see* **Note 1**).
2. Growth medium: Dulbecco's Modified Eagle's Medium (DMEM) with 10% (v/v) fetal bovine serum (FBS, heat-inactivated for 30 min at 56 °C) and 2 mM L-glutamine, 100 U/mL penicillin, and 100 μg/mL streptomycin.
3. Maintenance medium: DMEM with 5% (v/v) FBS and 2 mM L-glutamine, 100 U/mL penicillin, and 100 μg/mL streptomycin.
4. Six-well plates.
5. Full-length clone plasmid, e.g., pT7-ZEBOV (wt), pAmp-EBOV (wt), etc. (*see* **Note 2**).
6. Helper plasmids, e.g., pCAGGS-NP, pCAGGS-VP35, pCAGGS-VP30, pCAGGS-L, pCAGGS-eGFP, pCAGGS-T7 (*see* **Note 3**).
7. Transit-LT1 (or FuGENE 6) transfection reagent.
8. Opti-MEM.

2.2 Passage of Recombinant Viruses

1. Cells for passaging, e.g., Vero or Huh7 cells (*see* **Note 1**).
2. Growth medium.
3. Six-well plates.
4. T150 tissue culture flasks.
5. Cell scrapers.
6. FBS (heat-inactivated for 30 min at 56 °C).

3 Methods

Safety: This protocol results in the production of infectious EBOV, and as such any rescue attempts must be conducted according to approved institutional guidelines, which should include transfections for virus rescue being performed under BSL4 conditions. We also strongly recommend the use of designated BSL2 areas for full-length clone construction, propagation, and storage in order to keep these constructs physically separated from other plasmids involved in virus rescue (i.e., RNP protein expression plasmids), in order to eliminate any risk of contamination and accidental rescue of infectious viruses outside a BSL4 laboratory. In this context it is also important to mention that the term "infectious clone," which is sometimes used to refer to full-length clones, is somewhat misleading. In contrast to the situation with positive-strand RNA viruses, for negative-sense RNA viruses such as EBOV, these plasmids as well as the cDNA they encode are in fact noninfectious and thus can be safely handled outside of a BSL4 environment, as long as they are not combined with helper plasmids and transfected into mammalian cells.

3.1 Initial Transfection (For Rescue of Recombinant Virus)

1. Seed cells for transfection (p0 cells; *see* **Note 1**) in 2 mL growth medium per well into six-well plates such that they achieve ~50% confluence on the next day. Incubate cells overnight at 37 °C and 5% CO_2 in a humidified incubator.
2. At the time of transfection (i.e., approximately 24 h after cell splitting), combine the helper plasmids in a tube as shown in Table 1. Prepare a negative control rescue (−L; *see* **Note 4**) in addition to the positive rescue samples (+L) (*see* **Note 4**).
3. Add 100 μL Opti-MEM per sample to the DNA mixtures, vortex, and spin down the samples.
4. Transfer the cells, diluted helper plasmids, and an aliquot of full-length clone plasmid into a BSL4 laboratory.
5. Add 250 ng full-length plasmid to each of the diluted helper plasmid mixtures (Table 1).
6. Briefly vortex vial with Transit-LT1. Add 6 μL (*see* **Note 5**) Transit-LT1 per well to the mix, vortex, and spin down. Incubate the DNA to TransIT mixtures for 15 min at room temperature to allow transfection complexes to form.
7. After 15 min, add 100 μL of the transfection complexes dropwise to the cells ensuring that complexes are distributed over as much of the cell monolayer as possible.

Table 1
Transfection mixtures for virus rescue

	Positive rescue (+L)	Negative control (−L)
pCAGGS-NP	125 ng	125 ng
pCAGGS-VP35	125 ng	125 ng
pCAGGS-VP30	75 ng	75 ng
pCAGGS-L	1000 ng	–
pCAGGS-eGFP	–	1000 ng
pCAGGS-T7	250 ng	250 ng
Full-length plasmid (e.g., pAmp-EBOV (wt)	250 ng	250 ng
Total DNA mass	1825 ng	1825 ng

8. Rock plates back and forth and from side to side (*see* **Note 6**). Place the cells in a 37 °C and 5% CO_2 humidified incubator.
9. At approximately 24 h post-transfection, replace the medium with 4 mL maintenance medium (*see* **Note 7**). Return cells to the incubator.

3.2 Passage of Recombinant Viruses

1. At 7 days post-transfection, seed cells in six-well plates (p1 cells; *see* **Note 8**) in 3 mL growth medium so that they attain ~90% confluence on the next day. Incubate the cells overnight at 37 °C and 5% CO_2 in a humidified incubator.
2. On the next day, take the p1 cells into the BSL4 laboratory. Add 1 mL of supernatant from each p0 cells to a corresponding p1 cells (without removing the medium on those cells). Place the cells in a 37 °C and 5% CO_2 in a humidified incubator.
3. Freeze remaining p0 supernatants in 1 mL aliquots at −80 °C.
4. After 1 day, completely remove the supernatant from the p1 cells, and add 4 mL of maintenance medium.
5. From this point onward, check the p1 cells daily for the formation of cytopathic effect (CPE) (*see* **Note 9**). If rescuing reporter-expressing viruses (e.g., those expressing fluorescent proteins or luciferase), reporter activity can also be used as a readout.
6. Once CPE is clearly visible, passage 1 mL of p1 supernatant into a T150 flask with 90% confluent cells (p2 cells) in 60 mL growth medium to prepare a working stock. Alternatively, thaw 1 mL of p0 supernatant (from Subheading 3.2, **step 3**), and use this material to inoculate the T150 (p2 cells) for preparation of the working stock (*see* **Note 10**). Place the cells in a 37 °C and 5% CO_2 in a humidified incubator.
7. Check for CPE in p2 cells daily.

3.3 Harvest of Recombinant Viruses

1. Once p2 cells show clearly discernibly CPE, harvest the virus by scraping any cells that remain attached to the flask into the medium using a cell scraper.
2. Transfer the cell-containing supernatant into 2 × 50 mL Falcon tubes (i.e., approximately 30 mL each), and pellet the cells and cell debris by centrifugation for 10 min at 1000 × *g* at 4 °C.
3. Pour the clarified supernatant into new Falcon tubes each containing 3 mL of FBS, mix the samples by inversion, and aliquot the virus stocks as desired (usually 0.5 mL aliquots prove convenient).
4. Store the virus aliquots in liquid nitrogen.

3.4 Characterization of Newly Rescued Virus

Our experience has shown that rescued ebolaviruses frequently show unwanted mutations, and that while this can to some extent be mitigated by selection of a more suitable cell type (e.g., Huh7 cells, *see* **Note 1**), it is absolutely necessary to completely sequence the rescued viruses prior to use in further experiments. As a result of this phenomenon, it is also beneficial to perform several rescues in parallel, as recommended in this protocol, to obtain multiple virus clones, in order to increase the likelihood of obtaining one that is correct.

Further, we have observed that given the clonal nature of recombinant viruses rescued using reverse genetics (i.e., in contrast to the diverse quasispecies that exist in natural virus populations), the properties of clonal recombinant viruses can differ slightly from the natural isolates to which their sequences correspond. As such it is highly recommended to conduct growth kinetics and/or LD_{50} studies comparing any recombinant viruses to the corresponding natural isolate. Further, it makes it potentially very important to compare mutant viruses not only to the natural wild-type virus but to a recombinant clonal wild-type population as well (i.e., EBOV-wt vs. rEBOV-wt vs. rEBOV-mut).

4 Notes

1. Over the years several cell lines, or combinations thereof, have been used for the rescue of recombinant filoviruses. These include Vero cells, a 1:1 mixture of 293 and Vero cells, and more recently Huh-7 cells. The choice of cell line for virus rescue must take into account both the transfectability of the cells, which is critical for successful generation of the initial virus particles in the p0 cells, and their susceptibility to filoviruses infection, which allows these initial particles to amplify to detectable levels. As such both of these factors contribute to rescue efficiency. For example, Vero cells are difficult to transfect efficiently but highly susceptible to filovirus infection. In contrast, 293 and 293 T cells are highly transfectable, but not very susceptible to filovirus infection, although this can now be overcome by expressing the virus adhesion factor Tim-1 in these cells [21, 30]. As such, early protocols for filovirus rescue often used a mixture of these cells so that the 293/293T cells are transfected and produce the initial virus particles, which then infect the Vero cells to further amplify the recombinant virus. More recent work has shown that cell line selection can also influence the sequence fidelity of the recombinant viruses and that this is an important, but previously unappreciated, consideration. Specifically, filovirus rescue in Vero cells was found to result in a higher frequency of certain mutations, particularly A insertions in poly-A stretches, than other cells [31]. As such we now recommend Huh-7 cells for initial rescue (p0 cells), and either Huh-7 or Vero cells for further passaging (p1 cells).

2. A number of full-length EBOV clones have been described, and in principle any of these can be used for virus rescue. As noted, all existing filovirus full-length clones published to date have been based on T7-mediated transcription of the viral genome, but other systems (e.g., Pol I- and Pol II-driven constructs) are likely suitable as well and have in some cases been shown to be even more efficient for other virus families. However, not all systems work well for a given virus family, so if using alternate polymerase systems for EBOV rescue, it would be advisable to first validate the principle suitability of these systems, for instance, using minigenome systems. For Pol I this has already been done for Reston virus [7], and this system seems to present a suitable alternative, although this remains to be verified using full-length constructs. Further, if using different constructs the amount of full-length construct to be transfected might need to be adjusted to provide optimal rescue efficiency.
3. While we and many others currently base our helper plasmids on pCAGGS, this is by no means necessary. Other RNA Pol II expression constructs (e.g., those based on pCDNA3 or pI.18) as well as those based on T7 (e.g., pTM1) will almost certainly be suitable for use in virus rescue; however, expression levels will vary, and the amount of each plasmid used would need to be optimized accordingly. This is also true for helper plasmids for other Ebola virus species, which may have lower expression or exhibit different levels of protein activity. Alternatively, the helper plasmids for a highly efficient filovirus rescue system, such as that for EBOV, can be used for the initial generation of other recombinant filoviruses, which then lose these heterologous plasmid-encoded proteins during subsequent rounds of infection and passage.
4. Since rescue efficacy for most, if not all full-length clone systems, is <100%, and in some cases much less, it is highly advisable to transfect several replicate +L wells when attempting to rescue a virus. This is particularly the case if attempting to rescue recombinant viruses, where the introduced mutations may have an additional negative impact on stages of the virus life cycle that could further decrease rescue efficiency. One negative control well where pCAGGS-L is substituted with equal amount pCAGGS-eGFP should normally be sufficient to distinguish real CPE from cell death or changes to cell morphology developing, for instance, due to aging or overconfluence of the cells in the absence of virus.
5. Depending on the cell type used for virus rescue, and possibly experimental parameters, the amount of Transit-LT1 used may need to be adjusted to obtain optimal transfection efficacy. As a starting point for most cell types, including 293/293 T and

Huh7, a ratio of approximately 3 μL Transit-LT1 per 1 μg DNA is recommended and has produced good results. However, if using Vero cells only, it may be beneficial to increase this ratio to 6:1.

6. It is important to rock the plates front to back and side to side rather than swirling them in a circular fashion, as the latter results in transfection complexes being pushed to the edge of the well and produces uneven transfection of the monolayer.
7. It is extremely important at this stage to remove all transfection supernatant prior to adding fresh medium. To accomplish this we recommend to first removing the bulk of the supernatant using a serological pipette, and then removing any remaining supernatant using a 1000 μL pipette. This ensures that there are no remaining traces of the cDNA plasmid from the transfection, which can cause problems in downstream procedures, particularly PCR-based assays. Further, already at this point, it has to be considered that virus may already be being produced and thus standard practices to avoid cross-contaminating wells need to be implemented (i.e., different tips for medium exchange need to be used between different samples, including replicate +L samples, and between positive and negative samples).
8. Again the choice of cells can be varied. Most commonly used at this point are Vero cells; however, continuous passaging in these cells can lead to insertions in the GP gene editing site [31, 32].
9. The time of onset of CPE is highly dependent on the properties of the virus being rescued, as well as the rescue efficacy itself. As a rule of thumb, for EBOV clear CPE (as compared to the −L negative control cells) should start to be visible about 1 week after passaging. However, for other Ebola virus species, particularly Reston virus, where titers are lower, and/or CPE is less pronounced, this can take significantly longer (i.e., for Reston virus, only modest CPE forms and should occur by 14 days after passaging).
10. Using p0 material for infection of flasks for the production of virus stocks allows the researcher to reduce the number of passages required for preparation of the working stock from two to one. This in turn reduces the chance that unwanted mutations may occur. As noted, this can be a particular problem when using Vero cells. Importantly, this approach still allowing for the screening of a larger number of wells for successful virus rescue by storing the p0 material until the success of the rescue can be determined in the p1 samples using a smaller, and therefore more convenient and less expensive, six-well plate format.

Acknowledgments

The authors thank Thomas Hoenen for critical reading of the manuscript.

References

1. Hoenen T, Groseth A, de Kok-Mercado F, Kuhn JH, Wahl-Jensen V (2011) Minigenomes, transcription and replication competent virus-like particles and beyond: reverse genetics systems for filoviruses and other negative stranded hemorrhagic fever viruses. Antivir Res 91(2):195–208. doi:10.1016/j.antiviral.2011.06.003. PubMed PMID: 21699921; PubMed Central PMCID: PMCPMC3586226
2. Flatz L, Bergthaler A, de la Torre JC, Pinschewer DD (2006) Recovery of an arenavirus entirely from RNA polymerase I/II-driven cDNA. Proc Natl Acad Sci U S A 103(12):4663–4668. doi:10.1073/pnas.0600652103. PubMed PMID: 16537369; PubMed Central PMCID: PMCPMC1450228
3. Flick K, Hooper JW, Schmaljohn CS, Pettersson RF, Feldmann H, Flick R (2003) Rescue of Hantaan virus minigenomes. Virology 306(2):219–224. PubMed PMID: 12642095
4. Flick R, Flick K, Feldmann H, Elgh F (2003) Reverse genetics for crimean-congo hemorrhagic fever virus. J Virol 77(10):5997–6006. PubMed PMID: 12719591
5. Flick R, Pettersson RF (2001) Reverse genetics system for Uukuniemi virus (Bunyaviridae): RNA polymerase I-catalyzed expression of chimeric viral RNAs. J Virol 75(4):1643–1655. PubMed PMID: 11160662
6. Freiberg A, Dolores LK, Enterlein S, Flick R (2008) Establishment and characterization of plasmid-driven minigenome rescue systems for Nipah virus: RNA polymerase I- and T7-catalyzed generation of functional paramyxoviral RNA. Virology 370(1):33–44. doi:10.1016/j.virol.2007.08.008. PubMed PMID: 17904180; PubMed Central PMCID: PMCPMC2716073
7. Groseth A, Feldmann H, Theriault S, Mehmetoglu G, Flick R (2005) RNA polymerase I-driven minigenome system for ebola viruses. J Virol 79(7):4425–4433. PubMed PMID: 15767442
8. Pinschewer DD, Perez M, de la Torre JC (2003) Role of the virus nucleoprotein in the regulation of lymphocytic choriomeningitis virus transcription and RNA replication. J Virol 77(6):3882–3887. PubMed PMID: 12610166; PubMed Central PMCID: PMCPMC149515
9. Martin A, Staeheli P, Schneider U (2006) RNA polymerase II-controlled expression of antigenomic RNA enhances the rescue efficacies of two different members of the Mononegavirales independently of the site of viral genome replication. J Virol 80(12):5708–5715. doi:10.1128/JVI.02389-05. PubMed PMID: 16731909; PubMed Central PMCID: PMCPMC1472609
10. Wang J, Wang C, Feng N, Wang H, Zheng X, Yang S et al (2015) Development of a reverse genetics system based on RNA polymerase II for Newcastle disease virus genotype VII. Virus Genes 50(1):152–155. doi:10.1007/s11262-014-1137-x. PubMed PMID: 25384536
11. Beniac DR, Melito PL, Devarennes SL, Hiebert SL, Rabb MJ, Lamboo LL et al (2012) The organisation of ebola virus reveals a capacity for extensive, modular polyploidy. PLoS One 7(1):e29608. doi:10.1371/journal.pone.0029608. PubMed PMID: 22247782; PubMed Central PMCID: PMC3256159
12. Bharat TA, Noda T, Riches JD, Kraehling V, Kolesnikova L, Becker S et al (2012) Structural dissection of Ebola virus and its assembly determinants using cryo-electron tomography. Proc Natl Acad Sci U S A 109(11):4275–4280. doi:10.1073/pnas.1120453109. PubMed PMID: 22371572; PubMed Central PMCID: PMC3306676
13. Muhlberger E, Weik M, Volchkov VE, Klenk HD, Becker S (1999) Comparison of the transcription and replication strategies of marburg virus and Ebola virus by using artificial replication systems. J Virol 73(3):2333–2342. PubMed PMID: 9971816
14. Mehedi M, Hoenen T, Robertson S, Ricklefs S, Dolan MA, Taylor T et al (2013) Ebola virus RNA editing depends on the primary editing site sequence and an upstream secondary structure. PLoS Pathog 9(10):e1003677. doi:10.1371/journal.ppat.1003677. PubMed PMID: 24146620; PubMed Central PMCID: PMC3798607

15. Neumann G, Ebihara H, Takada A, Noda T, Kobasa D, Jasenosky LD et al (2005) Ebola virus VP40 late domains are not essential for viral replication in cell culture. J Virol 79(16):10300–10307. PubMed PMID: 16051823
16. Hoenen T, Volchkov V, Kolesnikova L, Mittler E, Timmins J, Ottmann M et al (2005) VP40 octamers are essential for Ebola virus replication. J Virol 79(3):1898–1905. PubMed PMID: 15650213
17. Volchkov VE, Volchkova VA, Muhlberger E, Kolesnikova LV, Weik M, Dolnik O et al (2001) Recovery of infectious Ebola virus from complementary DNA: RNA editing of the GP gene and viral cytotoxicity. Science 291(5510):1965–1969. PubMed PMID: 11239157
18. Neumann G, Feldmann H, Watanabe S, Lukashevich I, Kawaoka Y (2002) Reverse genetics demonstrates that proteolytic processing of the Ebola virus glycoprotein is not essential for replication in cell culture. J Virol 76(1):406–410. PubMed PMID: 11739705
19. Biedenkopf N, Hartlieb B, Hoenen T, Becker S (2013) Phosphorylation of Ebola virus VP30 influences the composition of the viral nucleocapsid complex: impact on viral transcription and replication. J Biol Chem 288(16):11165–11174. doi:10.1074/jbc.M113.461285. PubMed PMID: 23493393; PubMed Central PMCID: PMC3630872
20. Martinez MJ, Volchkova VA, Raoul H, Alazard-Dany N, Reynard O, Volchkov VE (2011) Role of VP30 phosphorylation in the Ebola virus replication cycle. J Infect Dis 204(Suppl 3):S934–S940. doi:10.1093/infdis/jir320. PubMed PMID: 21987772
21. Watt A, Moukambi F, Banadyga L, Groseth A, Callison J, Herwig A et al (2014) A novel life cycle modeling system for Ebola virus shows a genome length-dependent role of VP24 in virus infectivity. J Virol 88(18):10511–10524. doi:10.1128/JVI.01272-14. PubMed PMID: 24965473; PubMed Central PMCID: PMCPMC4178905
22. Hoenen T, Shabman RS, Groseth A, Herwig A, Weber M, Schudt G et al (2012) Inclusion bodies are a site of ebolavirus replication. J Virol 86(21):11779–11788. doi:10.1128/JVI.01525-12. PubMed PMID: 22915810; PubMed Central PMCID: PMCPMC3486333
23. Shabman RS, Hoenen T, Groseth A, Jabado O, Binning JM, Amarasinghe GK et al (2013) An upstream open reading frame modulates ebola virus polymerase translation and virus replication. PLoS Pathog 9(1):e1003147. doi:10.1371/journal.ppat.1003147. PubMed PMID: 23382680; PubMed Central PMCID: PMCPMC3561295
24. Groseth A, Marzi A, Hoenen T, Herwig A, Gardner D, Becker S et al (2012) The Ebola virus glycoprotein contributes to but is not sufficient for virulence in vivo. PLoS Pathog 8(8):e1002847. doi:10.1371/journal.ppat.1002847. PubMed PMID: 22876185; PubMed Central PMCID: PMCPMC3410889
25. Prins KC, Delpeut S, Leung DW, Reynard O, Volchkova VA, Reid SP et al (2010) Mutations abrogating VP35 interaction with double-stranded RNA render Ebola virus avirulent in guinea pigs. J Virol 84(6):3004–3015. doi:10.1128/JVI.02459-09. PubMed PMID: 20071589; PubMed Central PMCID: PMC2826052
26. Hoenen T, Groseth A, Callison J, Takada A, Feldmann H (2013) A novel Ebola virus expressing luciferase allows for rapid and quantitative testing of antivirals. Antivir Res 99(3):207–213. doi:10.1016/j.antiviral.2013.05.017. PubMed PMID: 23751367; PubMed Central PMCID: PMCPMC3787978
27. Towner JS, Paragas J, Dover JE, Gupta M, Goldsmith CS, Huggins JW et al (2005) Generation of eGFP expressing recombinant Zaire ebolavirus for analysis of early pathogenesis events and high-throughput antiviral drug screening. Virology 332(1):20–27. PubMed PMID: 15661137
28. Albarino CG, Uebelhoer LS, Vincent JP, Khristova ML, Chakrabarti AK, McElroy A et al (2013) Development of a reverse genetics system to generate recombinant Marburg virus derived from a bat isolate. Virology 446(1-2):230–237. doi:10.1016/j.virol.2013.07.038. PubMed PMID: 24074586
29. Enterlein S, Volchkov V, Weik M, Kolesnikova L, Volchkova V, Klenk HD et al (2006) Rescue of recombinant Marburg virus from cDNA is dependent on nucleocapsid protein VP30. J Virol 80(2):1038–1043. doi:10.1128/JVI.80.2.1038-1043.2006. PubMed PMID: 16379005; PubMed Central PMCID: PMCPMC1346851
30. Kondratowicz AS, Lennemann NJ, Sinn PL, Davey RA, Hunt CL, Moller-Tank S et al (2011) T-cell immunoglobulin and mucin domain 1 (TIM-1) is a receptor for Zaire Ebolavirus and Lake Victoria Marburgvirus. Proc Natl Acad Sci U S A 108(20):8426–8431. doi:10.1073/pnas.1019030108. PubMed PMID: 21536871; PubMed Central PMCID: PMC3100998

31. Tsuda Y, Hoenen T, Banadyga L, Weisend C, Ricklefs SM, Porcella SF et al (2015) An improved reverse genetics system to overcome cell-type-dependent Ebola virus genome plasticity. J Infect Dis. doi:10.1093/infdis/jiu681. PubMed PMID: 25810440

32. Volchkova VA, Dolnik O, Martinez MJ, Reynard O, Volchkov VE (2011) Genomic RNA editing and its impact on Ebola virus adaptation during serial passages in cell culture and infection of guinea pigs. J Infect Dis 204(Suppl 3):S941–S946. doi:10.1093/infdis/jir321. PubMed PMID: 21987773

Chapter 14

Luciferase-Expressing Ebolaviruses as Tools for Screening of Antivirals

Thomas Hoenen

Abstract

Ebolaviruses cause severe hemorrhagic fever with high case fatality rates. Despite recent progress, there is a continued need for the development of antivirals against these viruses. Reporter-expressing ebolaviruses, which can be generated using reverse genetics systems, are powerful tools for antiviral screening. While viruses expressing fluorescent reporters are amenable for this purpose and can be used for high-content imaging-type screens, as an alternative, luciferase-expressing reporter viruses have recently been developed and have the advantages of being extremely easy to use and having short assay times. Here we provide a detailed protocol for the use of such a luciferase-expressing reporter virus for antiviral screening in a 96-well format, with parallel assessment of cytotoxicity of the screened compounds.

Key words Ebolaviruses, Filoviruses, Reverse genetics, Reporter-expressing viruses, Luciferase, Drug screening

1 Introduction

The recent epidemic of ebolavirus disease in West Africa has highlighted the need for preventative as well as therapeutic countermeasures against ebolavirus infections. While there have been impressive advances over the last few years in terms of the development of such countermeasures, there are still no approved specific therapeutics available [1]. Reporter-expressing viruses present a promising strategy for the screening of antiviral compounds, as they have the potential for faster, more precise, and technically less complicated, and thus less error-prone, readouts than classical cytopathic effect-based readouts of antiviral activity [2]. Such reporter-expressing viruses can be produced using full-length clone systems, which allow the generation of custom-tailored, recombinant viruses (*see* Chapter 13 for a protocol). The first fluorescent protein-expressing ebolaviruses were developed some years ago [3–5], and are particularly amenable to high-content imaging [6], making them valuable tools for antiviral screening [7].

Thomas Hoenen and Allison Groseth (eds.), *Ebolaviruses: Methods and Protocols*, Methods in Molecular Biology, vol. 1628,
DOI 10.1007/978-1-4939-7116-9_14,

However, this approach requires highly specialized equipment and software and is therefore not a viable option for many groups. As an alternative approach, luciferase-expressing viruses have also recently become available [8]. These viruses allow a more rapid assessment of antiviral activity than fluorescent reporter-expressing viruses, as well as very sensitive measurements of viral gene expression in infected cells, making them valuable tools for basic research questions [9]. Analysis of luciferase expression also relies on comparatively very simple, inexpensive, and widely available equipment. However, these viruses have the disadvantage that they cannot be used for high-content imaging, and thus it is technically difficult to assess parameters other than antiviral effects (e.g., cell viability/cytotoxic effects of the drug or parameters indicating potential mechanisms of action) in an initial screen. Nevertheless, their extreme ease of use makes them an interesting alternative to other reporter-expressing viruses for initial screening of antivirals, and they are beginning to be exploited for this purpose [10]. Here, we present a protocol that allows screening for antiviral activity against infectious ebolaviruses in vitro in a 96-well format, with results available within 48 h postinfection, and only minimal work required in a biosafety level 4 (BSL4) laboratory. In parallel, cytotoxic effects of the assessed drugs can be determined using a commercial cell viability assay outside the BSL4 laboratory.

2 Materials

1. VeroE6 cells (ATCC CRL 1586) (*see* **Note 1**).
2. Dulbecco's Modified Eagle's Medium (DMEM) with 10% (v/v; $DMEM_{10\%}$), 5% (v/v; $DMEM_{5\%}$), or 0% ($DMEM_{0\%}$) fetal bovine serum (FBS, heat-inactivated 30 min at 56 °C) and 1% L-glutamine (Q, 2 mM) and 1% penicillin/streptomycin (PS, 100 U/mL/100 μg/mL).
3. Trypsin/EDTA solution for cell dissociation (0.025% trypsin and 0.01% EDTA).
4. Sterile reagent reservoirs for use with multichannel pipettes.
5. Flat clear 96-well plates, tissue-culture treated.
6. Multichannel pipette (12-channel recommended).
7. U-bottom 96-well plate.
8. Recombinant Ebola virus-expressing luciferase (rgEBOV-luc2 [8]).
9. GloLysis buffer (Promega).
10. Bright-Glo reagent (Promega).
11. Luminometer(s) capable of reading 96-well plates (one within the BSL4 laboratory and, if also performing cytotoxicity assessments, one outside the BSL4 laboratory).

12. White (or black) opaque 96-well plates (*see* **Note 2**).
13. CellTiter-Glo reagent (Promega).
14. Orbital shaker.

3 Methods

3.1 Drug Treatment of Cells

1. Remove the medium from a confluent T75 flask of VeroE6 cells (*see* **Notes 1** and **3**), wash once briefly with 2 mL trypsin (*see* **Note 4**), and then add 2 mL trypsin. Wait until cells are showing signs of rounding off, and then dislodge them by tapping the flask. Add 8 mL of $DMEM_{10\%}$, and resuspend cells by repeatedly rinsing the bottom of the flask (*see* **Note 4**).
2. For two 96-well plates (*see* **Note 5**), add 4 mL cell suspension to 26 mL $DMEM_{10\%}$, mix by inversion, pour into a reagent reservoir, and immediately add 100 μL of cell suspension into each well using a multichannel pipette (*see* **Note 6**). Incubate the plates for 24 h (*see* **Note 3**).
3. After 24 h, prepare the drugs by diluting them in $DMEM_{0\%}$ in a 96-well U-bottom plate at double the concentration you want to test. Prepare a final volume of 150 μL (*see* **Note 7**).
4. Remove the supernatant from the 96-well plates with the cells, and add 50 μL of $DMEM_{5\%}$ to each well.
5. Add 50 μL of diluted drugs to each of the duplicate plates (one plate will be used for infection (called plate A), the other to assess possible cytotoxic effects of the drugs (called plate B)) (*see* **Notes 7–9**), and return the plates to the incubator for 2 h (*see* **Note 3**).
6. After 2 h, add 50 μL $DMEM_{0\%}$ to all wells of plate B, and return it to the incubator. Take plate A into the BSL4 laboratory.

3.2 Infection and Assessment of Antiviral Activity

1. Inside the BSL4 laboratory, dilute rgEBOV-luc2 in $DMEM_{0\%}$ for a final concentration of 2×10^4 $TCID_{50}$/mL (i.e., 1×10^3 $TCID_{50}$ per 50 μL) in a final volume of at least 6 mL per plate. Mix thoroughly (by gentle vortexing and/or inversion).
2. Using a multichannel pipette, add 50 μL of diluted virus to each well of plate A. Add 50 μL of $DMEM_{0\%}$ to the not treated, not infected control wells (*see* **Note 7**). Return the plate to the incubator.
3. After 48 h, completely remove the supernatant from the plate, and add 100 μL of GloLysis buffer to each well. Incubate for 10 min at room temperature.
4. In the meantime, add 50 μL of Bright-Glo reagent to each well of an opaque 96-well plate (*see* **Note 10**).

5. After the 10 min incubation, transfer 50 μL of cell lysate to the opaque plate using a multichannel pipette (*see* **Note 11**), and measure it in a luminometer, using an integration time of 0.5–1 s per well.

3.3 Assessment of Cytotoxicity

1. Forty hours after addition of drugs, remove 100 μL of supernatant from plate B, and add 50 μL CellTiter-Glo Reagent.
2. Incubate the plate at room temperature for 2 min on an orbital shaker at approximately 60 rpm and then for another 10 min at room temperature without shaking.
3. After the incubation is over, transfer 80 μL of supernatant from plate B into an opaque 96-well plate, and measure it in a luminometer, using an integration time of 0.5–1 s per well.

4 Notes

1. Different cell lines can be used for infection and antiviral testing (as long as they are susceptible to infection with filoviruses). However, the amount of input virus might have to be optimized for each cell line to obtain an adequate dynamic range of the assay, i.e., the difference in reporter activity between infected and noninfected cells.
2. Both white and black plates function well in this assay. In our experience, signals are about 2 logs higher with white plates; however, since this is true for both negative and positive samples, the dynamic range of the assay does not change appreciably with the choice of plate color. However, the sensitivity and/or linear range of detection of the luminometer used in the assay should be considered for the decision whether to use white or black plates.
3. Cells should be maintained in a humidified incubator with 5% CO_2 at 37 °C.
4. Alternatively, PBS can be used for this step.
5. When testing drugs, one plate should be used to test efficacy (this plate will be infected inside the BSL4 laboratory), and the other plate should be treated with the same compounds to test for cytotoxic effects of those compounds.
6. If cells are not seeded right away, they can settle in the reservoir, resulting in an uneven cell number in the plate, which can skew results.
7. The exact layout of the plate will of course vary between experiments. However, wells should be reserved for the following controls: (a) not treated, not infected samples (add 50 μL $DMEM_{0\%}$ to those plates at the time of drug addition and

another 50 μL $DMEM_{0\%}$ at the time of infection) and (b) not treated, infected samples (add 50 μL $DMEM_{0\%}$ to those plates at the time of drug addiction). In addition, it is advisable to include controls with the chemical solvents used to dissolve the drugs (e.g., DMSO) at the same concentration as they are present in the diluted drug samples.

8. It is advisable to have three (or more) biological replicates for each sample—this can be achieved either by having triplicate wells with identical samples on one plate or by running three identical plates in parallel. In the latter case, it makes sense to increase the volume of diluted drugs prepared in the drug dilution plate such that it will suffice for all replicate plates. This will simplify the workflow and minimize well-to-well variation.
9. During infection of 96-well plates, we sometimes observe that the corner wells or the outside rows/columns show spontaneous cell death independent of infection; however, this is usually only observed during experiments running for longer times (i.e., 1 week or longer). Nevertheless, when first establishing the assay, it might be reasonable to infect a complete 96-well plate and check whether reporter activity after 48 hours is uniform across the entire plate. The decision regarding the plate layout in future experiments (and particularly whether to include those potentially problematic wells in the assay) should then be based on the results of this initial test.
10. It is important to ensure that the Bright-Glo reagent has reached room temperature prior to measurement.
11. In our experience, during prior steps involving pipetting with the multichannel pipette, one can sometimes avoid changing tips by using a well-planned plate layout (i.e., pipetting steps are carried out from uninfected toward infected samples and from low to high drug concentrations). However, at this step, pipette tips have to be changed between each step, since even a slight carryover between samples will significantly skew results, due to the high dynamic range (3–4 $\log_{10}$) of the assay.

References

1. Mendoza EJ, Qiu X, Kobinger GP (2016) Progression of Ebola therapeutics during the 2014–2015 outbreak. Trends Mol Med 22(2): 164–173. doi:10.1016/j.molmed.2015.12.005
2. Falzarano D, Groseth A, Hoenen T (2014) Development and application of reporter-expressing mononegaviruses: current challenges and perspectives. Antivir Res 103C: 78–87. doi:10.1016/j.antiviral.2014.01.003
3. Towner JS, Paragas J, Dover JE, Gupta M, Goldsmith CS, Huggins JW, Nichol ST (2005) Generation of eGFP expressing recombinant Zaire ebolavirus for analysis of early pathogenesis events and high-throughput antiviral drug screening. Virology 332(1): 20–27. doi:10.1016/j.virol.2004.10.048
4. Ebihara H, Theriault S, Neumann G, Alimonti JB, Geisbert JB, Hensley LE, Groseth A, Jones SM, Geisbert TW, Kawaoka Y, Feldmann H (2007) In vitro and in vivo characterization of recombinant Ebola viruses expressing enhanced green fluorescent protein. J Infect

Dis 196(Suppl 2):S313–S322. doi:10.1086/520590

5. Albarino CG, Wiggleton Guerrero L, Lo MK, Nichol ST, Towner JS (2015) Development of a reverse genetics system to generate a recombinant Ebola virus Makona expressing a green fluorescent protein. Virology 484:259–264. doi:10.1016/j.virol.2015.06.013
6. Panchal RG, Kota KP, Spurgers KB, Ruthel G, Tran JP, Boltz RC, Bavari S (2010) Development of high-content imaging assays for lethal viral pathogens. J Biomol Screen 15(7):755–765. doi:10.1177/1087057110374357
7. Panchal RG, Reid SP, Tran JP, Bergeron AA, Wells J, Kota KP, Aman J, Bavari S (2012) Identification of an antioxidant small-molecule with broad-spectrum antiviral activity. Antivir Res 93(1):23–29. doi:10.1016/j.antiviral.2011.10.011
8. Hoenen T, Groseth A, Callison J, Takada A, Feldmann H (2013) A novel Ebola virus expressing luciferase allows for rapid and quantitative testing of antivirals. Antivir Res 99(3):207–213. doi:10.1016/j.antiviral.2013.05.017
9. Watt A, Moukambi F, Banadyga L, Groseth A, Callison J, Herwig A, Ebihara H, Feldmann H, Hoenen T (2014) A novel life cycle modeling system for Ebola virus shows a genome length-dependent role of VP24 in virus infectivity. J Virol 88(18):10511–10524. doi:10.1128/JVI.01272-14
10. Nelson EA, Dyall J, Hoenen T, Barnes A, Zhou H, Liang JY, Michelotti J, Dewey WH, deWald LE, Bennet RS, Morris PJ, Guha R, Klumpp-Thomas C, McKnight C, Chen Y, Xu X, Wang A, Hughes E, Martin S, Thomas C,Jahrling PB, Hensley LE, Ollinger GO, White J (2017) The Phosphatidylinositol-3-phosphate 5-kinase inhibitor Apilimod blocks filoviral entry and infection. PLoS Negl Trop Dis (in press)

Chapter 15

Live-Cell Imaging of Filoviruses

Gordian Schudt, Olga Dolnik, and Stephan Becker

Abstract

Observation of molecular processes inside living cells is fundamental to a deeper understanding of virus-host interactions in filoviral-infected cells. These observations can provide spatiotemporal insights into protein synthesis, protein-protein interaction dynamics, and transport processes of these highly pathogenic viruses. Thus, live-cell imaging provides the possibility for antiviral screening in real time and gives mechanistic insights into understanding filovirus assembly steps that are dependent on cellular factors, which then represent potential targets against this highly fatal disease. Here we describe analysis of living filovirus-infected cells under maximum biosafety (i.e., BSL4) conditions using plasmid-driven expression of fluorescently labeled viral and cellular proteins and/or viral genome-encoded expression of fluorescently labeled proteins. Such multiple-color and multidimensional time-lapse live-cell imaging analyses are a powerful method to gain a better understanding of the filovirus infection cycle.

Key words Recombinant filoviruses, Fluorescent time-lapse imaging, Fluorescent tag, Fusion protein, Intracellular transport, Viral spread, Assembly

1 Introduction

Live-cell imaging techniques are used by an increasing number of investigators to provide critical insight into the fundamental dynamics of host and virus interaction. This is possible due to the rapid technological advances that are currently being made in the field. Among the most significant technical challenges for performing successful live-cell imaging experiments is to maintain the cells in a viable state on the microscope stage, allowing analysis over the whole intended time span, although they are infected and additionally illuminated in the presence of synthetic fluorescent proteins. Controlling the environment regarding physical parameters like temperature, atmospheric conditions (gas mixture and humidity), nutritional supplements, and pH of the culture medium is essential for successful experiments [5, 6].

Thomas Hoenen and Allison Groseth (eds.), *Ebolaviruses: Methods and Protocols*, Methods in Molecular Biology, vol. 1628,
DOI 10.1007/978-1-4939-7116-9_15, © Springer Science+Business Media LLC 2017

Monitoring filoviral infection in real time is even more challenging since all experiments need to be done under biosafety level 4 (BSL-4) conditions, as all members of the family *Filoviridae* (marburgviruses, ebolaviruses, and cuevaviruses) are classified as BSL-4 agents [7]. These highly pathogenic, non-segmented, single-stranded RNA viruses cause severe hemorrhagic fever diseases with often fatal outcome in humans. The filoviruses are transmissible during the symptomatic phase of the disease—indeed, almost all body fluids from patients with severe symptoms contain high amounts of infectious virus [8–10]. Since working time under BSL-4 conditions should be minimized, the different procedures need to be as easy and secure as possible. Further, the used equipment should be safe for the user to handle, all components should only need low maintenance, and the microscope should provide the possibility to be remote controlled to monitor and control ongoing experiments from outside of the BSL-4 lab [1–4].

Visualizing filoviral infection in living cells needs a reverse genetic system to allow integration of additional open reading frames (ORF) encoding fluorescent viral proteins into the filoviral genome [11–15].

Filoviruses with an additional ORF for GFP (e.g., MARV-GFP, EBOV-NP/35-eGFP, EBOV-VP30/24-eGFP) or viruses with genomes encoding additional viral fusion proteins (GP-mCherry, L-mCherry, RFP-VP40, VP30-GFP, and others) could be successfully rescued during recent years and were used for live-cell imaging [4, 15–19]. If, in addition to the virus-encoded fluorescent protein, cells transiently express fluorescently labeled host or viral fusion proteins, more information about intracellular localization as well as behavior of these proteins during infection and the dynamics of interactions and transport can also be obtained (*see* **Note 1**) [15, 19].

The frequent filoviral outbreaks, the absence of specific treatments approved for human use, and the lack of understanding of filoviral pathogenesis highlight the need for active and broad research on these neglected tropical diseases. Here we show the different steps necessary to perform live-cell imaging with highly pathogenic filoviruses under BSL4 conditions. By imaging a full replication cycle of Marburg or Ebola virus, we were able to visualize the entry and assembly processes, cell to cell spread, as well as intracellular transport and budding steps (Fig. 1).

2 Materials

2.1 Growth Media and Solutions

1. Cell culture medium: Dulbecco's modified Eagle medium (DMEM) supplemented with 10% fetal bovine serum (FBS), 2 mM L-glutamine (Q), 50 U/mL penicillin, and 50 μg/mL streptomycin (PS).

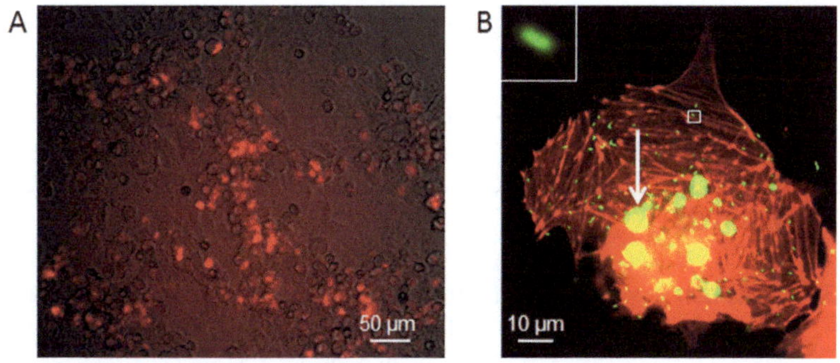

Fig. 1 Representative phenotypes of EBOV-infected cells. (**a**) Live-cell merged image of EBOV-infected VeroE6 cells 10 days post-infection acquired with low magnification (20×) in DIC and EPI fluorescence. Expression of a marker protein coded on the viral genome leads to red fluorescence in infected cells. (**b**) Live-cell merged image of EBOV-infected HuH-7 cells transiently expressing VP30-GFP and TagRFP-Actin, 24 h p.i. acquired with high magnification (63× oil) in dual-color EPI fluorescence. Cell is filled with inclusions (e.g., *white arrow*) and several 1 μm long rods, resembling viral nucleocapsids (*white inset*)

2. Transfection medium: Reduced serum medium, no phenol red (i.e., Opti-MEM I).
3. Imaging medium: Leibovitz's medium with 100 U/mL penicillin and 100 μg/mL streptomycin, 20% FBS, and 400 μM 6-hydroxy-2,5,7,8-tetramethylchromane-2-carboxylic acid (Trolox) (*see* **Note 2**).
4. PBS^{def}: 137 mM NaCl, 2,7 mM KCl, 10 mM Na_2HPO_4, 1,8 mM KH_2PO_4, pH = 7.4.

2.2 Cell Culture Dish

1. 35 mm μ-dish high wall (2 mL volume, 3.1 cm^2 growth area), polymer coverslip, tissue culture treated, sterilized (Ibidi) (*see* **Note 3**).

2.3 Microscope

1. Inverted fluorescence microscope (e.g., Leica DMI 6000 B) with additional opaque incubator chamber and heating unit (*see* **Note 4**). Motorized condenser, polarizer, camera port, DIC prism, filter wheels, filter cubes, 1.5× magnifying lens, objective turret, and sample stage (*see* **Note 5**). Low magnifying dry objective for cell culture flasks (e.g., 20× dry objective, NA 0.2). High magnifying objective (e.g., 63× Oil PL APO NA 1.4) (*see* **Note 6**). An at least 1.3 megapixel camera with high-sensitivity monochrome sensor and acquiring frame rate of >10 frames per second.

2.4 Computer and Software

1. Acquiring desktop computer including Internet/LAN connection for remote control.
2. Acquiring software (e.g., LAS AF).

3. Remote Desktop Connection (e.g., Microsoft Windows Remote Control).
4. Remote stage control software.

3 Methods

3.1 Infection

1. One day before infection, seed 2×10^4 well-suited, infectable cells (e.g., Huh-7) resuspended in 400 μL cell culture medium per μ-chamber (3.1 cm^2 growth area).
2. Grow cells in culture medium for at least 8 h at 37 °C, 5% CO_2.
3. Remove cell supernatant and rinse once with preheated (37 °C) PBS^{def}, and never let cell layer get dry.
4. Inoculate cells with preheated 400 μL stock virus suspension and infect cells with a multiplicity of infection (MOI) between 0.1 and 5 (*see* **Note 7**). During the infection period, incubate at 37 °C, 5% CO_2 for 1 h.
5. Remove virus suspension, rinse once with preheated PBS^{def} and add 2.0 mL preheated imaging medium.
6. Place infected cells in an appropriate chamber holder inside of the 37 °C preheated microscope incubator (*see* **Note 8**). Put one small drop of preheated immersion oil on oil objective, and also spread one drop around on the bottom of the cell culture chamber, if oil is desired (*see* **Note 9**).

3.2 Optional Step: Transfection

Perform any transfections after the infection step, since transfection might cause reduced infectivion of cells. Transfection might be performed with plasmids encoding (e.g., fluorescent viral fusion proteins or fluorescently labeled cellular proteins) and should be carried out in transfection medium according to the manufacturer's instructions (*see* **Note 10**).

3.3 Acquisition

Temporal and spatial resolution of acquisition is dependent on the processes that are of interest: Perform acquisition of infected cells between 0 h and 5 days post-infection (p.i.). For example, spread of infection in the cell layer can be preferably monitored at low magnification and at a low frequency frame rate of 1 frame per hour. Subviral structures might better be visualized at a high magnification, e.g., with a 63× oil objective. To follow the formation of inclusion bodies, acquisition between 3 and 22 h p.i. with frame rates lower than 1 frame per 5 min is suitable. Movement of subviral structures, e.g., capsids or vesicles, might be performed at 1 frame per 2 s and faster (*see* **Note 11**).

4 Notes

1. Fluorescence labeling of viral proteins can adversely influence protein function. The use of fluorescent fusion proteins needs adequate characterization in terms of intracellular processing and localization, protein-protein interactions, possible artificial phenotypes, and expression levels to avoid misinterpretation of the data. Sometimes a mixture of labeled and authentic protein is necessary to recover the wild-type phenotype. A titration of the DNA amounts of authentic to labeled protein might also be useful in this context.
2. Special inhibitors or chemicals, dissolved in the same volume of imaging medium, can be added carefully directly into the medium while the cell chamber is still placed on the microscope stage and acquisition is ongoing. Alternatively, the cell chamber can be removed from the microscope stage, inhibitors, or chemicals added, and infected cells are monitored after placement back onto the stage. To find particular cells again, special chambers with imprinted grids are available (e.g., from Ibidi).
3. There are a lot of different cell culture dishes available. They differ in material (glass or plastic bottom), number of wells (e.g., multi-well slides, like the μ-Slide 8 Well), or surface treatment. Also special mounts which fit regular microscope stages are available for the different culture dishes. For Ibidi 3.5 mm μ-dishes, a μ-dish microscopy rack with magnetic fixation, 127.5/85.5/19.5 mm (Ibidi), and appropriate magnetic fixation lids are available.
4. The best place to set up a live-cell microscope under BSL-4 conditions is on a special buffered anti-vibration table, far from any vibration-generating equipment (e.g., autoclaves, pumps, motors, etc.). The table should be placed underneath a hood to prevent contamination of the lab by aerosols. Enough space should be provided to store hardware components like heating unit, light source, computer, monitor, etc. outside the direct vicinity of the microscope to avoid vibrations. Regular maintenance of the BSL-4 lab equipment requires in most facilities an annual decontamination. In our experience a yearly paraformaldehyde fumigation did no harm to optical lenses or filters.
5. A programmable power outlet strip with LAN interface (e.g., EnerGenie EG-PM2) might be useful to turn individual hardware components (heating unit, light source, etc.) on or off via LAN remote control.
6. Due to physical limitations, the spatial resolution limit of a normal fluorescent microscope is approximately 350 nm in x and y and around 500 nm in z. Some of the viral structures, e.g., the filoviral capsid with a dimension of approximately

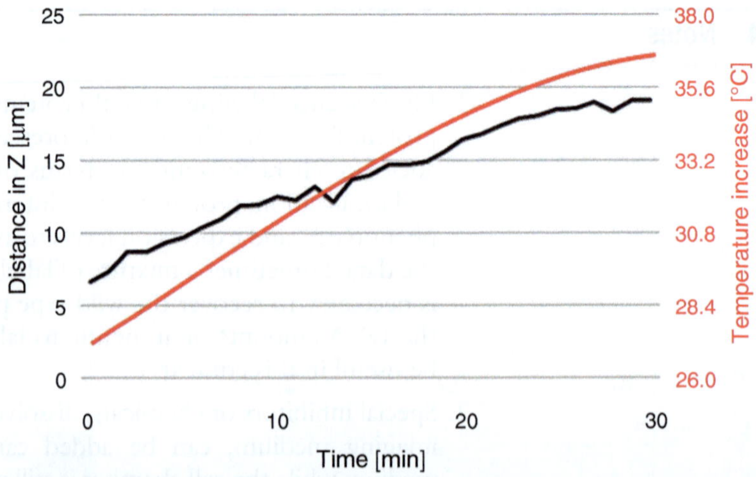

Fig. 2 Influence of increasing temperature on focus plane. During incubator heat-up, the *z*-value was adjusted every minute to be in the right focal plane

80 nm in diameter and 1000 nm in length, will be displayed in the fluorescent image as a structure of 300 nm × 1000 nm.

7. Infection of cells with a low MOI (lower than 1) is desirable to follow spread of infection from cell to cell. MOI values greater than 1 particularly in combination with plasmid-driven protein expression might lead to cell exhaustion and thus to artificial phenotypes.

8. It is a well-known challenge to acquire images for a certain time period in the correct focus plane. Physical or thermal effects like vibrations or fluctuations of temperature might negatively influence the image sharpness. As a solution, try to minimize the time in which the cell culture dish is not exposed to 37 °C. Also minimize the time in which the incubator chamber is open. The correct focus plane is dependent on temperature stability (Fig. 2). Sinusoidal variations of *z*-values over time suggest the need to improve the heat control in the incubation chamber. Recurring alterations of *z*-values in one direction suggest a problem with the *z*-drive in the microscope. A software-based autofocus which uses bright field images to detect an adequate focus plane is commonly available in acquisition software packages. Nevertheless a hardware-based auto focus might be a desirable advantage for acquiring with higher precision.

9. Preheating of immersion oil is recommended to minimize focus drift. Be aware that some immersion oils can cause damage to the plastic cell chamber bottom (Fig. 3). To avoid damage and contamination, use only approved oil, e.g., from Cargille, USA (type HF*, Cat. Nr: 16245).

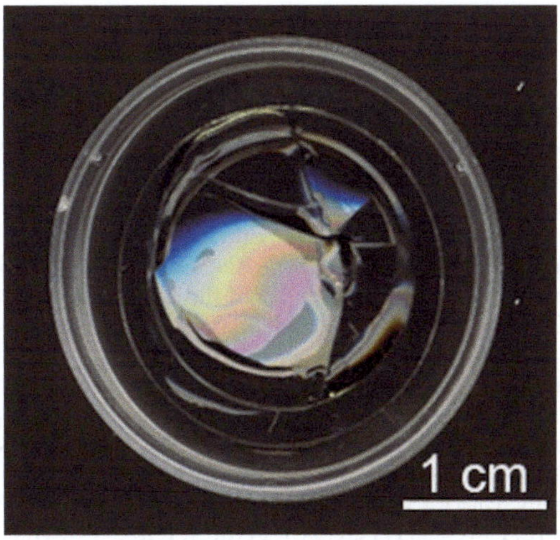

Fig. 3 Broken cell culture chamber. Leica immersion oil type F was used on the plastic bottom of a 35 mm μ-dish (Ibidi) for 68 h at 37 °C

10. For transfection use a FBS insensitive transfection reagent (e.g., Transit, Mirus). To reduce damage to the cells, the DNA amount should not exceed 1.5 μg per 3.1 cm^2 well.
11. Minimize exposure time to avoid unnecessary cell damage. Exposure times longer than 1 s and more than 50% light intensity cause notable irreversible bleaching of fluorophores over time. But also consider longer exposure times combined with lower light intensity is gentler for cells than short exposure time and brightest light intensity. Be aware that certain immersion oils, some cell chamber types, some cell culture medium, and almost all cell types can cause autofluorescence. These signals can easily be misinterpreted as a weak specific fluorescent signal. Investigators should test which combination of equipment, products, and cell types leads to the best signal-to-noise ratio.

References

1. Hoenen T, Groseth A, Callison J, Takada A, Feldmann H (2013) A novel Ebola virus expressing luciferase allows for rapid and quantitative testing of antivirals. Antivir Res 99(3):207–213. doi:10.1016/j.antiviral.2013.05.017
2. Kondratowicz AS, Maury WJ (2012) Ebolavirus: a brief review of novel therapeutic targets. Future Microbiol 7(1):1–4. doi:10.2217/fmb.11.110
3. Dolnik O, Stevermann L, Kolesnikova L, Becker S (2015) Marburg virus inclusions: a virus-induced microcompartment and interface to multivesicular bodies and the late endosomal compartment. Eur J Cell Biol 94(7–9):323–331. doi:10.1016/j.ejcb.2015.05.006
4. Schmidt KM, Schumann M, Olejnik J, Krahling V, Muhlberger E (2011) Recombinant Marburg virus expressing EGFP allows rapid screening of virus growth and real-time visualization of virus spread. J Infect Dis 204(Suppl 3):S861–S870. doi:10.1093/infdis/jir308
5. Parhamifar L, Wu L, Andersen H, Moghimi SM (2014) Live-cell fluorescent microscopy platforms for real-time monitoring of polyplex-cell interaction: basic guidelines. Methods 68(2):300–307. doi:10.1016/j.ymeth.2014.02.004

6. Ettinger A, Wittmann T (2014) Fluorescence live cell imaging. Methods Cell Biol 123:77–94. doi:10.1016/B978-0-12-420138-5.00005-7
7. Kuhn JH, Andersen KG, Bao Y, Bavari S, Becker S, Bennett RS, Bergman NH, Blinkova O, Bradfute S, Brister JR, Bukreyev A, Chandran K, Chepurnov AA, Davey RA, Dietzgen RG, Doggett NA, Dolnik O, Dye JM, Enterlein S, Fenimore PW, Formenty P, Freiberg AN, Garry RF, Garza NL, Gire SK, Gonzalez JP, Griffiths A, Happi CT, Hensley LE, Herbert AS, Hevey MC, Hoenen T, Honko AN, Ignatyev GM, Jahrling PB, Johnson JC, Johnson KM, Kindrachuk J, Klenk HD, Kobinger G, Kochel TJ, Lackemeyer MG, Lackner DF, Leroy EM, Lever MS, Muhlberger E, Netesov SV, Olinger GG, Omilabu SA, Palacios G, Panchal RG, Park DJ, Patterson JL, Paweska JT, Peters CJ, Pettitt J, Pitt L, Radoshitzky SR, Ryabchikova EI, Saphire EO, Sabeti PC, Sealfon R, Shestopalov AM, Smither SJ, Sullivan NJ, Swanepoel R, Takada A, Towner JS, van der Groen G, Volchkov VE, Volchkova VA, Wahl-Jensen V, Warren TK, Warfield KL, Weidmann M, Nichol ST (2014) Filovirus RefSeq entries: evaluation and selection of filovirus type variants, type sequences, and names. Virus 6(9):3663–3682. doi:10.3390/v6093663
8. McElroy AK, Spiropoulou CF (2014) Biomarkers for understanding Ebola virus disease. Biomark Med 8(9):1053–1056. doi:10.2217/bmm.14.75
9. Sanchez A, Geisbert T, Feldmann H (2007) Filoviridae – Marburg and Ebola viruses. In: Knipe D (ed) Fields virology, vol 1, 5th edn. Lippincott Williams and Wilkins, Philadelphia, PA, pp 1410–1448
10. Towner JS, Khristova ML, Sealy TK, Vincent MJ, Erickson BR, Bawiec DA, Hartman AL, Comer JA, Zaki SR, Stroher U, Gomes da Silva F, del Castillo F, Rollin PE, Ksiazek TG, Nichol ST (2006) Marburgvirus genomics and association with a large hemorrhagic fever outbreak in Angola. J Virol 80(13):6497–6516. doi:10.1128/JVI.00069-06
11. Krahling V, Dolnik O, Kolesnikova L, Schmidt-Chanasit J, Jordan I, Sandig V, Gunther S, Becker S (2010) Establishment of fruit bat cells (*Rousettus aegyptiacus*) as a model system for the investigation of filoviral infection. PLoS Negl Trop Dis 4(8):e802. doi:10.1371/journal.pntd.0000802
12. Enterlein S, Volchkov V, Weik M, Kolesnikova L, Volchkova V, Klenk HD, Muhlberger E (2006) Rescue of recombinant Marburg virus from cDNA is dependent on nucleocapsid protein VP30. J Virol 80(2):1038–1043. doi:10.1128/JVI.80.2.1038-1043.2006
13. Volchkov VE, Volchkova VA, Muhlberger E, Kolesnikova LV, Weik M, Dolnik O, Klenk HD (2001) Recovery of infectious Ebola virus from complementary DNA: RNA editing of the GP gene and viral cytotoxicity. Science 291(5510): 1965–1969. doi:10.1126/science.1057269
14. Kuhn JH, Bao Y, Bavari S, Becker S, Bradfute S, Brister JR, Bukreyev AA, Chandran K, Davey RA, Dolnik O, Dye JM, Enterlein S, Hensley LE, Honko AN, Jahrling PB, Johnson KM, Kobinger G, Leroy EM, Lever MS, Muhlberger E, Netesov SV, Olinger GG, Palacios G, Patterson JL, Paweska JT, Pitt L, Radoshitzky SR, Saphire EO, Smither SJ, Swanepoel R, Towner JS, van der Groen G, Volchkov VE, Wahl-Jensen V, Warren TK, Weidmann M, Nichol ST (2013) Virus nomenclature below the species level: a standardized nomenclature for natural variants of viruses assigned to the family Filoviridae. Arch Virol 158(1):301–311. doi:10.1007/s00705-012-1454-0
15. Schudt G, Kolesnikova L, Dolnik O, Sodeik B, Becker S (2013) Live-cell imaging of Marburg virus-infected cells uncovers actin-dependent transport of nucleocapsids over long distances. Proc Natl Acad Sci U S A 110(35):14402–14407. doi:10.1073/pnas.1307681110
16. Ebihara H, Theriault S, Neumann G, Alimonti JB, Geisbert JB, Hensley LE, Groseth A, Jones SM, Geisbert TW, Kawaoka Y, Feldmann H (2007) In vitro and in vivo characterization of recombinant Zaire ebolavirus expressing eGFP. J Infect Dis 196(Suppl 2):313–322
17. Hoenen T, Shabman RS, Groseth A, Herwig A, Weber M, Schudt G, Dolnik O, Basler CF, Becker S, Feldmann H (2012) Inclusion bodies are a site of ebolavirus replication. J Virol 86(21):11779–11788. doi:10.1128/JVI.01525-12
18. Dolnik O, Kolesnikova L, Welsch S, Strecker T, Schudt G, Becker S (2014) Interaction with Tsg101 is necessary for the efficient transport and release of nucleocapsids in marburg virus-infected cells. PLoS Pathog 10(10):e1004463. doi:10.1371/journal.ppat.1004463
19. Schudt G, Dolnik O, Kolesnikova L, Biedenkopf N, Herwig A, Becker S (2015) Transport of Ebolavirus nucleocapsids is dependent on actin polymerization: live-cell imaging analysis of Ebolavirus-infected cells. J Infect Dis 212(Suppl 2):S160–S166. doi:10.1093/infdis/jiv083

Chapter 16

Assessment of Inhibition of Ebola Virus Progeny Production by Antiviral Compounds

Darryl Falzarano

Abstract

Assessment of small molecule compounds against filoviruses, such as Ebola virus, has identified numerous compounds that appear to have antiviral activity and should presumably be further investigated in animal efficacy trials. However, despite the many compounds that are purported to have good antiviral activity in in vitro studies, there are few instances where any efficacy has been reported in nonhuman primate models. Many of the high-throughput screening assays use reporter systems that only recapitulate a portion of the virus life cycle, while other assays only assess antiviral activity at relatively early time points. Moreover, many assays do not assess virus progeny production. A more in-depth evaluation of small numbers of test compounds is useful to economize resources and to generate higher quality antiviral hits. Assessing virus progeny production as late as 5 days post-infection allows for the elimination of compounds that have initial antiviral effects that are not sustained or where the virus rapidly develops resistance. While this eliminates many potential lead compounds that may be worthy of further structure-activity relationship (SAR) development, it also quickly excludes compounds that in their current form are unlikely to be effective in animal models. In addition, the inclusion of multiple assays that assess both cell viability and cell cytotoxicity, via different mechanisms, provides a more thorough assessment to exclude compounds that are not direct-acting antivirals.

Key words Ebola virus, Antiviral, Cytotoxicity, Cell viability, Virus progeny, Assay, Drug screening

1 Introduction

High-throughput screening (HTS) of clinically approved compounds or chemical libraries [1–5] has identified potential lead compounds that may be useful antivirals against Ebola virus. While numerous compounds have provided protection in Ebola rodent models [3, 4, 6–10], only recently has a compound been identified that has sufficient antiviral activity to provide protection in the gold-standard macaque model [11]. While HTS is valuable for providing potential hits, they have not been particularly effective at predicting whether a compound will have in vivo activity. The discord between in vitro and in vivo studies can likely be attributed to a number of key characteristics: (1) the disease models use high

Thomas Hoenen and Allison Groseth (eds.), *Ebolaviruses: Methods and Protocols*, Methods in Molecular Biology, vol. 1628,
DOI 10.1007/978-1-4939-7116-9_16,

challenge doses (in terms of LD_{50}) and are essentially 100% lethal in a rapid time frame (5–10 days); (2) most antivirals assessed to date have relatively high effective concentration $(EC)_{50}$ values, and little to no information is provided on what the EC_{90}, EC_{99}, etc. values are; and (3) interpretation of in vitro testing results according to guidelines that in the past were useful for identifying viral inhibitors of chronic viral conditions (HIV, HBV, HCV) is not necessarily translatable to severe, acute conditions that require rapid intervention, such as Ebola virus infection. In addition, HTS uses either reporter-expressing viruses or various reporter assays that model parts of the virus life cycle to identify prospective antiviral compounds. Unfortunately, many of these assays do not assess the effect on virus progeny production and, in addition, only take into account early time points (i.e., 2 days or less) that do not account for the development of resistance. Ultimately, this means that many false positives must be further evaluated and it is important to exclude unpromising compounds as early as possible.

More in-depth in vitro analysis that assesses virus progeny production at multiple time points and includes sufficient assays to assess cytotoxic effects, which can also be aided by microscopic observation, can likely exclude many hits identified in current analyses. Unfortunately, these characteristics are not easily and economically amenable to HTS studies. The recent success in a nonhuman primate model of GS-5734 involved achieving concentrations of drug in vivo that exceed the EC_{90} [11]. Thus, it has been suggested that compounds that do not have an achievable EC_{90} likely need not be assessed in animal models.

Obtaining EC values calculated on progeny production instead of reduced reporter activity at relevant time points is likely a critical factor in assessing the antiviral activity of test compounds. Simultaneously, collecting data on viral loads in addition to virus titers can also be useful, but is not necessarily essential. Typically the earliest time point that is assayable yields the greatest number of hits, but for Ebola, past experience would indicate that compounds with small effects are insufficient in vivo. Thus, setting more stringent screen parameters may be useful for identifying hits that translate to animal models. Here we suggest using viral titers at day 5 post-infection as a better predictor of antiviral success.

The assay described in this protocol is not amenable to screening large numbers of compounds but instead provides more data points to make an assessment as to whether a lead compound or groups of compounds, potentially identified using HTS methods, should be further assessed. It provides both early and late time points and can assess the efficacy of test compounds over time compared to cytotoxicity over time. While using an Ebola virus with a reporter is suggested to make the readouts easier and quicker, it is not essential, and the data collected is not reporter dependent.

2 Materials

1. VeroE6 cells (ATCC# CRL-1586).
2. 96-well cell culture plates.
3. 24-well cell culture plates.
4. Plain DMEM: DMEM with no additives.
5. Complete DMEM: DMEM with 10% heat-inactivated FBS, 1% penicillin/streptomycin, 1% L-glutamine.
6. Maintenance DMEM: DMEM with 2% heat-inactivated FBS, 1% penicillin/streptomycin, 1% L-glutamine.
7. Test compounds.
8. Solvent to dissolve compounds (water, DMSO, ethanol, etc.).
9. Ebola virus. The use of a recombinant virus expressing enhanced green fluorescent protein (eGFP) or another reporter is not essential [12], but is convenient, if available.
10. CellTiter 96 AQueous One Solution Cell Proliferation Assay System (Promega).
11. LDH Cytotoxicity Assay Kit (Pierce).
12. CompuSyn software (www.combosyn.com).
13. Microplate reader capable of reading 490 and 680 nm (LDH assay) and 490 nm (CellTiter96 assay).
14. Fluorescent microscope capable of detecting green fluorescent protein (GFP) (optional).

3 Methods

3.1 General Methods

1. Approximately 24 h prior to initiating the assays, split the confluent VeroE6 cells 1:3 into 24-well plates (1 mL total volume/well) for the antiviral assay (*see* Subheading 3.2) and 1:3 into 96-well plates (100 μL total volume/well) in complete DMEM for the cell proliferation/cytotoxicity assays (see Subheadings 3.3 and 3.4). Incubate plates in a humidified atmosphere at 37 °C, 5% CO_2. This should result in near-confluent (>90%) cells the next day.
2. The following day, dissolve test compounds in the appropriate solvent (water, DMSO, ethanol, etc.). Selection of solvent must be determined based on the test compounds' known properties or empirically. Solvent concentration may affect cell viability and/or cause cytotoxicity.
3. Prepare a minimum of six dilutions of test compounds using 0.5 log dilutions if the concentration range to be assayed is unknown. Alternatively, if a concentration range is known,

two-fold dilutions over a minimum of five dilutions would be sufficient. All conditions should be performed in triplicate. Include solvent-only (no test compound) and mock (no solvent or test compound) controls. Dilutions of test compound should be performed in the solvent used in order to maintain the same concentration of solvent in all dilutions. Diluted test compounds are then added to maintenance DMEM.

3.2 Antiviral Assay

1. Ensure cells are >90% confluent by phase-contrast microscopy.
2. In biosafety level 4 (BSL4), thaw Ebola virus or recombinant Ebola virus eGFP stocks and dilute in plain DMEM to achieve a concentration of 100 focus-forming units (FFU)/250 μL.
3. Remove medium from the wells of the 24-well plates and add 250 μL of the virus inoculum prepared in the previous step to each well. Incubate plates in a humidified atmosphere at 37 °C, 5% CO_2 for 1 h.
4. During incubation add diluted test compounds to maintenance DMEM (approximately 4 mL of medium per test compound dilution is required to perform all three assays in triplicate).
5. Following the incubation remove the virus inoculum from the wells and discard. Add 500 μL of maintenance medium containing test compounds. Incubate plates in a humidified atmosphere at 37 °C, 5% CO_2 for 24 h.
6. Collect supernatant into cryotubes (add 140 μL of supernatant to a separate tube containing AVL if performing viral load assay (*see* Subheading 3.6)). Store collected samples at −80 °C until ready to process.
7. Replace with 500 μL/well fresh maintenance medium containing test compounds.
8. Repeat **steps 6** and **7** on days 3 and 5 post-infection. Depending on the stability of the test compounds, a stock solution may be maintained as appropriate, but dilutions should be made fresh on days 1, 3, and 5 post-infection.
9. Cells can be viewed by fluorescence microscopy on days 3 and 5 post-infection to determine if eGFP expression (as a readout for the amount of virus) is reduced or absent at given concentrations of test compound compared to mock-treated cells (*see* **Note 1**).
10. Following collection of supernatant on day 5 post-infection, discard plates as per institutional protocols.

3.3 Cell Viability Using CellTiter 96 AQueous One Solution Cell Proliferation Assay

1. Split VeroE6 cells 1:3 into 96-well plates (100 μL/well). Incubate cells overnight (humidified 37 °C, 5% CO_2). Set up one set of plates for each time point when samples are collected in the antiviral assay (e.g., day 1, 3, 5) (*see* **Notes 2** and **3**).

2. Remove and discard supernatant. Using the same concentrations as in the antiviral assay, add 100 μL of test compounds dissolved in solvent to wells in triplicate (include solvent-only and mock controls).
3. Incubate one set of plates for each time period (e.g., 1, 3, and 5 days) analyzed in the antiviral assay at 37 °C, 5% CO_2. Remove supernatant and replace with fresh maintenance medium containing test compounds as in the antiviral assay on days 1 and 3 post-infection (with the exception of the plate being assayed on that day).
4. Thaw CellTiter 96 AQueous One Solution reagent (Promega) at room temperature.
5. Add 20 μL of solution to each well of the 96-well plate containing the samples in 100 μL of culture medium. Return the plate to the incubator (37 °C, 5% CO_2) for 1 h.
6. Read the plate in a microplate reader at A_{490}. If necessary, 25 μL of 10% SDS can be added to the well to stop the reaction, and the plate can be read later.
7. Establish a standard curve. Using the same starting concentration of cells that will be used for the antiviral assay, make 1:2 dilutions of cells using a minimum of five dilutions and seed these in triplicate into 96-well plates. Incubate the cells for 6–24 h. Add CellTiter 96 Aqueous One Solution as described above. Plates should be read after 1 h. If absorbance is plotted on the *Y*-axis versus the number of cells per well, the curve should be linear. The standard curve generated is used to quantify cell viability at all time points.

3.4 Cell Cytotoxicity Using the Pierce LDH Cytotoxicity Assay

1. Split VeroE6 cells 1:3 into 96-well plates (100 μL/well). Incubate cells overnight (humidified 37 °C, 5% CO_2). Set up one set of plates for each time point when samples are collected in the antiviral assay (e.g., day 1, 3, 5) (*see* **Notes 2** and **3**).
2. Remove supernatant and discard. Using the same concentrations as in the antiviral assay, add 100 μL of test compounds dissolved in solvent to wells in triplicate. Include the following control wells in triplicate: solvent only, mock, spontaneous LDH activity (add 10 μL of sterile ultrapure water), and maximum LDH activity.
3. Incubate one set of plates for the same time periods (e.g., 1, 3, and 5 days) as the antiviral assay at 37 °C, 5% CO_2. To the maximum LDH activity control wells, add 10 μL of 10× lysis buffer and mix. Incubate the plate at 37 °C, 5% CO_2 for 45 min. Remove supernatant and replace with fresh maintenance medium containing test compounds as in the antiviral assay on days 1 and 3 post-infection (with the exception of the plate being assayed on that day).

4. Transfer 50 μL of sample to be assayed to a 96-well flat bottom plate in triplicate. Add 50 μL of reaction mixture to each sample well and mix using a multichannel pipette.
5. Incubate the plate at room temperature for 30 min protected from light.
6. Add 50 μL of stop solution to each sample well and mix by gentle tapping.
7. Measure the A_{490} and A_{680}. To determine the LDH activity, subtract the A_{680} from the A_{490}.
8. To calculate % cytotoxicity, subtract the LDH activity of the spontaneous LDH release control from the compound treated sample LDH activity and divide by the total LDH activity:

$$\%\,cytotoxicity = \frac{compound\ treated\ LDH\ activity - spontaneous\ LDH\ activity}{maximum\ LDH\ activity - spontaneous\ LDH\ activity} \times 100.$$

3.5 Titration of Samples (See Note 4)

1. Split VeroE6 cells 1:3 into 96-well plates (100 μL/well). Incubate cells overnight (humidified 37 °C, 5% CO_2).
2. In BSL4, thaw samples from the antiviral assay and prepare serial tenfold dilutions from neat to 10^{-6} in plain DMEM in a 96-well round bottom plate.
3. Remove medium from cells in 96-well plates and discard.
4. Add 50 μL of diluted samples to each well. Return the plate to incubator and incubate for 1 h.
5. Remove virus inoculum and replace with fresh maintenance DMEM.
6. On day 1 post-infection, check cells microscopically to ensure there is no contamination.
7. On days 5–7 post-infection, view cells by fluorescence microscopy to determine the dilution(s) where eGFP is no longer detectable. Calculate the $TCID_{50}$ based on the calculations of Reed and Muench [13] for each replicate of test compounds.

3.6 Viral Load Determination (See Note 4)

1. Thaw collected supernatants and add them to AVL (Qiagen), with further inactivation according to institutional protocols and then RNA extraction.
2. The extracted RNA can be subsequently quantified by real-time qRT-PCR. Previously described [14] primers and probe targeting Ebola virus NP are used with the Rotor-Gene Probe RT-PCR Kit (Qiagen). A tenfold dilution series of viral RNA based on $TCID_{50}$ equivalents can be used as a standard.

3.7 Calculation of Effective/Cytotoxic Dose

1. Determine the effective concentration that reduces virus replication by 50% and 90% (EC_{50}, EC_{90}) using ComboSyn (www.combosyn.com) [15]. ComboSyn is a free program that uses mass action to calculate EC*x* values for both single and combination compound treatments (*see* **Note 5**).
2. Determine the cytotoxic concentration that reduces cell viability by 50% or increases cell cytotoxicity by 50% using ComboSyn as described above.

4 Notes

1. A dose-dependent reduction in eGFP expression, which can be observed microscopically with a microscope capable of detecting GFP, should be noted on the antiviral plate if a compound has antiviral activity against Ebola virus. Caution should be used in the interpretation of this data, as test compounds that inhibit either eGFP expression or fluorescence, as well as compounds that are cytotoxic, could also have the same effect. This is in part why subsequent viral loads or titers are used as the principle readout.
2. Multiple assays that assess cell viability and/or cytotoxicity are not necessarily required. This should be assessed on a compound-by-compound (or class of compounds) basis. Correlation with microscopic findings also plays a key role assessing toxicity of compounds. In our experience we have found compounds that lead to little to no change in Nicotinamide adenine phosphate (NADPH/NADH) based assays while causing detachment of up to 50% of cells. Conversely, we have also observed 50% decrease in activity in NADPH/NADH based assays in cells that appear microscopically intact, can be subcultured, and can be maintained in a tenfold increase in the test compound.
3. Some compounds do not induce noticeable effects on cell viability or cytotoxicity at early (1-day post-infection) time points [16]; however, typically by 3 days post-infection if compounds are cytotoxic, this is noted. Thus, late time points can be highly valuable.
4. It is not necessarily essential to determine both viral load and viral titer from samples collected from the antiviral assay. This method describes both. The samples collected on day 1 post-infection should be analyzed for viral load only as readily quantifiable levels of infectious virus should not have been produced at this time point.
5. Alternatively GraphPad Prism can be used to determine EC_x values using the inhibitor function in the Analysis Toolbox.

References

1. McMullan LK, Flint M, Dyall J, Albarino C, Olinger GG, Foster S, Sethna P, Hensley LE, Nichol ST, Lanier ER, Spiropoulou CF (2016) The lipid moiety of brincidofovir is required for in vitro antiviral activity against Ebola virus. Antivir Res 125:71–78. doi:10.1016/j.antiviral.2015.10.010
2. Long J, Wright E, Molesti E, Temperton N, Barclay W (2015) Antiviral therapies against Ebola and other emerging viral diseases using existing medicines that block virus entry. F1000 Res 4:30. doi:10.12688/f1000research.6085.2
3. Huggins J, Zhang ZX, Bray M (1999) Antiviral drug therapy of filovirus infections: S-adenosylhomocysteine hydrolase inhibitors inhibit Ebola virus in vitro and in a lethal mouse model. J Infect Dis 179(Suppl 1):S240–S247. doi:10.1086/514316
4. Aman MJ, Kinch MS, Warfield K, Warren T, Yunus A, Enterlein S, Stavale E, Wang P, Chang S, Tang Q, Porter K, Goldblatt M, Bavari S (2009) Development of a broad-spectrum antiviral with activity against Ebola virus. Antivir Res 83(3):245–251. doi:10.1016/j.antiviral.2009.06.001
5. Johansen LM, DeWald LE, Shoemaker CJ, Hoffstrom BG, Lear-Rooney CM, Stossel A, Nelson E, Delos SE, Simmons JA, Grenier JM, Pierce LT, Pajouhesh H, Lehar J, Hensley LE, Glass PJ, White JM, Olinger GG (2015) A screen of approved drugs and molecular probes identifies therapeutics with anti-Ebola virus activity Sci Transl Med 7 (290):290ra289. doi:10.1126/scitranslmed.aaa5597
6. Warren TK, Wells J, Panchal RG, Stuthman KS, Garza NL, Van Tongeren SA, Dong L, Retterer CJ, Eaton BP, Pegoraro G, Honnold S, Bantia S, Kotian P, Chen X, Taubenheim BR, Welch LS, Minning DM, Babu YS, Sheridan WP, Bavari S (2014) Protection against filovirus diseases by a novel broad-spectrum nucleoside analogue BCX4430. Nature 508(7496):402–405. doi:10.1038/nature13027
7. Warren TK, Warfield KL, Wells J, Enterlein S, Smith M, Ruthel G, Yunus AS, Kinch MS, Goldblatt M, Aman MJ, Bavari S (2010) Antiviral activity of a small-molecule inhibitor of filovirus infection. Antimicrob Agents Chemother 54(5):2152–2159. doi:10.1128/AAC.01315-09
8. Panchal RG, Reid SP, Tran JP, Bergeron AA, Wells J, Kota KP, Aman J, Bavari S (2012) Identification of an antioxidant small-molecule with broad-spectrum antiviral activity. Antivir Res 93(1):23–29. doi:10.1016/j.antiviral.2011.10.011
9. Bray M, Driscoll J, Huggins JW (2000) Treatment of lethal Ebola virus infection in mice with a single dose of an S-adenosyl-L-homocysteine hydrolase inhibitor. Antivir Res 45(2):135–147
10. Garrison AR, Giomarelli BG, Lear-Rooney CM, Saucedo CJ, Yellayi S, Krumpe LR, Rose M, Paragas J, Bray M, Olinger GG Jr, McMahon JB, Huggins J, O'Keefe BR (2014) The cyanobacterial lectin scytovirin displays potent in vitro and in vivo activity against Zaire Ebola virus. Antivir Res 112:1–7. doi:10.1016/j.antiviral.2014.09.012
11. Warren TK, Jordan R, Lo MK, Ray AS, Mackman RL, Soloveva V, Siegel D, Perron M, Bannister R, Hui HC, Larson N, Strickley R, Wells J, Stuthman KS, Van Tongeren SA, Garza NL, Donnelly G, Shurtleff AC, Retterer CJ, Gharaibeh D, Zamani R, Kenny T, Eaton BP, Grimes E, Welch LS, Gomba L, Wilhelmsen CL, Nichols DK, Nuss JE, Nagle ER, Kugelman JR, Palacios G, Doerffler E, Neville S, Carra E, Clarke MO, Zhang L, Lew W, Ross B, Wang Q, Chun K, Wolfe L, Babusis D, Park Y, Stray KM, Trancheva I, Feng JY, Barauskas O, Xu Y, Wong P, Braun MR, Flint M, McMullan LK, Chen SS, Fearns R, Swaminathan S, Mayers DL, Spiropoulou CF, Lee WA, Nichol ST, Cihlar T, Bavari S (2016) Therapeutic efficacy of the small molecule GS-5734 against Ebola virus in rhesus monkeys. Nature 531(7594):381–385. doi:10.1038/nature17180
12. Ebihara H, Theriault S, Neumann G, Alimonti JB, Geisbert JB, Hensley LE, Groseth A, Jones SM, Geisbert TW, Kawaoka Y, Feldmann H (2007) In vitro and in vivo characterization of recombinant Ebola viruses expressing enhanced green fluorescent protein. J Infect Dis 196(Suppl 2):S313–S322. doi:10.1086/520590
13. Reed LJ, Muench H (1938) A simple method of estimating fifty percent endpoints. Am J Hygiene 27(3):493–497
14. Marzi A, Ebihara H, Callison J, Groseth A, Williams KJ, Geisbert TW, Feldmann H (2011) Vesicular stomatitis virus-based Ebola vaccines with improved cross-protective efficacy. J Infect Dis 204(Suppl 3):S1066–S1074. doi:10.1093/infdis/jir348
15. Chou TC, N. M (2005) CompuSyn for drug combinations and for general dose-effect analysis. ComboSyn Inc., Paramus, NJ
16. Falzarano D, Safronetz D, Prescott J, Marzi A, Feldmann F, Feldmann H (2015) Lack of protection against ebola virus from chloroquine in mice and hamsters. Emerg Infect Dis 21(6):1065–1067. doi:10.3201/eid2106.150176

Chapter 17

Analysis of the Cellular Stress Response During Ebola Virus Infection by Immunofluorescence

Emily V. Nelson and Kristina M. Schmidt

Abstract

In this chapter, the use of immunofluorescence analysis as a tool to examine stress granule (SG) formation in Ebola virus (EBOV)-infected cells is described. The following protocol focuses on the process of inducing and analyzing the cellular stress response, including treatment of cells with inducers and inhibitors of the SG formation, and also describes EBOV infection, DNA transfection, and the usage of different cell lines.

Key words *Zaire ebolavirus*, Ebola virus, Filoviruses, Stress granules, Cellular stress response, Immunofluorescence, Microscopy, Antibodies, Sodium arsenite

1 Introduction

Exogenous environmental stress such as heat, nutrient deprivation, ultraviolet radiation, and viral infection can trigger the cellular stress response, which acts as a form of mRNA triage to prioritize the translation of those mRNAs that are essential for cell survival. This response culminates in stress granule (SG) formation, which occurs concomitantly with a global decrease in cellular protein translation [1–4]. This widespread translational repression is unsustainable for viral replication, which requires the host translational machinery to synthesize viral proteins. Therefore, many viruses have evolved strategies to antagonize or avoid the stress response, which is increasingly becoming appreciated as an antiviral response. Few viruses have been shown to coexist with fully formed SGs over the entire course of infection, and many studies indicate that viruses employ a range of different mechanisms to prevent or exploit SGs and their components (reviewed in [5–7]). The increasing number of studies demonstrating an interaction between SGs and viruses highlights the significance of this response during viral infection and emphasizes the complex and unique interactions between them. For Ebola virus (EBOV), it has been shown that SG

Thomas Hoenen and Allison Groseth (eds.), *Ebolaviruses: Methods and Protocols*, Methods in Molecular Biology, vol. 1628, DOI 10.1007/978-1-4939-7116-9_17, © Springer Science+Business Media LLC 2017

formation is not induced during infection [8] and that the virus employs several mechanisms to interfere with the cellular antiviral responses in different ways (reviewed in [9]). Here, we focus on a method for examining SG formation in EBOV-infected cells or cells expressing the viral protein (VP) 35, which has been shown to disrupt SG formation [8].

The induction of SGs has been best characterized via the activation of four cytoplasmic kinases that sense various forms of stress and initiate the signaling cascade that eventually leads to a state of translational arrest (Fig. 1). These kinases include protein kinase R (PKR), heme-regulated inhibitor kinase (HRI), PKR-like endoplasmic reticulum

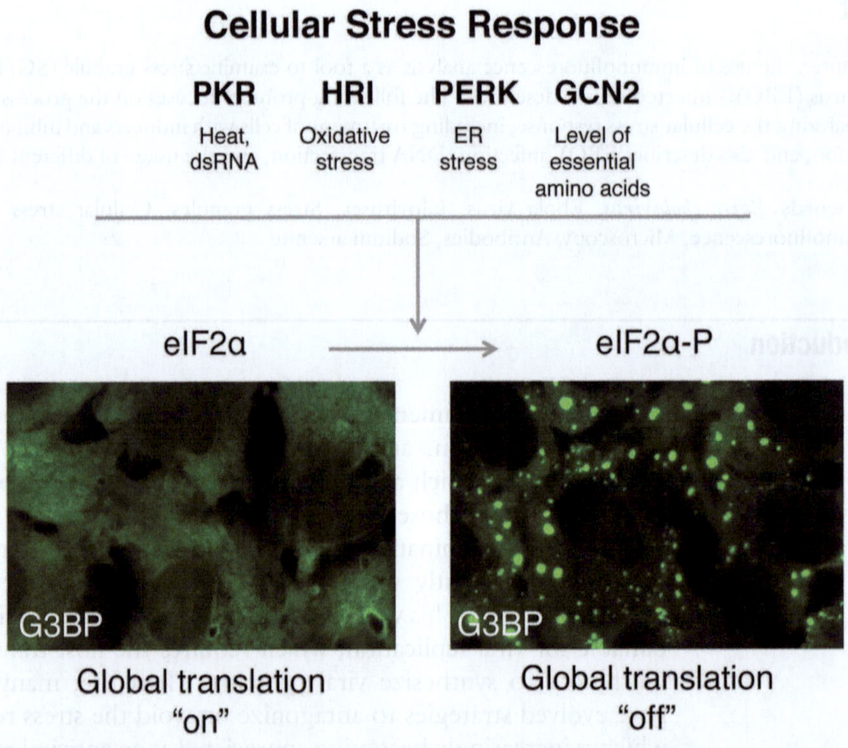

Fig. 1 Overview of the induction of SG formation. Environmental stress such as heat, virus infection (which can lead to dsRNA production), oxidative stress, ER stress, and amino acid deprivation are sensed by at least four different cellular kinases, which include protein kinase R (PKR), heme-regulated inhibitor kinase (HRI), PKR-like endoplasmic reticulum kinase (PERK), and general control nonderepressible 2 (GCN2). Upon activation, these kinases phosphorylate the alpha (α) subunit of eukaryotic translation initiation factor 2 (eIF2), thereby preventing the assembly of translation pre-initiation complexes, causing global translational arrest. These stalled translation initiation complexes accumulate on mRNAs and are visible as stress granules (SGs). A phospho-eIF2α (eIF2α-P)-independent stimulus leading to the formation of SGs can also occur via changes in the expression level of specific translation initiation factors and canonical SG components, e.g., eIF4A, eIF4G, and G3BP. Depending on the stressor, SGs can consist of different components. However, the components found consistently across all SG types include the small (40S) ribosomal subunit, a number of early transcription initiation factors, mRNAs, and a subset of RNA binding proteins

kinase (PERK), and general control nonderepressible 2 (GCN2), which sense dsRNA, oxidative stress, ER stress, and amino acid deprivation, respectively [2, 4]. The activation of these kinases leads to the phosphorylation of their downstream target eIF2α, which can be detected by Western blot analysis. However, this is not a direct measure of SG formation. In order to fully determine whether SGs are induced, immunofluorescence analysis or microscopy techniques are recommended. Because of the complex nature of SGs and their continuous recycling of different cellular components, the examination of these transient structures is technically challenging. Furthermore, the isolation or purification of intact SG complexes has thus far been unsuccessful. Therefore, the most useful and widely accepted way to observe and study SGs and their dynamics is through microscopy and immunofluorescence-based techniques [10–13].

SGs are non-membranous granules that form exclusively in the cytoplasm and are typically 1–2 μm in diameter [14]. The protein composition of SGs can vary depending on the type of stress present, but many SG proteins, such as Ras GTPase-activating protein-binding protein 1 (G3BP), T-cell-restricted intracellular antigen 1 (TIA-1), and eukaryotic translation initiation factor 4G (eIF4G), are found across all SG types and are, therefore, considered canonical SG components or SG marker proteins [3, 12]. In addition, SGs also contain mRNAs and translation factors, as well as scaffold proteins, RNA helicases, RNA-binding proteins involved in mRNA stability, and components of other signaling pathways [15]. The localization of the various SG marker proteins can be observed using fluorescently labeled protein-specific antibodies or fluorescently tagged proteins that are either introduced exogenously (via transfection, transduction, etc.) or stably expressed in modified immortalized cell lines.

Here, we outline the method of immunofluorescence analysis to examine the dynamics of SG formation during EBOV infection. We also expand the same technique to examine cells that have been transfected with VP35. This protocol can be adapted to address a number of questions involving the interaction between EBOV and SG proteins, or other cellular proteins of interest. Furthermore, through the use of SG-inducing chemicals or drugs (here we use the drug sodium arsenite (Ars); others are outlined in Table 1 and reviewed in [13]), it is possible to address whether EBOV interferes with or disrupts these pathways and how it accomplishes this. To further characterize and differentiate canonical SGs from other cytoplasmic RNA granules, cycloheximide (CHX), an inhibitor of protein synthesis, can also be used. CHX blocks polysome disassembly and leads to the dissolution of SGs but not other cytoplasmic granules [16, 17]. EBOV has a wide cellular tropism, which is advantageous for this assay because it allows for the use of cell lines that are optimal for immunofluorescence and cellular imaging. This protocol details the detection of SG marker protein

Table 1
Commonly used treatments/drugs to examine SGs

Treatment	Pathway	Effects and tested doses
Sodium arsenite (Ars)	eIF2α dependent via HRI activation	SGs induced at concentrations of 0.2–2 mM Ars for 30 min (*see* **Note 10**)
Heat shock	eIF2α dependent via PKR activation	SGs induced at 44 °C for 1 h
Thapsigargin	eIF2α dependent via PERK activation	SGs induced (not tested)
Nutrient starvation	eIF2α dependent via GCN2 activation	SGs induced (not tested)
Carbonyl cyanide-*p*-trifluorometh oxyphenylhydrazone (FCCP)	eIF2α independent, inhibits eIF4A helicase activity	SGs induced (not tested)
Pateamine A (Dr. Jerry Pelletier, McGill University [20])	eIF2α independent, inhibits eIF4A helicase activity	SGs induced (not tested)
Hippuristanol (Dr. Jerry Pelletier, McGill University [21])	eIF2α independent, inhibits eIF4A helicase activity	SGs induced at concentrations of 1–1.5 μM for 30 min
Cycloheximide (CHX)	Polysomes stabilized and assembly blocked	SGs dissolved at concentrations of 100 μg/mL for 30 min
Emetine	Polysomes stabilized and assembly blocked	SGs dissolved (not tested)

localization in EBOV-infected human osteosarcoma cells (U2OS), or DNA-transfected human hepatocellular carcinoma cells (Huh7), which are optimal for microscopy-based techniques due to their flat cell morphology as well as large cytoplasm and overall size.

2 Materials

2.1 Plating Cells for Transfection or Infection (See Note 1)

1. Human hepatocellular carcinoma cell line (Huh7; *see* **Note 2**).
2. Human osteosarcoma epithelial cell line (U2OS, ATCC HTB-96).
3. DMEM 10%: Dulbecco's Modified Eagle Medium (DMEM) supplemented with 10% fetal bovine serum (FBS), penicillin (50 units/mL), streptomycin (50 mg/mL), and L-glutamine (200 mM).
4. Phosphate buffered saline (PBS), Mg^{2+}-, and Ca^{2+}-free (e.g., from Lonza).
5. 0.025% Trypsin-EDTA.
6. Sterile glass coverslips or 8-well chamber slides.
7. 6-well cell culture plate.

2.2 Transfection and Infection of Cells

2.2.1 Transfection of Cells

1. pCAGGS-VP35-HA plasmid, with the HA tag located at the N-terminus of the protein [8].
2. Nuclease-free water.
3. DMEM 10%.
4. Lipofectamine LTX with PLUS Reagent (*see* **Note 3**).
5. Opti-MEM I Reduced Serum Medium.
6. Sterile 1.5 mL microcentrifuge tubes.

2.2.2 Infection of Cells (*See* **Note 4**)

1. Ebola virus (EBOV), Kikwit 1995 isolate (Genbank accession: KR867676.1) grown in Vero E6 cells with virus titers determined by plaque assay.
2. DMEM 0%: DMEM supplemented with penicillin (50 units/mL), streptomycin (50 mg/mL), and L-glutamine (200 mM).
3. DMEM 3%: DMEM supplemented with 3% FBS, penicillin (50 units/mL), streptomycin (50 mg/mL), and L-glutamine (200 mM).

2.3 Sodium Arsenite Treatment for Stress Granule Induction

1. 100 mM Ars stock solution: 0.129 g $NaAsO_2$, 10 mL H_2O, or PBS (for other stress inducers, *see* Table 1); sodium arsenite is toxic and must be handled and disposed of accordingly (*see* **Notes 5** and **6**).
2. Thermal packs, pre-warmed to 37 °C.

2.4 Fixation of Transfected/Infected Cells

1. PBS.
2. 4% paraformaldehyde (PFA) in DMEM 0% (paraformaldehyde is toxic and must be handled and disposed of accordingly).

2.5 Immunofluore scence Analysis

1. 12-well cell culture plates.
2. PBS.
3. 0.1% Triton X-100 in PBS.
4. 0.1 M glycine in PBS.
5. Blocking reagent: 20 g bovine serum albumin (BSA), 2 mL Tween-20, 30 mL glycerin, 10 mL sodium azide (NaN_3, 5% solution), and 1 L sterile PBS.
6. Parafilm.
7. Cell culture plate lid.
8. Plastic nontransparent/lightproof box for antibody incubation.
9. Rabbit anti-eIF4G primary antibody (Santa Cruz Biotechnology, Inc. sc-11373, used at 1:100).
10. Mouse anti-HA tag primary antibody (Covance MMS-101P, used at 1:50).

11. Mouse anti-VP35 primary antibody (used at 1:1000, *see* **Note 2**).
12. Goat anti-rabbit Alexa Fluor488 secondary antibody (e.g., Invitrogen).
13. Donkey anti-mouse Alexa Fluor594 secondary antibody (e.g., Invitrogen).
14. 4′,6-diamidino-2-phenylindole (DAPI) nuclear stain.
15. Mounting reagent (Calbiochem) (optional nail polish to seal).
16. Ultrapure H_2O.
17. Fluorescence microscope.

3 Methods

3.1 Plating of Cells onto Coverslips or in Chamber Slides

1. For transfection experiments, seed 3×10^5 cells per well of a 6-well plate in a volume of 2 mL DMEM 10% per well onto three glass coverslips per well (*see* **Note** 7).
2. For infection experiments, seed 1.25×10^4 U2OS cells per well of an 8-well chamber slide in a volume of 400 μL DMEM 10%.
3. Incubate cells at 37 °C and 5% CO_2 in an incubator overnight.

3.2 Transfection of Cells (See Note 8)

1. Prepare aliquots of the desired concentrations of plasmid(s) of interest in 1.5 mL microcentrifuge tubes using nuclease-free water to dilute DNA. Here we used 500 ng of pCAGGS-VP35-HA per well of a 6-well plate.
2. Dilute the DNA for each transfection in 500 μL OptiMEM.
3. Add 2.5 μL PLUS reagent to each DNA sample.
4. Mix by vortexing each tube briefly three times.
5. Incubate the diluted DNA/PLUS reagent mixture at room temperature for 5 min.
6. Add 6.25 μL Lipofectamine LTX to each sample.
7. Mix by vortexing each tube briefly three times.
8. Incubate the DNA/PLUS reagent/Lipofectamine LTX mixture at room temperature for 30 min.
9. After 25 min, remove the cells from incubator (they should be 70–80% confluent) and wash once gently with 1 mL Opti-MEM.
10. Add 1 mL fresh Opti-MEM per well.
11. Once the DNA/PLUS reagent/LTX mix has incubated for 30 min, add the mixture dropwise to cells and rock the plate gently to ensure proper mixing into the medium.
12. Return cells to incubator.

13. Add 1 mL DMEM 10% to each well between 16 and 24 h post transfection.
14. Continue to incubate cells until the desired level of protein expression is achieved (we generally incubate for a total of 48 h post transfection for EBOV VP35).

3.3 Infection of Cells

1. Prepare virus dilution in DMEM 0% for infection at a multiplicity of infection (MOI) of 1 focus-forming unit (ffu) per cell.
2. Add 100 μL of the virus dilution per well of the 8-well chamber slide.
3. Incubate for 1 h at 37 °C.
4. Replace the virus dilution with 300 μL of DMEM 3% and incubate the cells for desired length of time for analysis (we generally incubate the cells for 24 h).

3.4 Induction of SG Formation

1. Heat thermal packs to 37 °C in an incubator or water bath (*see* **Note 9**).
2. At the desired time points postinfection (24 h) or post transfection (48 h), carefully remove cell culture plates from the incubator and place them on top of the warmed thermal packs.
3. Add Ars (*see* **Note 10**) dropwise to the cells to achieve a final 0.5 mM concentration. Mock-treated cells should receive the same volume of H_2O or PBS, added dropwise to cells and gently swirl the medium (*see* **Note 11**).
4. Return the cells to the incubator and incubate at 37 °C for 30 min (*see* **Note 12**).

3.5 Fixation of Transfected Cells

1. At the appropriate times post transfection (*see* **Note 13**), and after Ars treatment (as described in Subheading 3.4), remove medium from cells (*see* **Note 14**).
2. Carefully wash cells with enough PBS to cover the cells (about 1 mL).
3. With all PBS removed, add 1 mL 4% PFA in DMEM 0%, or enough to fully submerge cells.
4. Incubate cells at room temperature for at least 20 min, or at 4 °C overnight.

3.6 Inactivation and Fixation of EBOV-Infected Cells

1. After an appropriate infection period and incubation time with Ars (as described in Subheading 3.4), remove medium from cells (*see* **Note 14**).
2. Wash cells once with enough PBS to cover the cells (about 1 mL).
3. Add freshly prepared 4% PFA in DMEM 0% for inactivation of EBOV and fixation of the cells, with incubation times and fixative changes performed according to the institutional BSL4 inactivation protocols (*see* **Note 15**).

3.7 Immunofluorescence Assay

1. Remove 4% PFA from cells (PFA is toxic and must be disposed of according to institutional regulations).
2. Wash coverslips or chamber slides 3× with PBS (*see* **Note 16**).
3. To permeablize the cells, add 0.1% Triton X-100 in PBS, adding enough to fully submerge and cover the cells.
4. Incubate for 5–10 min at room temperature.
5. Remove 0.1% Triton X-100 and wash 3× with PBS.
6. Add enough 0.1 M glycine to cover the cells, and incubate for 5–10 min at room temperature.
7. Remove 0.1 M glycine and wash 3× with PBS.
8. After washing, add enough blocking reagent to cover the cells, and incubate at room temperature for a minimum of 10 min.
9. Prepare the antibody incubation chamber by placing a wet paper towel inside a box or container (this should be opaque to protect the contents from light) large enough to accommodate the lid of a cell culture plate. Create a clean and smooth surface by stretching a piece of parafilm around the lid of a cell culture plate, making sure that there are no ridges, creases, or large pockets of air (*see* **Note 17**). Place this lid on top of the moist paper towel, and ensure that the container is still able to fully close.
10. If using chamber slides, remove the chamber from the glass coverslip as recommended by the manufacturer (*see* **Note 18**).
11. Incubate the coverslips or slides with the desired primary antibodies diluted in blocking solution (25 μL total volume per coverslip and 300 μL total volume per chamber slide). To do so, pipette the appropriate volume of primary antibody dilution onto the parafilm and place the coverslips or chamber slides with the cell side facing down, in direct contact with the antibody dilution. Ensure air bubbles do not become trapped while inverting the coverslips/chamber slides onto the antibody droplet.
12. Close the incubation chamber.
13. Incubate the primary antibody for at least 1 h at room temperature or at 4 °C overnight (*see* **Note 19**).
14. After incubation with the primary antibodies, wash coverslips or chamber slides 3× with PBS.
15. Prepare secondary antibody dilutions (*see* **Note 20**).
16. Replace the parafilm on top of the cell culture plate lid with a new piece.
17. Incubate the cells with the secondary antibody as described above in steps 11–13, making sure that the coverslips or slides are well protected from light at this and subsequent steps to avoid bleaching of the secondary antibody fluorophore.

18. Incubate for 1 h at room temperature or at 4 °C overnight (*see* **Note 21**).
19. Wash coverslips or chamber slides 3× in PBS.
20. Dip coverslips or chamber slides gently in ultrapure water to rinse (*see* **Note 22**).
21. Mount coverslips cell side down onto microscope slides using about 10 μL of mounting reagent, making sure to avoid introducing air bubbles.
22. For chamber slides, mount a rectangular microscope slide coverslip on top of the chamber slide using a few drops of mountain reagent, again making sure to avoid introducing air bubbles.
23. Let mounted coverslips and slides sit protected from light for at least 1 h to dry before examining them (*see* **Note 23**).
24. Analyze the mounted coverslips or chamber slides using a fluorescent microscope and appropriate imaging software (Figs. 2 and 3).

4 Notes

1. The cell line used can vary depending on their intended use. Huh7 cells were used in this protocol for transfection experiments because they can be efficiently transfected and are still useful for immunofluorescence analysis. U2OS cells were used in this protocol for infection experiments because they can be easily infected with EBOV and are beneficial for general use in immunofluorescence analysis because they are typically flat cells with a large cytoplasm. Importantly, both Huh7 and U2OS cells are sensitive to Ars treatment. There are a number of U2OS-based cell line alternatives that overexpress different SG proteins that are fluorescently tagged (typically with GFP). These cell lines are useful as well because they allow for the visualization of SG formation without antibody staining. They also can be useful when an antibody is unreliable or uncommon.
2. The Huh7 cell line used in this protocol was kindly provided by J. Alonso, Texas Biomedical Research Institute, San Antonio, TX. The VP35 antibody was a kind gift from C. F. Basler, Georgia State University, Institute for Biomedical Sciences, Atlanta GA. Other EBOV-specific antibodies are commercially available and are appropriate for use in immunofluorescence analysis.
3. Generally, transfections can be done using a variety of commercially available transfection reagents. Here we use the LTX transfection reagent because it yielded high transfection

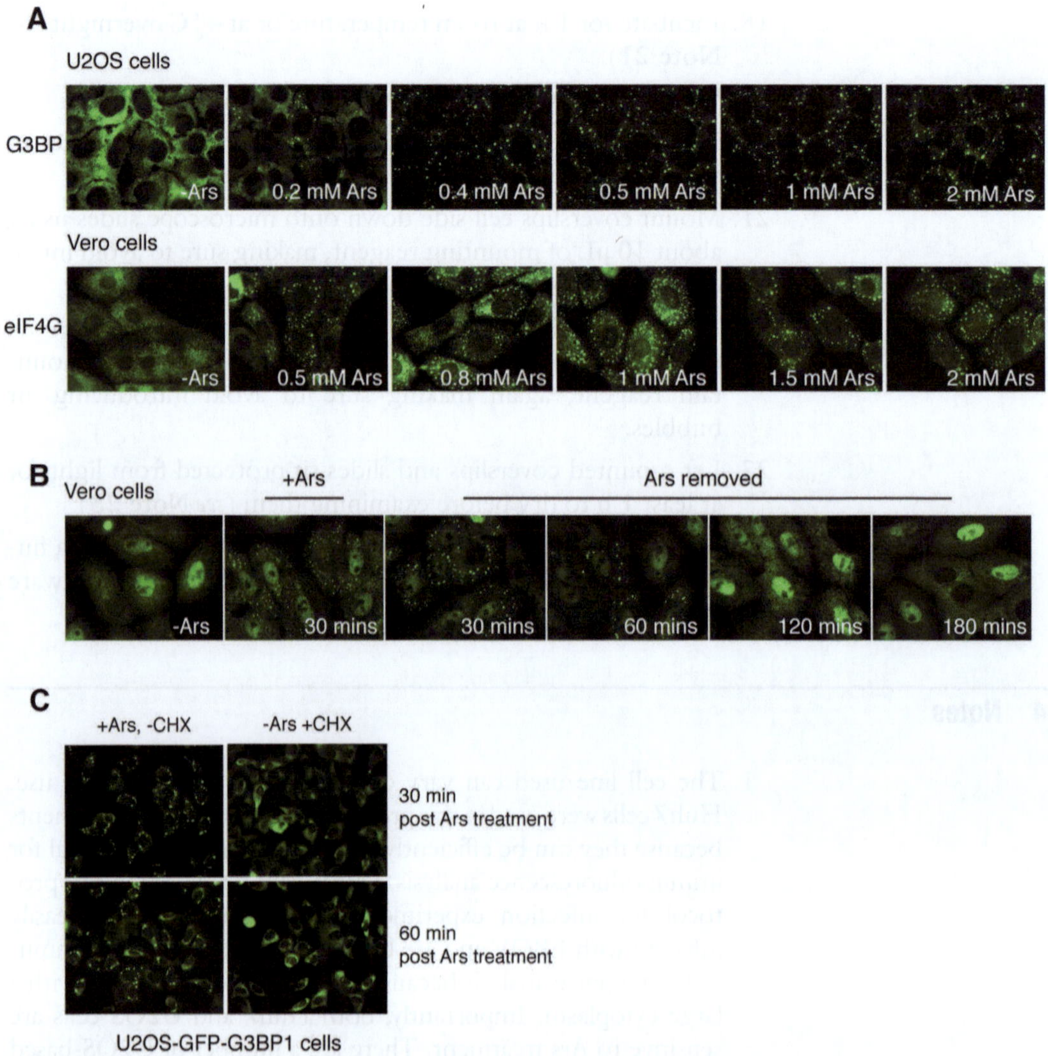

Fig. 2 Induction and examination of SG dynamics in different cell lines. (**a**) U2OS (*top row*) or Vero E6 (*bottom row*) cells were treated with increasing concentrations of Ars for 30 min and analyzed by immunofluorescence analysis (IFA) for the SG marker protein G3BP and eIF4G, respectively. Results show increasing SG formation correlating to higher concentrations of Ars used for the treatment. The lowest concentration of Ars required for SG formation in nearly 100% of cells depends on the cell type. (**b**) Vero cells were treated with 0.8 mM Ars for 30 min, at which point Ars was removed. Cells were then fixed at the indicated time points post removal of arsenite. Cells were subjected to immunofluorescence analysis and examined for SG formation by staining for the SG marker protein HuR (*green*). SG formation is reversible, and therefore when Ars is removed, SGs dissolve over time. (**c**) U2OS-GFP-G3BP1 cells (described in [8]) were treated with 0.2 mM Ars for 30 min. After 30 min, medium was replaced with medium containing 0.2 mM Ars alone (+Ars, −CHX) or medium containing CHX alone (−Ars, +CHX). SG formation was visualized by GFP-G3BP1 fluorescence via live cell imaging at 30 or 60 min after the medium was exchanged. Ars treatment in the absence of CHX leads to canonical SG formation, as observed by the aggregation of GFP-G3BP1 (*left panels*). Treatment of cells with CHX after Ars treatment leads to the dissolution of SGs

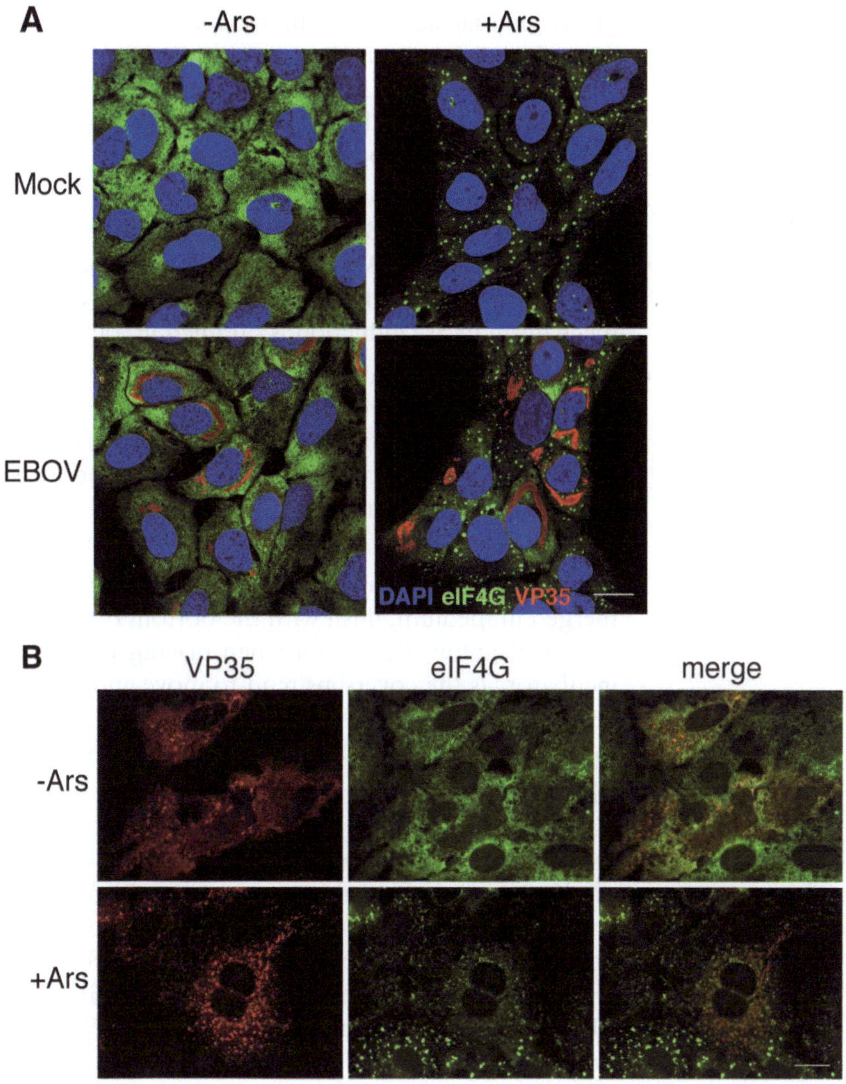

Fig. 3 Examination of the interaction of SGs with EBOV proteins. (**a**) U2OS cells were infected with EBOV at an MOI of 1. At 24 h postinfection, cells were either mock-treated or treated with 0.5 mM Ars. After 30 min of incubation, cells were fixed and examined by immunofluorescence analysis, staining for the SG marker protein eIF4G (*green*) and the EBOV protein VP35 (*red*). Blue DAPI staining indicates the cell nuclei. Scale bar = 20 μm. (**b**) Huh7 cells were transfected with 500 ng of pCAGGS-VP35-HA. Two days post transfection cells were either mock-treated or treated with 0.5 mM Ars for 30 min. After Ars treatment, cells were fixed and examined by IFA. SGs were visualized by staining for eIF4G (*green*), while VP35 was visualized using an HA tag-specific antibody (*red*). Ars-treated cells expressing VP35 show disrupted SG formation. Scale bar = 20 μm. Further data and a more detailed analysis of SG formation during EBOV infection is published in [8]

efficiency in the cell lines used, which are more difficult to transfect. However, many other commercially available transfection reagents are also acceptable.

4. Experiments performed using EBOV must be done under biosafety level (BSL) 4 conditions following specific and approved

protocols that adhere to the established institutional regulations. Samples must be inactivated prior to removal from the BSL4 laboratory, again according to approved institutional procedures. Specific BSL4 regulations and protocols are not outlined in this protocol.

5. Ars induces SG formation via the kinase heme-regulated inhibitor kinase (HRI) [18]. There are a number of drugs that can be used in place of Ars that induce SG formation via kinases other than HRI. A list of commonly used drugs and their mechanism of SG induction is outlined in Table 1.
6. Ars is highly toxic, and it is therefore essential to wear appropriate personal protective equipment (PPE). For handling Ars, it is recommended to wear a lab coat, goggles, face mask or face shield, and double gloves. It is also important to make sure lab coat sleeves are not dragging on surfaces and that there is no exposed skin. Prepare the solution in the fume hood.
7. It is important to make sure that the coverslips are fully submerged in medium, flush with the bottom of the well, and are not overlapping. Be careful when placing the plates into the incubator, as the coverslips tend to move and can slide underneath one another.
8. Many viruses encode proteins that directly inhibit or interfere with SG production (reviewed in [6, 19]). Therefore, it is useful to study these proteins in isolation, out of the context of viral infection, to determine more directly if and by which mechanism(s) they are able to prevent or disrupt SG formation and to determine if there is a direct interaction with SG components. For this purpose, transfection of cells with plasmids encoding these different proteins can be informative. For EBOV, it was found that the viral protein VP35 is able to disrupt SG formation at high levels of expression [8].
9. When adding Ars to cells, the cell culture plates must be removed from the incubator. The temperature of the cells can have an effect on SG formation. For optimal SG induction, keep the cells warm during this step when they are out of the incubator by keeping them on top of a thermal pack or similar heat block warmed to 37 °C. However, be careful when doing so because the plate will not be entirely stable on top of a thermal pack.
10. A preliminary immunofluorescence analysis examining a titration of Ars concentrations in the desired cell line of usage is recommended. Some cell lines require more Ars to induce SG formation in ~100% of cells (Fig. 2a). U2OS cells remain highly sensitive to 0.5 mM Ars and at this concentration show SG formation in nearly all cells. The passage number of the cell line can also make a difference in regard to SG formation and sensitivity to Ars or other stress-inducing agents. Cells that are

higher in passage may stop responding to Ars treatment altogether. It is therefore recommended that cells be used at a lower passage number.

11. As a control, add the same amount of either water or PBS (whichever was used to dissolve the Ars stock solution). This can be accomplished by simply adding Ars or mock treatment dropwise to cells directly or by replacing the cell culture medium with fresh medium supplemented with the appropriate final concentration of Ars.
12. Thirty minutes is the recommended time to treat cells with Ars and induces SGs in nearly 100% of the treated cells (Fig. 2b). Treating for shorter periods of time will not induce SGs in the majority of cells, and treating for longer periods of time is often lethal for the cells. If Ars is removed, SGs will gradually decrease in size and number over time (Fig. 2b).
13. The length of time for transfection can vary depending on the desired level of protein expression. Typically, examining the cells at 2 days post transfection provides enough time for adequate protein expression with minimal cell death.
14. Remember to be careful when handling the cell culture plates here as they contain Ars, which is highly toxic. Be sure to collect, dispose of, and label waste accordingly.
15. Once inactivated, all samples are treated as noninfectious and can be further processed and analyzed under BSL2 conditions.
16. If using coverslips, it is often easier to transfer each cover slip into an individual well of a 24-well tissue culture plate after fixation. Ensure that the coverslips are face up in the well after transfer, and be sure to clearly label the wells.
17. Before covering the plate lid with parafilm, it is useful to label which antibodies are being used and which coverslips these correspond to directly on the plate lid.
18. Removing the chamber can be difficult. Take extra care when doing this, as the slides can easily break at this point. It is still possible to stain using broken slides; however, it is much more difficult to image after staining (as these fragments will not fit in a standard slide holding apparatus). Further, care must be taken to avoid injury due to the sharp edges.
19. Depending on the antibody used, incubation times, particularly for primary antibodies, can vary substantially. Check with the manufacturer to determine the recommended starting point for optimization.
20. It is important to test the desired antibody combinations to ensure that they are specific for the protein of interest and do not cross react with other proteins. For the detection of multiple

proteins in one sample, it is crucial to use primary antibodies generated in different species. Furthermore, it is important to test that the secondary antibodies distinguish between and do not cross react with the primary antibodies. Additionally, the fluorophores should be chosen carefully to prevent bleed-through between channels, e.g., Alexa Fluor 488 and 647 are well separated, so that bleed-through can be excluded in this combination.

21. Although secondary antibody incubation can take place at 4 °C overnight, in our experience the best results are often obtained after a 1 h incubation at room temperature. This often reduces the amount of background fluorescence signal.

22. It is useful to gently dab the edge of the coverslips or chamber slides on a paper towel to remove excess water after this step.

23. Sometimes it helps to seal the edges of the mounted coverslips or chamber slides. To do so, a thin layer or nail polish around the edges is sufficient. However, be careful not to cover too much of the coverslip with polish, as this will obstruct the field of view in the microscope. Make sure that there are no air bubbles under the mounted coverslip before sealing. It is also advisable to wait about 10 min before sealing with nail polish so that the mounting reagent has time to harden slightly. This prevents the coverslip from moving around when trying to seal the edges.

Acknowledgments

The work was performed in the lab of Dr. Elke Mühlberger at the Boston University School of Medicine, Department of Microbiology, and was funded by the National Institute of Allergy and Infectious Disease (NIAID) of the National Institutes of Health (NIH) under award numbers U01-AI082954 and UC06AI058618. The authors are also indebted to Dr. Nancy Kedersha at Brigham and Women's Hospital, Department of Rheumatology, Immunology and Allergy, for her expertise on stress granules and valuable technical advice.

References

1. Kedersha N, Stoecklin G, Ayodele M, Yacono P, Lykke-Andersen J, Fritzler MJ, Scheuner D, Kaufman RJ, Golan DE, Anderson P (2005) Stress granules and processing bodies are dynamically linked sites of mRNP remodeling. J Cell Biol 169:871–884. doi:10.1083/jcb.200502088
2. Anderson P, Kedersha N (2009) Stress granules. Curr Biol 19:R397–R398. doi:10.1016/j.cub.2009.03.013
3. Anderson P, Kedersha N (2008) Stress granules: the Tao of RNA triage. Trends Biochem Sci 33:141–150. doi:10.1016/j.tibs.2007.12.003
4. Buchan JR, Parker R (2009) Eukaryotic stress granules: the ins and outs of translation. Mol Cell 36:932–941. doi:10.1016/j.molcel.2009.11.020
5. Reineke LC, Lloyd RE (2013) Diversion of stress granules and P-bodies during viral infection. Virology 436:255–267. doi:10.1016/j.virol.2012.11.017

6. Montero H, Trujillo-Alonso V (2011) Stress granules in the viral replication cycle. Virus 3:2328–2338. doi:10.3390/v3112328
7. Beckham CJ, Parker R (2008) P bodies, stress granules, and viral life cycles. Cell Host Microbe 3:206–212. doi:10.1016/j.chom.2008.03.004
8. Nelson EV, Schmidt KM, Deflubé LR, Doğanay S, Banadyga L, Olejnik J, Hume AJ, Ryabchikova E, Ebihara H, Kedersha N, Ha T, Mühlberger E (2016) Ebola virus does not induce stress granule formation during infection and sequesters stress granule proteins within viral inclusions. J Virol 90:7268–7284. doi:10.1128/JVI.00459-16
9. Messaoudi I, Amarasinghe GK, Basler CF (2015) Filovirus pathogenesis and immune evasion: insights from Ebola virus and Marburg virus. Nat Rev Microbiol 13:663–676. doi:10.1038/nrmicro3524
10. Brielle S, Gura R, Kaganovich D (2015) Imaging stress. Cell Stress Chaperones 20:867–874. doi:10.1007/s12192-015-0615-y
11. Thomas MG, Loschi M, Desbats MA, Boccaccio GL (2011) RNA granules: the good, the bad and the ugly. Cell Signal 23:324–334. doi:10.1016/j.cellsig.2010.08.011
12. Anderson P, Kedersha N (2009) RNA granules: post-transcriptional and epigenetic modulators of gene expression. Nat Rev Mol Cell Biol 10:430–436. doi:10.1038/nrm2694
13. Panas MD, Kedersha N, McInerney GM (2015) Methods for the characterization of stress granules in virus infected cells. Methods. doi:10.1016/j.ymeth.2015.04.009
14. Souquere S, Mollet S, Kress M, Dautry F, Pierron G, Weil D (2009) Unravelling the ultrastructure of stress granules and associated P-bodies in human cells. J Cell Sci 122:3619–3626. doi:10.1242/jcs.054437
15. Kedersha N, Ivanov P, Anderson P (2013) Stress granules and cell signaling: more than just a passing phase? Trends Biochem Sci 38:494–506. doi:10.1016/j.tibs.2013.07.004
16. Kedersha N, Cho MR, Li W, Yacono PW, Chen S, Gilks N, Golan DE, Anderson P (2000) Dynamic shuttling of TIA-1 accompanies the recruitment of mRNA to mammalian stress granules. J Cell Biol 151:1257–1268
17. Simpson-Holley M, Kedersha N, Dower K, Rubins KH, Anderson P, Hensley LE, Connor JH (2011) Formation of antiviral cytoplasmic granules during orthopoxvirus infection. J Virol 85:1581–1593. doi:10.1128/JVI.02247-10
18. McEwen E, Kedersha N, Song B, Scheuner D, Gilks N, Han A, Chen J-J, Anderson P, Kaufman RJ (2005) Heme-regulated inhibitor kinase-mediated phosphorylation of eukaryotic translation initiation factor 2 inhibits translation, induces stress granule formation, and mediates survival upon arsenite exposure. J Biol Chem 280:16925–16933. doi:10.1074/jbc.M412882200
19. White JP, Lloyd RE (2012) Regulation of stress granules in virus systems. Trends Microbiol 20:175–183. doi:10.1016/j.tim.2012.02.001
20. Bordeleau M-E, Matthews J, Wojnar JM, Lindqvist L, Novac O, Jankowsky E, Sonenberg N, Northcote P, Teesdale-Spittle P, Pelletier J (2005) Stimulation of mammalian translation initiation factor eIF4A activity by a small molecule inhibitor of eukaryotic translation. Proc Natl Acad Sci U S A 102:10460–10465. doi:10.1073/pnas.0504249102
21. Bordeleau M-E, Mori A, Oberer M, Lindqvist L, Chard LS, Higa T, Belsham GJ, Wagner G, Tanaka J, Pelletier J (2006) Functional characterization of IRESes by an inhibitor of the RNA helicase eIF4A. Nat Chem Biol 2:213–220. doi:10.1038/nchembio776

Chapter 18

Analyzing Apoptosis Induction and Evasion in Ebola Virus-Infected Cells

Judith Olejnik and Emily V. Nelson

Abstract

In this chapter, we describe a Western blot assay for the successful detection of apoptosis in Ebola virus (EBOV)-infected cells. The protocol includes all steps from cell culture, infection of cells, generation of lysates, and analysis using Western blot to detect caspase cleavage as marker of apoptosis.

Key words *Zaire ebolavirus*, Ebola virus, Filoviruses, Apoptosis, Caspases, Western blot, Antibodies

1 Introduction

Viruses rely on a functional cellular machinery for replication and propagation, and, therefore, cell death is an essential host defense mechanism for eliminating virus-infected cells. However, many viruses have evolved multiple mechanisms to interfere with cell death signaling to ensure efficient viral replication [1]. Apoptosis is the best-characterized form of programmed cell death and is regulated by a complex signaling network. Induction of apoptosis can be mediated via extracellular signals through the binding of apoptosis-inducing ligands to their respective receptors or mediated by intracellular mitochondrial changes [2, 3]. A hallmark of apoptosis is the activation of cysteine aspartate-specific proteases, the caspases [2–4]. Activation of caspases occurs via the cleavage of the inactive precursor protein into the active form. Different initiator caspases are associated with death receptor (caspase 8, 10) or mitochondrial signaling (caspase 9), but ultimately they activate the same downstream effector caspases, including caspase 3 [3–6].

Here we focus on the use of Western blot analysis to detect activated caspases in Ebola virus (EBOV)-infected cells. Detection of caspase cleavage by Western blot analysis is a commonly used method to assess apoptosis in cell culture [7, 8]. There are multiple

Thomas Hoenen and Allison Groseth (eds.), *Ebolaviruses: Methods and Protocols*, Methods in Molecular Biology, vol. 1628, DOI 10.1007/978-1-4939-7116-9_18, © Springer Science+Business Media LLC 2017

antibodies commercially available for the detection of different caspase proteins. Using antibodies that recognize both the active cleaved form and the inactive precursor is recommended, as this provides a quantifiable comparison of the amount of cleaved caspase versus precursor protein. The ability to use lysates generated from EBOV-infected cells to detect multiple caspases performing multiple Western blot analysis is advantageous given the difficulties and limitations associated with work using EBOV under high containment/biosafety level (BSL) 4 conditions. After preparation of the cell lysates and inactivation, the analysis can then be performed under BSL2 conditions using standard laboratory equipment for Western blot analysis.

The disadvantage of using Western blot analysis to detect apoptosis in virus-infected cells is the inability to discern if active caspases are present in infected or noninfected cells (*see* also ref. 7). Therefore, it is essential to either have all cells that are analyzed infected to ensure correct interpretation of the data or to use a second method of apoptosis detection to analyze infected cells directly. This can be performed by immunofluorescence analysis by co-staining for the virus and apoptosis markers, for example, caspase activation or annexin V translocation [7–9] (*see* **Note 1**). EBOV readily infects a variety of cell types (reviewed in [10]), and the described apoptosis assays can be adapted for use in the cell line/cell type of interest as needed. It is essential to include apoptosis-inducing controls, which will differ depending on the cells used and/or caspase analyzed. This protocol describes the detection of caspase 3 activation in Vero cells, which are commonly used to generate EBOV viral stocks. As a control, infection with vesicular stomatitis virus (VSV), a known inducer of apoptosis, is also described [11].

2 Materials

2.1 Cells

1. DMEM-2: Dulbecco's Modified Eagle Medium (DMEM) supplemented with 2% fetal bovine serum (FBS), penicillin (50 units/mL), L-glutamine (200 mM), and streptomycin (50 mg/mL).
2. DMEM-10: Dulbecco's Modified Eagle Medium (DMEM) supplemented with 10% fetal bovine serum (FBS), penicillin (50 units/mL), L-glutamine (200 mM), and streptomycin (50 mg/mL).
3. Phosphate buffered saline (PBS), Mg^{2+}- and Ca^{2+}-free (e.g., Lonza).
4. 0.025%Trypsin-EDTA.
5. Vero cells (ATCC CCL-81).

6. Bright-field microscope.
7. 6-well tissue culture plates.
8. Neubauer cell counting chamber.

2.2 Infection of Cells

1. DMEM-2.
2. EBOV virus stock.
3. VSV virus stock.

2.3 Generation of Western Blot Lysates

1. PBS.
2. Lysis buffer: 3-[(3-cholamidopropyl)-dimethylammonio]-1-propanesulfonate (CHAPS) buffer (Cell Signaling Technology) supplemented with 5 mM dithiothreitol (DTT, Cell Signaling Technology) and 1 mM cOmplete protease inhibitor mix (Roche).
3. 2× sodium dodecyl sulfate (SDS) sample buffer (e.g., Laemmli buffer (Sigma)).
4. Cell scrapers.

2.4 Western Blot Analysis

1. Vertical gel electrophoresis chamber with gel casting stand and glass plates (e.g., Mini-PROTEAN Electrophoresis System (Bio-Rad)).
2. Thirty percent acrylamide/bisacrylamide (37.5:1) mix.
3. Ten percent ammonium persulfate (APS): 100 mg APS, 1 mL H_2O.
4. Tetramethylethylenediamine (TEMED).
5. Ten percent SDS: 10 g SDS, 100 mL H_2O.
6. 1.5 M Tris [pH 6.8]: 181.71 g Tris, 1 L H_2O, pH adjusted to 6.8 with HCl.
7. 1.5 M Tris [pH 8.8]; 181.71 g Tris, 1 L H_2O, pH adjusted to 8.8 with HCl.
8. 100% ethanol (EtOH).
9. Heat block.
10. 1× SDS Running buffer: 100 mL 10× SDS running buffer (Boston BioProducts), 900 mL H_2O.
11. Prestained Western blot marker (e.g., PageRuler Plus Prestained Protein Ladder (Fermentas)).
12. Protein transfer chamber (e.g., Fastblot B34 (Biometra)).
13. Transfer anode buffer I: 36.34 g Tris, 200 mL EtOH, 1 L H_2O.
14. Transfer anode buffer II: 3.06 g Tris, 200 mL EtOH, 1 L H_2O.
15. Transfer cathode buffer: 6.25 g E-amino-caproic acid, 3.03 g Tris, 200 mL EtOH, 1 L H_2O.

16. Whatman paper (0.34 mm thickness) cut into 6 × 9 cm pieces, nine pieces per gel.
17. Polyvinylidene fluoride (PVDF) membrane for use with fluorescent detection cut into 6 × 9 cm pieces (e.g., Millipore Immobilon-FL).
18. 100% methanol.
19. Tris-buffered saline (TBS): 100 mL 10× TBS (e.g., Boston BioProducts), 900 mL H_2O.
20. TBS + 0.1% Tween (TBS/T): 1 mL Tween-20, 1 L TBS.
21. Odyssey blocking buffer (LI-COR).
22. Rabbit anti-human caspase 3 antibody (Cell Signaling Technology).
23. Mouse anti-β-actin antibody (Abcam).
24. Goat anti-rabbit IRDye800 conjugated antibody (LI-COR).
25. Goat anti-mouse IRDye680 conjugated antibody (LI-COR).
26. Odyssey scanner and Image Studio software (LI-COR).
27. Plastic tweezers.
28. Plastic box for membrane incubation, preferably opaque.
29. Aluminum foil (if an opaque box is not used).
30. Shaker or rocker.

3 Methods

3.1 Plating of Cells (See Note 2)

1. Remove cell culture medium from Vero cells grown in DMEM-10 to 80–100% confluency in a 75 cm^2 cell culture flask using sterile plastic pipettes.
2. Gently add 5–10 mL of PBS to the flask and gently swirl flask to wash (*see* **Note 3**).
3. Remove PBS using sterile plastic pipettes.
4. Repeat **steps 2** and **3** for a total of two washes.
5. Add 1.5 mL of trypsin to the cells.
6. Incubate the cells with trypsin until cells are detached. Monitor cell detachment using a bright-field microscope (*see* **Note 4**).
7. Add 8.5 mL of DMEM-2 to detached cells.
8. Thoroughly resuspend detached cells (*see* **Note 5**).
9. Count cells using a Neubauer cell counting chamber (*see* **Note 6**).
10. Seed 1 × 10^5 cells per well of a 6-well cell culture plate in 2 mL of DMEM-2 (*see* **Note 7**).
11. Incubate cells at 37 °C, 5% CO_2 until ready for infection the following day.

3.2 Infection of Cells (See Note 8)

1. Infection will be performed 1 day after seeding cells as described above in Subheading 3.1.
2. Prepare the virus inoculum by mixing 0.5 mL DMEM-2 and the appropriate amount of EBOV or VSV (positive control) for each well to be infected. Prepare 0.5 mL DMEM-2 without any virus as a negative control (*see* **Note 9**).
3. Carefully remove the medium from cells in a 6-well cell culture plate using sterile plastic pipettes.
4. Add 0.5 mL EBOV inoculum or controls as prepared in **step 1** to each well to be infected.
5. Incubate cells at 37 °C, 5% CO_2 for 1 h (*see* **Note 10**).
6. Add 1 mL DMEM-2 to each well of the 6-well plate (*see* **Note 11**).
7. Incubate cells at 37 °C, 5% CO_2 until ready for preparation of Western blot lysates (*see* **Note 12**).

3.3 Generation of Western Blot Lysates (See Note 8)

1. Prepare 50 μL of lysis buffer per sample (*see* Subheading 2.3, **item 2**) (*see* **Note 13**).
2. Analyze cells under bright-field microscope to determine the presence and extent of detached cells (*see* **Note 14**).
3. Scrape cells into cell culture medium using a cell scraper and transfer to 2 mL microcentrifuge tubes (*see* **Note 14**).
4. Spin cells down at 20,000 × *g* for 3 min at 4 °C in a tabletop centrifuge (*see* **Note 15**).
5. Remove supernatants carefully by pipetting.
6. Add 1 mL cold PBS to the tube (*see* **Note 16**).
7. Centrifuge at 20,000 × *g* for 2 min at 4 °C in a tabletop centrifuge.
8. Remove PBS carefully by pipetting.
9. Repeat **steps 5–8** for a total of two washes.
10. Resuspend pellet in 50 μL cell extraction buffer by pipetting up and down several times.
11. Vortex sample (*see* **Note 17**).
12. Transfer samples to −80 °C and freeze sample completely (*see* **Note 18**).
13. Transfer samples to room temperature (RT) and thaw completely (*see* **Note 18**).
14. Vortex sample (*see* **Note 17**).
15. Repeat **steps 11–13** twice for a total of three freeze/thaw cycles.
16. Centrifuge at 20,000 × *g* for 10 min at 4 °C.

17. Transfer supernatants to new microcentrifuge tubes containing SDS sample buffer (*see* **Note 19**).
18. Inactivate samples and remove from BSL4 laboratory according to institutionally approved procedures (*see* **Note 19**).
19. Store samples at −20 °C until analysis by Western blot is performed.

3.4 Western Blot Analysis

1. Assemble glass plates in a gel casting chamber according to the manufacturer's instructions (*see* **Note 20**).
2. Prepare the solution for the separating gel, as outlined in Table 1 (*see* **Note 21**).
3. Add the separating gel solution in between the glass plates in gel cast chamber (*see* **Note 22**).
4. Overlay carefully with 1–2 mL 100% EtOH.
5. Wait for separating gel to solidify, and then remove the ethanol overlay (*see* **Note 23**).
6. Prepare the solution for the stacking gel, as outlined in Table 1.
7. Add the stacking gel solution on top of separating gel and insert a well comb (*see* **Note 24**).
8. Once the stacking gel is set, transfer the gel to an electrophoresis chamber filled with 1× SDS running buffer and remove the comb (*see* **Note 25**).
9. Thaw samples prepared in Subheading 3.3 on ice (*see* **Note 26**).

Table 1
Preparation of SDS-polyacrylamide gels (PAGE). Template of ingredients for preparation of SDS-PAGE gels, including both separating and stacking gels. The listed amounts will yield two gels

	Stacking gel	Separating gel		
	4%	10%	12%	15%
H_2O	2.9 mL	4 mL	3.3 mL	2.3 mL
30% acrylamide/ bisacrylamide mix (37.5:1)	750 μL	3.3 mL	4 mL	5 mL
10% SDS	50 μL	100 μL	100 μL	100 μL
1.5 M Tris, pH 8.8	–	2.5 mL	2.5 mL	2.5 mL
1.5 M Tris, pH 6.8	1.25 mL	–	–	–
10% APS	50 μL	100 μL	100 μL	100 μL
TEMED	5 μL	5 μL	5 μL	5 μL

10. Prior to loading samples into the wells of the SDS-PAGE gel, incubate the samples in heating block for 3–5 min at 95 °C (*see* **Note 27**).
11. Load 15 μL of samples per well by pipetting sample into wells. Include at least one well loaded with a prestained protein ladder (*see* **Note 28**).
12. Attach the lid to the gel electrophoresis chamber, and run at 100–180 V until sufficient separation of the proteins has occurred, using the prestained protein ladder as a measure of separation.
13. Soak three pieces of precut Whatman paper per gel in each of the transfer buffers (anode buffer I, anode buffer II, and cathode buffer, i.e., nine pieces total per gel).
14. Activate the PVDF membrane by soaking it for 30 s in 100% methanol (*see* **Note 29**), and then equilibrate in anode buffer II for a few minutes.
15. To create the protein transfer sandwich, place the three pieces of Whatman soaked in anode buffer I onto the bottom of the semidry protein transfer chamber.
16. On top of this, place the three pieces of Whatman paper soaked in anode buffer II, followed by the PVDF membrane. Following this step, it is important to work quickly so that the PVDF membrane does not dry out.
17. Carefully remove the electrophoresis gel from the electrophoresis chamber and glass plates. Carefully remove the top part of gel containing the wells, and place remaining gel down on top of the PVDF membrane.
18. Finally, place the Whatman paper soaked in cathode buffer on top (*see* **Note 30** and Fig. 1).

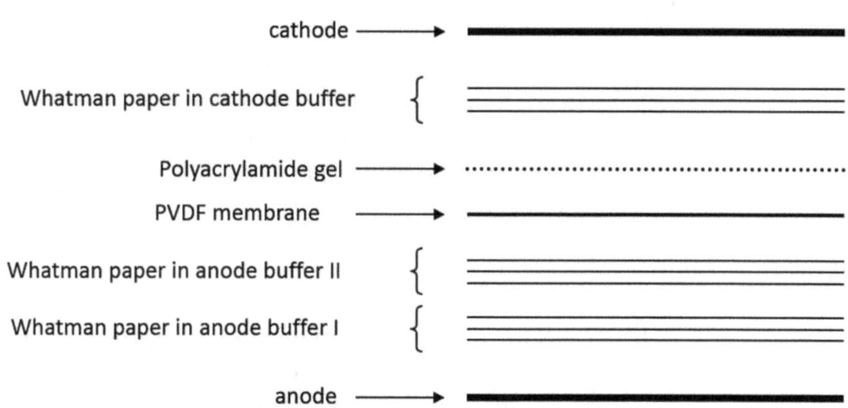

Fig. 1 Assembly of Western blot transfer sandwich. Schematic of transfer sandwich setup described in Subheading 3.4, **steps 15–18**

19. Place the lid on the semidry transfer chamber, connect to a power supply, and run at 30 V for 30 min (*see* **Note 31**).
20. After the transfer, remove the PVDF membrane carefully from the transfer sandwich using plastic tweezers, and incubate in blocking reagent for 1 h at RT on a shaker (*see* **Note 32**). Work quickly so that the PVDF membrane does not dry out.
21. Remove blocking buffer and add TBS/T. Incubate for 10 min at RT on a shaker (*see* **Note 32**).
22. Repeat **step 20** (remove TBS/T) twice for a total of three washes.
23. Remove TBS/T and add caspase 3 and β-actin antibodies diluted in Odyssey blocking buffer supplemented with 0.1% Tween, and incubate at 4 °C overnight on a shaker (*see* **Note 33**).
24. Remove the primary antibody solution and add TBS/T (*see* **Note 32**). Incubate for 10 min at RT on a shaker.
25. Repeat **step 20** (remove TBS/T) twice for a total of three washes.
26. Remove TBS/T, and add IRDye-conjugated secondary antibodies diluted in Odyssey blocking buffer supplemented with 0.1% Tween, for 1 h at RT (*see* **Note 34**). Make sure to protect the antibodies from light by covering the incubation container with foil or using an opaque container.
27. Remove the secondary antibody solution and add TBS/T (*see* **Note 32**). Incubate for 10 min at RT on a shaker while still protecting the membrane from light.
28. Repeat **step 23** (remove TBS/T) twice for a total of two washes.
29. Add TBS and incubate for 10 min at RT on a shaker.
30. Remove TBS, add fresh TBS, and incubate for 10 min at RT on a shaker for a total of two washes.
31. Scan the membrane using an Odyssey Infrared Imaging system, using the Image Studio software (*see* **Note 35**).
32. Analyze the caspase 3 banding pattern to determine if caspase cleavage has occurred, and quantify the results, if desired, using the Image Studio software (*see* **Notes 35** and **36** and Fig. 2) [12].

4 Notes

1. To ensure the most accurate data interpretation when using Western blot analysis to detect apoptosis, it is highly recommended that an immunofluorescence analysis using an EBOV-specific antibody is performed in parallel, to detect the relative level of infection in the cells. In addition, staining of apoptosis

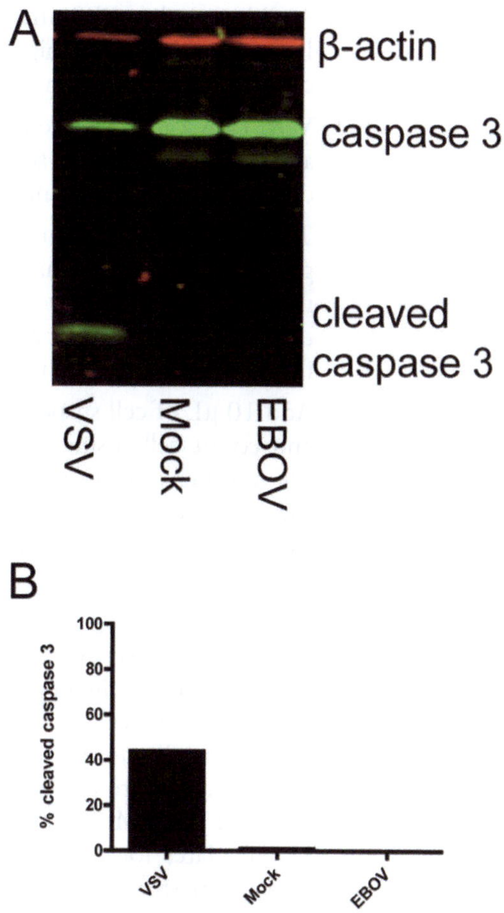

Fig. 2 Analysis of caspase 3 cleavage in EBOV- and VSV-infected cells. (**a**) Representative image of expected Western blots result using the described protocol. Cleaved caspase 3 (17 kDa) can only be detected in VSV-infected cells, not in EBOV-infected cells or in noninfected control cells (Mock), whereas the precursor protein (caspase 3, 35 kDa) is present in all samples. (**b**) Quantification of caspase 3 cleavage based on results presented in (**a**). *See* **Note 35** for details about the quantification

markers can also be performed prior to the fixation of cells under BSL4 conditions following all necessary safety precautions. Other assays that have been successfully used to analyze apoptosis in EBOV-infected cells include fluorochrome-labeled inhibitors of caspases (FLICA) staining to detect active caspases or annexin V staining to analyze annexin V translocation [9].

2. Cell culture work must be performed in a biosafety cabinet with correct sterile technique. Cell culture medium, PBS, and trypsin are stored at 4 °C. Do not use cold reagents on cells. Warm PBS and medium in a 37 °C water bath or at room temperature before use. Trypsin should only be warmed at room temperature.

3. Avoid pipetting PBS and trypsin directly onto the cell layer. Instead, add them by pipetting slowly onto the side of the cell culture flask and rocking to distribute the liquid over the cells.
4. Vero cells can be placed into the 37 °C incubator after the addition of trypsin for several minutes if needed. Monitor the detachment of cells using a bright-field microscope.
5. The cell monolayer can be detached easily by hitting the flask gently against your hand while avoiding splashing of the medium. Resuspend the cells by pipetting up and down several times and ensure that all cells have been removed from the flask and a single cell suspension is achieved.
6. Add 10 μL of cell suspension to Neubauer counting chamber, and count cells using a bright-field light microscope. Count the number of cells in all four big grid squares, and calculate the mean cell number for one square. This number represents the total cell number in 0.1 μL of the cell suspension.
7. Cells for apoptosis detection should be seeded sub-confluently to avoid inducing cell death signaling via cell crowding. Seeding 10^5 cells per well of a 6-well cell culture plate will yield 30–40% cell confluency 1 day after seeding cells. Determine the amount of cell suspension needed using the information about the cell number present in 0.1 μL as described above (*see* **Note 6**). Prepare a master mix of cell suspension mixed with 2 mL of DMEM-2 per well for all wells to be seeded. Add to 6-well plates for 2 mL total volume per well. Calculate and prepare master mix to include at least one additional well to account for pipetting errors.
8. All work with infectious EBOV must be performed under BSL4 conditions. This protocol will not describe specific safety details. However, all work must be performed adhering to regulations in place at the BSL4 facility following approved protocols. Samples can only be removed from the BSL4 laboratory after inactivation for safe processing at BSL2 by following approved institutional procedures and regulations, which are not addressed in this protocol.
9. Infections should be performed with a high multiplicity of infection (MOI) to ensure complete infection of all cells. A MOI of 5 for EBOV and MOI of 1 for VSV are recommended, but it is advisable to test the appropriate amount of virus to be used beforehand by performing infections using multiple MOIs and confirming infection rates by immunofluorescence. Prepare a master mix for all wells to be infected. Calculate for at least one additional well when preparing master mix to account for pipetting errors.
10. Optimal distribution of virus can be achieved by carefully rocking the plate gently multiple times during the incubation.

11. Virus inoculum is not removed in this protocol to increase the infection rate.
12. Analysis of a range of time points for apoptosis induction, starting at a few hours after infection up to multiple days, is recommended. It is important to test the optimal time point for the apoptosis control. Caspase 3 cleavage after VSV infection can be detected between 10 and 24 h after infection.
13. Lysis buffer should be prepared for at least one additional sample to account for pipetting errors. Lysis buffer needs to be kept on ice or at 4 °C until used.
14. Apoptosis induction will lead to cell rounding and detachment. Therefore, it is essential to harvest all cells for analysis including detached cells. Therefore, cells are scraped into the cell culture medium, and subsequent washes and lysis step are performed in microcentrifuge tubes instead of in the tissue culture plate.
15. Centrifugation of samples should be performed at a minimum of $15{,}000\times g$.
16. To minimize protein degradation, all reagents should be kept cold on ice or at 4 °C.
17. Vortexing for about 10 s is sufficient to optimize resuspension and cell lysis.
18. Freeze thaw cycles are used for cell lysis. Complete freezing of the 50 μL sample will take about 5 min at −80 °C. Freezing can also be performed using a −20 °C freezer with a longer incubation time. Check if the sample is frozen by inverting the tube. To thaw sample, place the tubes at RT for several minutes. Thawing will be faster than freezing. To avoid protein degradation, samples should be checked frequently and processed as soon as they are completely thawed.
19. In order to process samples at lower biosafety levels, inactivation must be performed adhering to the regulations in place at the BSL4 facility where the work is performed and following approved protocols to ensure complete inactivation of EBOV. This might include different final SDS concentrations in the sample and/or different boiling times or temperatures. *See* also **Note 8**.
20. Follow instructions for the gel chamber and gel casting devices used if gels are prepared from reagents as described. Alternatively, commercially available precast gels can be used. If using precast gels, skip ahead in the protocol to **step 8**.
21. For the detection of caspase cleavage, preparation of a 15% separating gel is recommended. The listed amounts will be sufficient to prepare two separate gels. Acrylamide solution is a hazardous chemical and toxic when in solution; ensure that

you are following appropriate safety measures while handling this substance.

22. Use a plastic pipette to add solution between the glass plates, and avoid air bubbles. The solution should fill about ¾ of the space. This ensures sufficient protein separation while leaving enough room for addition of the stacking gel and comb.

23. Remove EtOH by pouring off or soaking up with Whatman paper. If using Whatman paper, be careful not to touch the surface of the set separating gel. Ensure complete removal of ethanol as residual ethanol will lead to disrupted solidification of the stacking gel.

24. Use a well comb with sufficient space to load desired sample amount. If quantification of samples using the LI-COR Odyssey will be performed, it is recommended that all samples are loaded on the same gel to minimize experimental differences.

25. Solidified gel residue can be present in the wells. Therefore, it is recommended that the wells are washed out before loading samples. Pipette SDS running buffer repeatedly into the wells to do so.

26. To avoid protein degradation, thaw samples on ice and keep cold (on ice or at 4 °C) until use.

27. Three minutes of sample boiling is sufficient to disrupt protein secondary structures prior to loading of samples. As most inactivation procedures for the removal of Western blot samples from BSL4 involve heating/boiling, reducing the amount of additional time the samples are heated might increase the protein signal.

28. The amount of sample loaded per well will need to be optimized depending on the cell line and experimental conditions used. Generally loading at least 10 μL is necessary to achieve good protein detection using the described protocol. Choose a prestained protein marker with appropriately sized bands to identify full length caspase 3 (35 kDa), cleaved caspases 3 (17 kDa), and β-actin (42 kDa). Loading of the SDS gel can be performed using a Hamilton syringe, specific gel-loading pipette tips, or standard 100 μL pipette tips depending on user preference.

29. Work with 100% methanol under a fume hood, and dispose of it according to local regulations for hazardous waste. Methanol can be reused several times for membrane activation if kept in a sealed container to avoid evaporation. The PVDF membrane will change color from bright white to semitranslucent when activated. Check for white spots that resist hydration (e.g., from improper handling or residues on the membrane). These spots will not efficiently be able to bind protein, and the use of a fresh membrane is advised. Transfer the membrane into a

container with anode buffer II in the fume hood before transporting it to the area where the blot will be assembled.

30. To ensure correct protein transfer, air bubbles need to be avoided when assembling the transfer sandwich. To remove air from the assembled transfer sandwich, use a plastic pipette, and gently roll it over the top of the assembled transfer sandwich. Do not use too much pressure while rolling to ensure that the Whatman papers are still soaked with the corresponding buffers. Remove excess liquid with paper towels without disturbing the transfer sandwich.
31. Transfer settings will need to be adapted depending on the protein transfer chamber used. To ensure efficient protein transfer, check that the protein bands from the included marker are transferred from the gel onto the membrane. It is especially useful to check the marker within the size range of proteins to be detected to confirm protein transfer.
32. Use of the described blocking buffer (*see* 2.4.21) is recommended for use with the corresponding LI-COR secondary antibodies. The buffer contains low concentrations of sodium azide, which must be collected and disposed of according to local hazardous waste regulations. Alternatively, blocking solution can be prepared using skim milk powder or bovine serum albumin (BSA) at concentrations between 5 and 10%, if desired Ponceau staining (follow the manufacturer's instructions) for general protein detection can be performed before blocking of the membrane.
33. Antibody amounts should be optimized for different experimental conditions. Use of a 1:1000 dilution for the caspase 3 (*see* 2.4.22) and 1:40,000 (*see* 2.4.23) for the β-actin antibody has been used successfully in this protocol. The described antibodies are cross-reactive between human and monkey and can be successfully used in Vero cells and human cell lines. Detection of β-actin is used to probe for a cellular protein unaffected early in apoptosis as a protein loading control, to ensure that sufficient protein was loaded in cases caspase 3 staining is undetectable. Any other cellular protein can be used instead if its expression is not influenced by EBOV or VSV infection (or other apoptosis-inducing controls that are used).
34. Antibody amounts should be optimized for different experimental conditions. Use of 1:20,000 dilution for the described secondary antibodies (2.4.24 and 2.4.25) has been used successfully in this protocol. This protocol describes the use of fluorescently labeled secondary antibodies for use with the LI-COR Odyssey system, which are useful for quantification of the resulting protein bands. Protection from light during incubation and subsequent washes is, therefore, essential to avoid loss

of fluorescence. Alternatively peroxidase-conjugated secondary antibodies can be used in combination with detection using film exposure or other compatible imaging technologies.

35. Refer to instruction manual of the LI-COR Odyssey scanner for setup and scanning of the membrane. Briefly, this involves placing the membrane on the Odyssey scanner, covering it with the included silicone mat, and then scanning using the Image Studio software to generate a digital image.

36. *See* Fig. 2 for representative results. To quantify the amount of cleaved caspase, refer to Image Studio software instructions for Western blot quantification. Briefly, this involves selecting the area of the image desired to be quantified and choosing an appropriate background setup (refer to the instruction manual of the LI-COR Odyssey scanner). Select bands for caspase 3 and cleaved caspase 3 for quantification and export the signal intensity values to Excel or another other program for further analysis and calculations. The percentage of cleaved caspase 3 is equal to the intensity of cleaved caspase 3 divided by the sum of the intensities of cleaved caspase 3 and uncleaved caspase 3 (quantification strategy as demonstrated for RIPK1 cleavage during apoptosis) [12].

Acknowledgments

The authors are grateful to J.R. Pacheco (Boston University) for critical reading of the manuscript. The work was performed in the lab of Dr. Elke Mühlberger at the Boston University School of Medicine, Department of Microbiology and was funded by the National Institute of Allergy and Infectious Disease (NIAID) of the National Institutes of Health (NIH) under award numbers UC6 AI058618 and U01 AI082954.

References

1. Best SM (2008) Viral subversion of apoptotic enzymes: escape from death row. Annu Rev Microbiol 62:171–192. doi:10.1146/annurev.micro.62.081307.163009
2. Duprez L, Wirawan E, Vanden Berghe T, Vandenabeele P (2009) Major cell death pathways at a glance. Microbes Infect 11:1050–1062.doi:10.1016/j.micinf.2009.08.013.S1286-4579(09)00204-4 [pii]
3. Kiraz Y, Adan A, Kartal Yandim M, Baran Y (2016) Major apoptotic mechanisms and genes involved in apoptosis. Tumor Biol 37(7):8471–8486. doi:10.1007/s13277-016-5035-9
4. Degterev A, Boyce M, Yuan J (2003) A decade of caspases. Oncogene 22:8543–8567. doi:10.1038/sj.onc.1207107
5. Guicciardi ME, Gores GJ (2009) Life and death by death receptors. FASEB J 23:1625–1637. doi:10.1096/fj.08-111005
6. Riedl SJ, Salvesen GS (2007) The apoptosome: signalling platform of cell death. Nat Rev Mol Cell Biol 8:405–413. doi:10.1038/nrm2153
7. Galluzzi L, Aaronson SA, Abrams J et al (2009) Guidelines for the use and interpretation of assays for monitoring cell death in higher eukaryotes. Cell Death Differ 16:1093–1107. doi:10.1038/cdd.2009.44
8. Krysko DV, Vanden T, Herde KD et al (2008) Apoptosis and necrosis: detection, discrimination and phagocytosis. Methods

44:205–221. doi:10.1016/j.ymeth.2007.12.001

9. Olejnik J, Alonso J, Schmidt KM et al (2013) Ebola virus does not block apoptotic signaling pathways. J Virol 87:5384–5396. doi:10.1128/JVI.01461-12
10. Olejnik J, Ryabchikova E, Corley RB, Mühlberger E (2011) Intracellular events and cell fate in filovirus infection. Virus 3:1501–1531. doi:10.3390/v3081501
11. Gadaleta P, Vacotto M, Coulombie F (2002) Vesicular stomatitis virus induces apoptosis at early stages in the viral cycle and does not depend on virus replication. Virus Res 86:87–92
12. van Raam BJ, Ehrnhoefer DE, Hayden MR, Salvesen GS (2013) Intrinsic cleavage of receptor-interacting protein kinase-1 by caspase-6. Cell Death Differ 20:86–96. doi:10.1038/cdd.2012.98

Chapter 19

Electron Microscopy of Ebola Virus-Infected Cells

Takeshi Noda

Abstract

Ebola virus (EBOV) replicates in host cells, where both viral and cellular components show morphological changes during the process of viral replication from entry to budding. These steps in the replication cycle can be studied using electron microscopy (EM), including transmission electron microscopy (TEM) and scanning electron microscopy (SEM), which is one of the most useful methods for visualizing EBOV particles and EBOV-infected cells at the ultrastructural level. This chapter describes conventional methods for EM sample preparation of cultured cells infected with EBOV.

Key words Transmission electron microscopy, Scanning electron microscopy, Ultrathin sectioning

1 Introduction

Ebola virus (EBOV) virions are filamentous and have a uniform diameter of approximately 80 nm [1]. After their entry into host cells via macropinocytosis, nucleocapsids are released from virions into the cytoplasm. Transcription and replication of the viral genome occur in cytoplasmic inclusion bodies that hold a large number of nucleocapsids. Later, newly synthesized nucleocapsids, each containing a viral genome, are transported from the inclusion bodies to the plasma membrane, where assembly and budding of progeny virions occurs [2]. Because all these steps are accompanied by morphological changes in both the viral and cellular components, ultrastructural analysis of EBOV-infected cells can provide us with important insights into the replication mechanisms of EBOV [3–8].

Filoviruses, including EBOV, Marburg virus, and Cueva virus, are classified as Biosafety Level 4 (BSL4) pathogens, meaning that they must be handled in BSL4 facilities. For this reason, if electron microscopes are not installed in a BSL4 facility, EBOV-infected cells to be subjected to EM must be removed from the facility following prescribed sample inactivation procedures. Among disinfectants that can be used in inactivation procedures, glutaraldehyde is recommended. It is commonly used as a fixative in the preparation of

Thomas Hoenen and Allison Groseth (eds.), *Ebolaviruses: Methods and Protocols*, Methods in Molecular Biology, vol. 1628, DOI 10.1007/978-1-4939-7116-9_19, © Springer Science+Business Media LLC 2017

both TEM and SEM samples and preserves the ultrastructures of biological specimens. Once the inactivated samples are removed from the BSL4 facility, they can be prepared for EM using conventional methods.

2 Materials

2.1 General

1. Use reagents of EM grade or analytical grade, if available.
2. Prepare all reagents using ultrapure water.

2.2 Buffers

1. 0.1 M Cacodylate buffer, pH 7.2: Prepare a 0.2 M stock solution of sodium cacodylate ($(CH_3)_2AsO_2Na$) in distilled water. For a 100 mL stock solution, dissolve 4.28 g of $(CH_3)_2AsO_2Na{\cdot}3H_2O$ in 90 mL of distilled water, add 0.2 M HCl as necessary to adjust the pH, and add distilled water to obtain a final volume of 100 mL.

2.3 Fixatives

1. 2.5% Glutaraldehyde fixative (*caution*: toxic; wear gloves and handle in a fume hood): Prepare a 2.5% glutaraldehyde solution in 0.1 M cacodylate buffer. For 10 mL of fixative, dilute 1 mL of a 25% glutaraldehyde solution (EM grade) in a mixture of 5 mL of 0.2 M cacodylate buffer and 4 mL of distilled water.
2. Two percent osmium tetroxide fixative (*caution*: very toxic; wear gloves and handle in a fume hood): Prepare a 2% osmium tetroxide solution in 0.1 M cacodylate buffer from a 4% osmium tetroxide solution by making a 1:1 dilution using the 0.2 M cacodylate buffer.

2.4 Dehydration for Ultrathin Section TEM

1. Ethanol (50, 70, 90%) in distilled water.
2. 99.5% or 100% ethanol.
3. Propylene oxide (*caution*: carcinogenic substance).

2.5 Embedding Resins for Ultrathin Section TEM

1. Epoxy resin (TAAB 812 resin kit, TAAB Laboratories Equipment Ltd.):

   TAAB Epon 812: 9.6 mL.

   Dodecenylsuccinic anhydride (DDSA): 5.5 mL.

   Methyl nadic anhydride (MNA): 4.5 mL.

   2,4,6-Tri(dimethylaminoethyl)phenol (DMP-30): 0.4 mL.
2. Silicone capsule (silicone embedding board, e.g., Dosaka EM Co. Ltd.).

2.6 Poststaining for Ultrathin Section TEM

1. Two percent uranyl acetate in 70% ethanol.
2. Lead citrate: For a 50 mL stock solution, dissolve 1.33 g of lead nitrate [$Pb(NO_3)_2$] and 1.76 g of sodium citrate [$Na_3(C_6H_5O_7){\cdot}2H_2O$] in 30 mL of distilled water that has

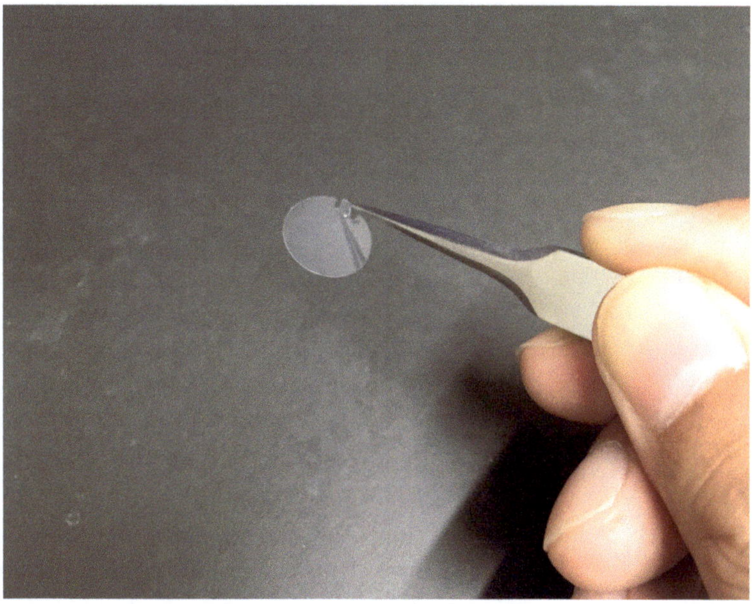

Fig. 1 A plastic disk with a tab for cell culture

been boiled and quickly chilled. Shake vigorously for 1 min, and continue intermittent shaking for 30 min. Add 8 mL of 1 N NaOH, at which point the solution should become clear. Add boiled and quickly chilled distilled water to obtain a final volume of 50 mL.

2.7 Freeze-Drying for SEM

1. Ethanol (50, 70, 90%) in distilled water.
2. Tertiary butanol.
3. Vacuum freeze dryer (e.g., VFD-30, Vacuum Device Inc.).
4. SEM specimen stubs (e.g., Ted Pella Inc.).
5. SEM carbon stickers (e.g., Ted Pella Inc.).
6. Silver paste (Fast-drying silver paint, e.g., Ted Pella Inc.).

2.8 Vacuum Evaporation Coating for SEM

1. Osmium coater (e.g., HPC-1SW, Vacuum Device Inc.).

2.9 Cell Culture Substrates for SEM

1. Thin-plastic cell-culture disks (Cell desk LF1, Sumitomo Bakelite Co. Ltd.) (Fig. 1).

3 Methods

General remarks: Throughout the process, take care not to let the specimens dry out. Handle them gently to prevent physical damage.

3.1 Sample Preparation for Ultrathin Section TEM

1. Grow EBOV-infected cells on a suitable cell-culture substrate (e.g., in 12-well plates).
2. Gently aspirate the culture medium, and add 1 mL of prechilled (4 °C) 2.5% glutaraldehyde in 0.1 M cacodylate buffer at the desired time point(s) post-infection (*see* **Note 1**).
3. Fix at 4 °C for 20 min.
4. Gently scrape off the fixed cells using a cell scraper, centrifuge the cell/fixative mixture at 4 °C at 13,000 × *g* for 10 min in a polypropylene reaction tube, and fix for an additional 30 min at 4 °C (*see* **Note 2**). If the sample is to be removed from a BSL4 facility, fix at 4 °C for additional time according to the prescribed institutional inactivation procedures before removal.
5. Remove the fixative, and replace with 1 mL of prechilled (4 °C) 0.1 M cacodylate buffer.
6. Transfer the pellet and buffer into a petri dish. Cut the pellet into ~1 mm cubes in a pool of the buffer (*see* **Note 3**).
7. Transfer the specimen cubes into cacodylate buffer in a polypropylene reaction tube (*see* **Note 4**). Replace the buffer three times, each for 10 min at 4 °C, with 1 mL of prechilled (4 °C) cacodylate buffer.
8. Remove the buffer and postfix the cubes with 100 μL of 2% osmium tetroxide for 1 h at 4 °C (*see* **Note 5**). Do not allow this fixation step to exceed 2 h.
9. Rinse three times, each for 10 min at 4 °C, in 1 mL of distilled water.
10. Dehydrate the specimen cubes successively in 1 mL of 50, 70, and 90% ethanol, each for 5 min at 4 °C, and then three times, each for 5 min at room temperature, in 1 mL of 99.5% or 100% ethanol. Ensure that the specimen cubes do not dry out.
11. Remove the excess ethanol (~800 μL) using a pipette, but do not allow the sample to become completely dry. Add 1 mL of propylene oxide, and replace it twice, holding the specimen in propylene oxide for 5 min each time at room temperature. Propylene oxide is highly volatile, so ensure that it does not evaporate, which may result in the specimen cubes drying out. While **steps 10** and **11** are being performed, prepare pure epoxy resin and a 1:1 mixture of propylene oxide/epoxy resin in a disposable 15 mL polypropylene tube (*see* **Note 6**).
12. Remove most of the propylene oxide using a pipette, and dispense 1 mL of the 1:1 mixture over the specimen cubes. Gently shake or rotate the specimen in a capped polypropylene reaction tube for 1.5 h at room temperature.
13. Uncap the polypropylene reaction tube to allow evaporation of the propylene oxide, and gently shake the sample for an additional 1.5 h at room temperature.

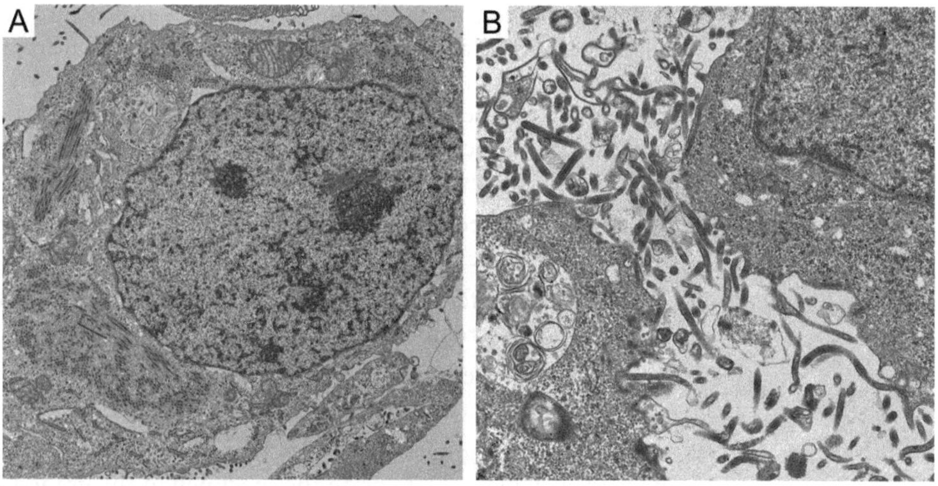

Fig. 2 TEM images of Ebola virus-infected Vero E6 cells at lower (*left panel*) and higher (*right panel*) magnification

14. Pour a small amount of pure epoxy resin (~200 μL) into each hole of a silicone capsule. Using forceps or a toothpick, gently transfer individual specimen cubes into each hole of the silicone capsule. Leave for at least 30 min at room temperature until the specimen cubes completely sink to the bottom.
15. Gently fill each hole with pure epoxy resin, and transfer the silicone capsule to a 60 °C oven for 2 days for complete polymerization of the resin. These resin blocks can then be subjected to ultrathin sectioning followed by heavy metal staining (*see* **Note 7**).
16. Prepare 50–100 nm thick ultrathin sections on grids (*see* **Note 8**). Dry them in a 50 °C oven for 30 min.
17. Place a drop of uranyl acetate on a piece of Parafilm for each grid (*see* **Note 9**). Transfer the grids onto the drops using clean forceps, with the section side down, and leave them for 5 min.
18. Fill three small petri dishes with distilled water. After the uranyl acetate staining, dip the grids in the first petri dish for 3 min. Repeat the process in the other two petri dishes to remove excess uranyl acetate.
19. Float the grids on a drop of lead citrate on Parafilm (*see* **Note 9**), with the section side down, and leave them for 5 min.
20. After the lead citrate staining, wash three times with distilled water as in *step 18*. Dry the grids completely in a 50 °C oven for more than 30 min before TEM observation (Fig. 2).

3.2 Sample Preparation for SEM

1. Grow EBOV-infected cells on thin plastic disks in 24-well plates [2]. Plastic disks with tabs are highly recommended for easy handling using forceps.

2. Aspirate the culture medium, gently add 1 mL of pre-chilled (4 °C) 2.5% glutaraldehyde in 0.1 M cacodylate buffer at the desired time point(s) post-infection, and fix at 4 °C for 1 h. If the sample will be removed from a BSL4 facility, fix at 4 °C for additional time according to the prescribed institutional inactivation procedures before removal.
3. Gently rinse three times, 10 min each at 4 °C, with 1 mL of 0.1 M cacodylate buffer.
4. Remove the buffer, and postfix with 200–300 μL of 1% buffered osmium tetroxide at 4 °C for 1 h (*see* **Note 10**).
5. Gently rinse three times, each for 10 min at 4 °C, with 1 mL of distilled water.
6. Prepare a 1 mL graded ethanol series of 50, 70, and 90% in a new 24-well plate. Dehydrate the specimen in the graded ethanol series (increasing concentration), with 5 min at room temperature for each step in the series. Ensure that the cells on the disks do not dry out.
7. Remove the ethanol, add 1 mL of tertiary butanol, and replace the tertiary butanol three times. Hold the specimen in the tertiary butanol for 15 min at 37 °C each time.
8. Pour tertiary butanol into the holders of a vacuum freeze dryer so that the EBOV-infected cells on the disks are soaked. Transfer the specimen disks in the tertiary butanol to the holders, and place the holders in a freezer until the tertiary butanol is frozen (*see* **Note 11**).
9. Quickly transfer the holders into a precooled chamber of the vacuum freeze dryer (e.g., VFD-30, Vacuum Device Inc.), and allow the tertiary butanol to dry completely (*see* **Note 12**).
10. Place and fix the specimen disks on SEM specimen stubs using SEM carbon stickers and silver paste. Leave at room temperature for at least 1 h to dry out the silver solvent.
11. Transfer the specimen stubs into the osmium-coater chamber (e.g., HPC-1SW, Vacuum Device Inc.), and coat the specimen surfaces with osmium. These samples are ready for SEM observation (Fig. 3).

4 Notes

1. Proper incubation time after virus infection should be determined for each experimental condition. In our experience, the time when cytopathic effects start to be visible is good for fixation to observe the virus particles budding from the cells and the nucleocapsids in the cytoplasm.

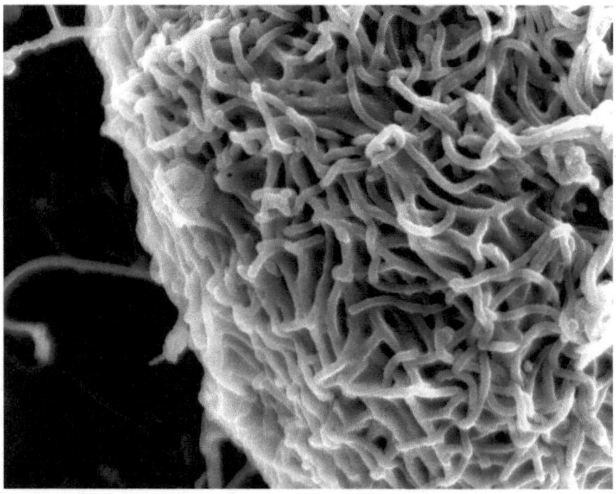

Fig. 3 A SEM image of Ebola virus particles budding from cell surface

2. After centrifugation, the pellet must be visible. The pellet size does not matter as long as you can handle it.
3. We use a razor blade.
4. We pick up the specimen cube within the buffer by surface tension between tips of forceps. Alternatively, you can use a 1 mL pipette with a wide orifice tip to convey the specimen cubes. Only a few cubes are enough to be processed to the next step.
5. The specimen cubes become black within 10 min.
6. More than ten blocks can be prepared from 20 mL of the epoxy resin. We make a few blocks for each sample.
7. Resin blocks can be stored at room temperature and be used for more than 10 years.
8. The thickness of ultrathin sections affects the quality of the images. More than 100 nm thick sections are not recommended for TEM with 80 kV.
9. Fifty microliter is enough for each staining drop.
10. Seal the lid of the 24-well plate with scotch tape. Otherwise, this step should be done in a fume hood.
11. Usually this takes less than 15 min.
12. While dependent on the amount of t-butanol you add, this usually takes more than 1 h.

Acknowledgments

This work is supported by JST PRESTO and by the Research Program on Emerging and Re-emerging Infectious Diseases from the Japan Agency for Medical Research and Development.

References

1. Geisbert TW, Jahrling PB (1995) Differentiation of filoviruses by electron microscopy. Virus Res 39:129–150
2. Noda T, Ebihara H, Muramoto Y et al (2006) Assembly and budding of Ebolavirus. PLoS Pathog 2:e99
3. Ellis DS, Simpson DIH, Francis DP et al (1978) Ultrastructure of Ebola virus particles in human liver. J Clin Pathol 31:201–208
4. Ellis DS, Stamford S, Lloyd G et al (1979) Ebola and Marburg viruses: I. Some ultrastructural differences between strains when grown in Vero cells. J Med Virol 4:201–211
5. Baskerville A, Fisher-Hoch SP, Neild GH et al (1985) Ultrastructural pathology of experimental Ebola hemorrhagic fever virus infection. J Pathol 147:199–209
6. Watanabe S, Watanabe T, Noda T et al (2005) Production of novel Ebola virus-like particles from cDNAs: an alternative to Ebola virus generation by reverse genetics. J Virol 78:999–1005
7. Neumann G, Ebihara H, Takada A et al (2005) Ebola virus VP40 late domains are not essential for viral replication in cell culture. J Virol 79:10300–10307
8. Halfmann O, Kim JH, Ebihara H et al (2008) Generation of biologically contained Ebola viruses. Proc Natl Acad Sci U S A 105: 1129–1133

Chapter 20

Validating the Inactivation Effectiveness of Chemicals on Ebola Virus

Elaine Haddock and Friederike Feldmann

Abstract

While viruses such as Ebola virus must be handled in high-containment laboratories, there remains the need to process virus-infected samples for downstream research testing. This processing often includes removal to lower containment areas and therefore requires assurance of complete viral inactivation within the sample before removal from high-containment. Here we describe methods for the removal of chemical reagents used in inactivation procedures, allowing for validation of the effectiveness of various inactivation protocols.

Key words High-containment, Inactivation, Validation, Dialysis, Detergent removal, Ebola virus

1 Introduction

Effective and reliable inactivation of samples containing infectious viruses, such as Ebola virus, is critical for the safe operation of high-containment laboratories, as well as for processing of potentially infectious field and clinical samples. In the past, such specimens have been inactivated based on historical experience rather than on well-documented and validated protocols [1–3]. There are multiple ways in which infectious material may be treated for downstream testing. These include, but are not limited to, irradiation, heat, chaotropic agents for RNA extraction, protein assay buffers containing sodium dodecyl sulfate (SDS) or other detergents, and fixatives such as formalin, paraformaldehyde and glutaraldehyde solutions. While the use of radiation and heat may pose little difficulty to downstream infectivity validation testing, chemical additives used in some inactivation processes may leave the sample toxic in downstream infectivity testing involving cell culture or animal model validation assays. One method employed has been dilution of inactivated samples such that the toxic reagents no longer negatively impact the infectivity assays [4], but this also allows for the potential complication of diluting out the infectious

Thomas Hoenen and Allison Groseth (eds.), *Ebolaviruses: Methods and Protocols*, Methods in Molecular Biology, vol. 1628, DOI 10.1007/978-1-4939-7116-9_20, © Springer Science+Business Media LLC 2017

particles to a level below detection limits. Centrifugation of samples to physically separate the virus pellet from liquid chemical agent through multiple washes has also been utilized [5, 6], but this might lead to physical damage of infectious particles rendering them potentially noninfectious and thus interfering with infectivity assays. A third option is the complete removal of the inactivating reagents using physical methods before employing validation testing, the procedure for which may vary greatly depending on the chemical additives involved. To us this seems to be the most reliable methodology for downstream infectivity testing.

Here we describe dialysis and detergent-removal protocols to successfully remove chaotropic agents, fixatives, and detergents from inactivated viral specimens, such as Ebola-infected materials, in a way that inactivation procedures may be confirmed in cell culture or animal validation assays. These methods may be utilized with a variety of other inactivating reagents or virus families.

2 Materials

2.1 Dialysis

1. Dialysis membrane/cassettes, 10 kDa molecular weight cutoff (*see* **Note 1**), as well as appropriate preparatory and equilibration buffers for dialysis products and/or appropriate needles/syringes to load and unload dialysis products.
2. Dialysis buffer: sterile Dulbecco's phosphate-buffered saline (DPBS) with calcium and magnesium: pH 7.2–7.6 (NaCl 8 g/L, KCl 0.2 g/L, Na_2HPO_4 1.15 g/L, KH_2PO_4 0.2 g/L, $CaCl_2{\cdot}2H_2O$ 0.1 g/L, $MgCl_2{\cdot}6H_2O$ 0.1 g/L).
3. Dialysis container(s), Nalgene, 2–4 L.
4. Sterile stir bar(s).
5. Stir plate at 4 °C.
6. Waste containers to store and decontaminate used dialysis buffer.
7. Micro-Chem Plus or other disinfectant for decontamination of dialysis buffer and membranes.

2.2 Detergent Removal

1. Detergent-removal columns (*see* **Note 2**).
2. Equilibration buffer: DPBS with calcium and magnesium, pH 7.2–7.6.
3. 15 mL conical tubes.
4. Swinging bucket centrifuge with adapters for 15 mL conical tubes.
5. Waste containers to store and decontaminate waste.
6. Micro-Chem Plus or other disinfectant for decontamination of flow through and columns.

3 Methods

For an overview of the procedure, *see* Fig. 1a.

3.1 Dialysis

1. Following appropriate contact times for inactivation of virus-infected samples (*see* **Note 3**) or control samples (*see* **Note 4**) with non-detergent containing inactivating reagents (*see* **Note 5**), transfer the full volume of each sample to a pre-wetted dialysis product.
2. Suspend dialysis products in DPBS in a container sized to allow all cassettes to move freely above a stir bar without impeding the bar.
3. Put the DPBS container on a stir plate at 4 °C and adjust stir speed such that fluid flows easily but without causing a vortex (*see* **Note 6**).
4. Dialyze samples with five total buffer changes (*see* **Note 7**), transferring the used dialysis buffer to a separate container and decontaminating before discarding after each change (*see* **Note 8**).
5. Remove dialyzed samples from the dialysis products and decontaminate and discard the dialysis membrane.
6. Use dialyzed samples, with all appropriate dialyzed controls, to infect cell culture or animal models used in your validation testing (*see* **Note 9**).

3.2 Detergent Removal

1. Following appropriate contact times for inactivation of virus-infected samples (*see* **Note 10**) or control samples with detergent-containing inactivating reagents, transfer the full volume of each test sample to a pre-equilibrated detergent removal column (*see* **Note 11**).
2. Change column to clean 15 mL conical tube from that used for equilibration.
3. Spin sample through column following manufacturer's recommended procedure and collect sample flow-through.
4. If necessary, flush column with additional DBPS and spin again (*see* **Note 12**), adding this flow-through to the original sample flow-through.
5. Use collected sample from columns, with all appropriate controls, to infect cell culture or animal models used in your validation testing (*see* **Note 13**).

3.3 Data Analyses

1. Depending upon your infectivity assay, results may be tabulated as residual titer, animal survival, number of cell culture flasks positive for cytopathic effect, etc. *See* Fig. 1b for two examples of inactivation results.

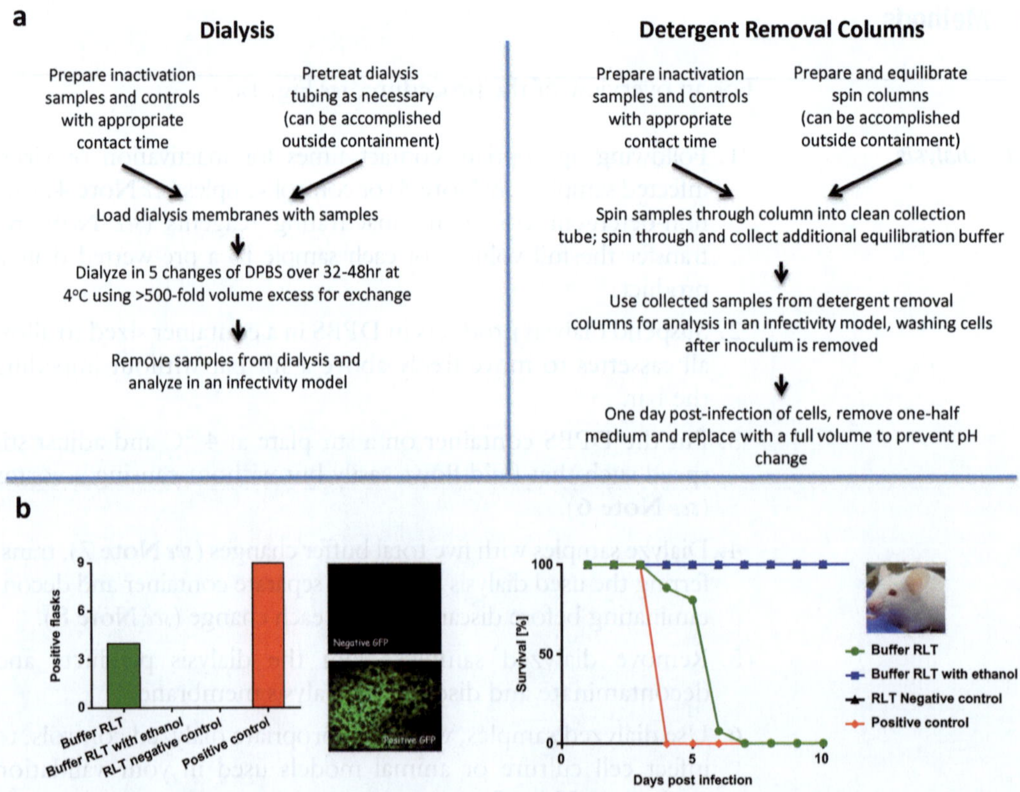

Fig. 1 (**a**) The basic steps of reagent removal by either dialysis or detergent removal columns. (**b**) Infectivity results from cell culture and animal infectivity assay [9]. Infectivity in cell culture was determined by the number of cell flasks positive for cytopathic effect and/or fluorescence after Buffer RLT (Qiagen) inactivation of wild-type EBOV expressing enhanced green fluorescent protein (EBOV-eGFP). Triplicate samples were dialyzed and each split to infect triplicate flasks, with appropriate positive and negative controls. Infectivity in an animal model was determined by survival of BALB/c mice after Buffer RLT inactivation of mouse-adapted EBOV (MA-EBOV). Triplicate samples were dialyzed and each split to infect five mice, with appropriate positive and negative controls

4 Notes

1. The choice of dialysis product is critical to the work in several ways. Standard dialysis tubing and clips may not only lead to leakage and sample loss, but may be difficult to handle in high-containment positive pressure suits. Dialysis cassettes such as the Thermo Fisher Slide-A-Lyzers are easy to use, with a large dialysis surface area and only require a 2 min equilibration in dialysis buffer; however, these are loaded with a needle/syringe combination that could be a safety concern in high-containment. Similar cassettes, the Thermo Fisher Slide-A-Lyzer G2 cassettes, allow for pipette-loading as an alternative

to needle loading. Another option is the Spectra/Por Float-A-Lyzer G2 tubes. These are dialysis tubing with prefitted end caps rather than clips, which allow the sample to be loaded by pipette with no needle puncture risk and a large surface area for dialysis. As with the Slide-A-Lyzers, these are available in a range of volume sizes. While the Float-A-Lyzer tubes are safe and easy to work with, they do require more pre-wetting per manufacturer's instructions: 10 min filled and submerged in 10% isopropanol or ethanol, followed by flushing with DI water and 15–20 min filled and submerged in DI water, flushing and pre-wetting in dialysis buffer. The pre-use handling of these tubes can be accomplished outside high containment to minimize containment work, with the tubes carried in already submerged in dialysis buffer.

2. Detergent-removal systems such as the G Biosciences DetergentOUT columns are highly efficient at removing unbound detergent from protein samples. G Biosciences provides an assortment of column sizes. We found the DetergentOUT system to be sufficient in our testing of SDS or Tween-20 and Triton X-100 treated specimens up to 1% total detergent concentration when using 1 mL inactivated sample with use of the GBS10-5000 columns, which handles volumes up to 1.25 mL.
3. Infected samples may be of a variety of sample types, including virus stock or supernatant, virus-infected cells, virus-infected tissue, etc. We suggest that virus stocks or virus-infected cells be harvested in bulk, titrated and properly stored for later use to ensure consistency in multiple validation studies. Virus-infected tissues may be treated (e.g., by formalin fixation) and then homogenized in DPBS before dialysis. Our procedure for this is to treat the tissue fully, and then to remove an internal portion no larger than 150 mg, and to transfer this to a 2 mL tube with 1 mL sterile DBPS and a 5 mm stainless steel bead. This is then homogenized in a Qiagen Tissue Lyser II at 30 Hz for 10 min. The homogenate is then dialyzed to remove residual chemical reagents.
4. Both positive and negative control samples should be tested for every validation procedure. Positive controls use equivalent viral samples with DPBS substituted for inactivating reagents. Negative controls use equivalent inactivating reagents with DPBS substituted for viral samples.
5. As discussed above, there is a variety of choice for dialysis membrane. We have successfully dialyzed Buffers AVL and RLT (Qiagen), TRIzol (Invitrogen), 10% formalin and 2% paraformaldehyde and glutaraldehyde with Slide-A-Lyzers and Float-A-Lyzers. We have determined that Slide-A-Lyzers are more durable when working with TRIzol, but that switching

the sample to a second Slide-A-Lyzer cassette or Float-A-Lyzer tube midway through the dialysis process does help to minimize dialysis membrane degradation by the caustic reagent.

6. Dialysis cassettes or tubes do not actually need to be moving within the container. The flow of liquid across them is sufficient for dialysis to proceed. Increasing the speed such that the cassettes spin may cause a vortex that pulls them down and increases the likelihood of failure of the dialysis tubing.
7. To ensure complete dialysis, we have used >500-fold exchange volumes with five buffer changes over 32–48 h. Additional buffer containers may be used to separate controls from test samples.
8. To ensure complete decontamination of DPBS, we added the used dialysis buffer directly to Micro-Chem Plus in a secondary container for a final concentration >5% disinfectant, with a room temperature contact time of no less than 1 h prior to disposal.
9. Your validation model may vary with your work. For Ebola virus validation we utilized VeroE6 cells at 80% confluency in 25 cm^2 flasks. Each sample, following dialysis, was brought to 3 mL in volume and split equally into three flasks, with a 1 h infection, followed by removal of the inoculum and addition of 6 mL DMEM with 2% FBS, L-glutamine (20 mM), Penicillin (500 U/mL) and Streptomycin (500 μg/mL). Mock infected cells were incubated as an infection control. Infection did not require a final wash step with the exception of TRIzol and glutaraldehyde samples, which were washed once with 6 mL DPBS before addition of medium. Validation studies have been published [5, 7, 8] with subsequent passage of virus to amplify low levels of virus and increase cytopathic effect (CPE). Fluorescent viruses, if available, may reduce this necessity.
10. Again, infected samples may be of a variety of sample types, including virus stock or supernatant, virus-infected cells, and virus-infected tissue. We have utilized DetergentOUT columns for virus stock, infected cells and tissue homogenates with SDS, Tween 20, and Triton X-100 to a total detergent concentration of 1%.
11. DetergentOUT columns can be prepared following the manufacturer's recommended instructions, and equilibrated with sterile DPBS. For the GBS10-5000 columns, the procedure is to centrifuge the column closed at 200 × *g* for 30 s to pellet the resin, then opening the column (top and bottom) and placing in a 15 mL collection tube for an additional 1 min spin at 200 × *g*. Flow-through is discarded, the column is loaded with 2 mL of equilibration buffer, centrifuged and flow-through discarded. Equilibration is repeated twice before transferring the column to a clean collection tube, loading the column with

sample and centrifuging 2 min at 200 × *g*. To minimize time in high-containment, columns may be capped after the last equilibration buffer is added, and carried to the high-containment laboratory for the final spin before sample addition.

12. A final spin of the columns with an additional 1–2 mL of DPBS should ensure that the viral particles are flushed into the flow-through tube with minimal loss of titer. This can be verified, if needed, based on the positive control.
13. Again, your validation model may vary with your work. For our Ebola virus validation, samples following detergent removal were brought to ~3 mL in volume and split equally over three flasks of 80% confluent VeroE6 cells, with a 1 h infection, followed by removal of inoculum and addition of 6 mL DMEM with 2% FBS, L-glutamine (20 mM), penicillin (500 U/mL), and streptomycin (500 μg/mL). Mock infected cells were incubated as an infection control. Infection required a final single wash step with 6 mL DPBS before addition of medium. In addition, we found that some portion of flasks underwent an extreme pH drop at around 2 days post-infection, which could be avoided by removing half the medium (3 mL) at day 1 post-infection and replacing it with 6 mL in all flasks.

Acknowledgment

This work was supported by the Intramural Research Program of the National Institute of Allergy and Infectious diseases (NIAID), National Institutes of Health (NIH).

References

1. Elliott LH, McCormick JB, Johnson KM (1982) Inactivation of Lassa, Marburg, and Ebola viruses by gamma irradiation. J Clin Microbiol 16(4):704–708
2. Lupton HW (1981) Inactivation of Ebola virus with 60Co irradiation. J Infect Dis 143(2):291
3. Mitchell SW, McCormick JB (1984) Physicochemical inactivation of Lassa, Ebola, and Marburg viruses and effect on clinical laboratory analyses. J Clin Microbiol 20(3):486–489
4. Blow JA, Dohm DJ, Negley DL, Mores CN (2004) Virus inactivation by nucleic acid extraction reagents. J Virol Methods 119(2):195–198. doi:10.1016/j.jviromet.2004.03.015
5. Smither SJ, Weller SA, Phelps A, Eastaugh L, Ngugi S, O'Brien LM, Steward J, Lonsdale SG, Lever MS (2015) Buffer AVL alone does not inactivate Ebola virus in a representative clinical sample type. J Clin Microbiol 53(10):3148–3154. doi:10.1128/JCM.01449-15
6. Moller L, Schunadel L, Nitsche A, Schwebke I, Hanisch M, Laue M (2015) Evaluation of virus inactivation by formaldehyde to enhance biosafety of diagnostic electron microscopy. Virus 7(2):666–679. doi:10.3390/v7020666
7. Kumar M, Mazur S, Ork BL, Postnikova E, Hensley LE, Jahrling PB, Johnson R, Holbrook MR (2015) Inactivation and safety testing of Middle East respiratory syndrome coronavirus. J Virol Methods 223:13–18. doi:10.1016/j.jviromet.2015.07.002
8. Chepurnov AA, Bakulina LF, Dadaeva AA, Ustinova EN, Chepurnova TS, Baker JR Jr (2003) Inactivation of Ebola virus with a surfactant nanoemulsion. Acta Trop 87(3):315–320
9. Haddock E, Feldmann F, Feldmann H (2016) Effective chemical inactivation of Ebola virus. Emerg Infect Dis 22(7):1292–1294. doi:10.3201/eid2207.160233

Chapter 21

Visualizing Ebolavirus Particles Using Single-Particle Interferometric Reflectance Imaging Sensor (SP-IRIS)

Erik P. Carter*, Elif Ç. Seymour*, Steven M. Scherr, George G. Daaboul, David S. Freedman, M. Selim Ünlü, and John H. Connor

Abstract

This chapter describes an approach for the label-free imaging and quantification of intact Ebola virus (EBOV) and EBOV viruslike particles (VLPs) using a light microscopy technique. In this technique, individual virus particles are captured onto a silicon chip that has been printed with spots of virus-specific capture antibodies. These captured virions are then detected using an optical approach called interference reflectance imaging. This approach allows for the detection of each virus particle that is captured on an antibody spot and can resolve the filamentous structure of EBOV VLPs without the need for electron microscopy. Capture of VLPs and virions can be done from a variety of sample types ranging from tissue culture medium to blood. The technique also allows automated quantitative analysis of the number of virions captured. This can be used to identify the virus concentration in an unknown sample. In addition, this technique offers the opportunity to easily image virions captured from native solutions without the need for additional labeling approaches while offering a means of assessing the range of particle sizes and morphologies in a quantitative manner.

Key words Biosensing, Interferometric imaging, Virus detection, Ebola, Viral hemorrhagic fevers, Label-free, High throughput

1 Introduction

Imaging of individual viruses has been performed using a variety of techniques such as fluorescence microscopy, electron microscopy, and surface plasmon resonance imaging. However, these techniques often require labeling, complicated sample preparation steps, and/or bulky and expensive optical setups. An alternative approach to virus detection, referred to as single-particle interferometric reflectance image sensing (SP-IRIS), has recently been developed. SP-IRIS is an approach that allows for the detection

*These authors contributed equally to this work.

Thomas Hoenen and Allison Groseth (eds.), *Ebolaviruses: Methods and Protocols*, Methods in Molecular Biology, vol. 1628,
DOI 10.1007/978-1-4939-7116-9_21, © Springer Science+Business Media LLC 2017

and analysis of nanoparticles on a silicon/silicon dioxide (Si/SiO_2) surface. Because assembled virus particles are themselves nanoparticles, there has been significant interest in determining how well SP-IRIS functions to image and count unlabeled virus particles [1, 2]. Initial experiments have been promising, as it has recently been used for the sensitive and multiplexed detection of whole virus particles from serum and blood samples without the need for labeling or significant sample preparation [1].

Interferometric reflectance imaging sensor (IRIS) was first introduced as a label-free biosensor that can detect biomass accumulation on a Si/SiO_2 substrate in a high-throughput microarray format [3, 4]. IRIS quantifies biomass accumulation on the sensor surface by measuring the optical path differences that cause a shift in spectral reflectivity. Optical path differences are then used to calculate the film thicknesses. SP-IRIS is a modified version of an earlier reflectance imaging technology with changes to the optical setup and the substrate [5]. These changes enable the detection of nanoscale particles that are captured near the silicon surface. Recently, SP-IRIS has been adapted into a microfluidic platform to allow real-time visualization of virus particles as they bind to the antibody capture probes on the sensor surface [6]. SP-IRIS has great potential for nanoparticle and protein detection and quantification, and we anticipate many other applications for this technology going forward.

There are several basic steps in the SP-IRIS imaging approach when starting from a pre-spotted chip (for additional information on chip spotting, *see* [7]). The experimental flow following pre-scanning is illustrated in Fig. 1 and involves sample incubation, washing, and final scanning. For highest sensitivity in this assay, it is preferable to pre-scan the chip that will be used to detect virus particles prior to use (non-virus particles that are detected on the antibody spot surface prior to incubation with sample will be subtracted from the post-scan data). Following scanning, the output data from this experiment includes the baseline (pre-) and endpoint (post-) particle counts illustrating the differential binding between different antibodies on the microarray, the average density of particles on

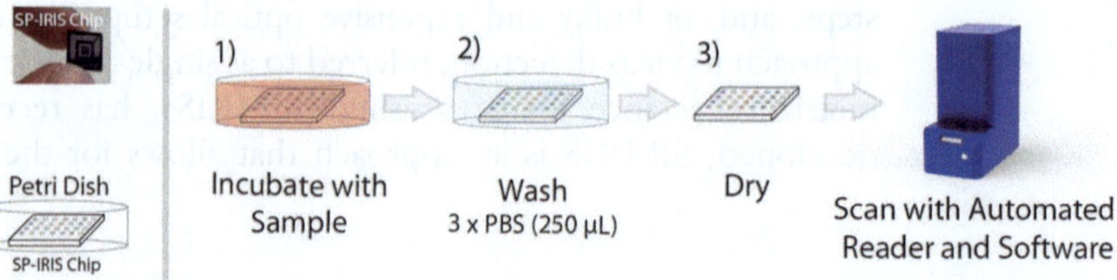

Fig. 1 SP-IRIS procedure flow. (*1*) Antibody-spotted SP-IRIS patterned chip is incubated with sample of interest. (*2*) Chip is washed and (*3*) dried. Finally, the chip is scanned. For procedural details, refer to Subheadings 3.2–3.4

each spot, and an image of each spot. The count data can be used to determine the average number of particles bound to antibody spots of a particular specificity and, among other applications, allows for the determination of virus titer or number of intact virus particles per unit volume by comparing particle counts against a standard curve (described in Subheading 3.6). The image data is useful in determining the range of virus sizes and morphologies. This information is also particularly useful for quality confirming preparations of inactivated virus, or virus-like particles (VLPs), where there is no biological readout (i.e., plaque assay) to confirm the presence of particles. These and other features make SP-IRIS a versatile research tool for the characterization of both infectious and noninfectious virus particles.

Here, we describe the methods for carrying out specific detection and quantification of intact viruses from serum samples in both high- and low-volume sample binding assays using incubation in a multi-well plate. We illustrate a simple virus particle detection assay using antibody microarray chips that are compatible with largely automated scanning and data analysis. We also describe a method by which the titer of an unknown virus sample can be determined by using a standard curve derived from serial tenfold dilutions of a stock of virus with a known concentration. The methods described here highlight the usefulness of this technology for studying and detecting virus particles in a specific manner through direct visualization as well as the generation of relative size and shape information, allowing for the quantitative evaluation of their morphologies.

2 Materials

2.1 Sample Incubation, Washing, and Scanning

1. SP-IRIS A1000 Reader (nanoView Diagnostics Inc., cat. no. NVDX10).
2. Antibody-spotted SP-IRIS patterned silicon chip (nanoView Diagnostics Inc., cat. no. NVDX 10-F6).
3. 24-well plates.
4. 12-well plates.
5. 10-cm petri dishes.
6. 1× phosphate-buffered saline (PBS) (e.g., Thermo Fisher Scientific, cat. no. 10010023) (*see* **Note 1**).
7. Forceps (*see* **Note 2**).
8. Deionized, reverse osmosis water.

2.2 Standard Curve Preparation for Virus Titration

1. Assay diluent: 1× PBS with 7.5 mM EDTA and 7.5 mM EGTA.
2. Heat-inactivated fetal bovine serum (e.g., Sigma-Aldrich, cat. no. F2442, maintain at 56 °C for 30 min in a heated water bath or purchase inactivated).

3. VSV ΔG-EBOV GP (or other pseudotyped virus of known titer).
4. 1.5 mL microcentrifuge tubes.

3 Methods

3.1 Scanning the IRIS Chip Using SP-IRIS Acquisition 2015.05 (V20)

1. Open the SP-IRIS Acquisition 2015.05 (V20) program. The program window is shown in Fig. 2.
2. Place the chip onto the SP-IRIS stage, and make sure that the corner of the chip, indicated by the white star in Fig. 3, aligns with the back left corner of the SP-IRIS stage. This is important to ensure proper alignment of the chip before scanning. Sensor should be connected to the in-house vacuum via a port on the back of the machine. Turn on the vacuum supply. This will keep the chip in place on the stage.

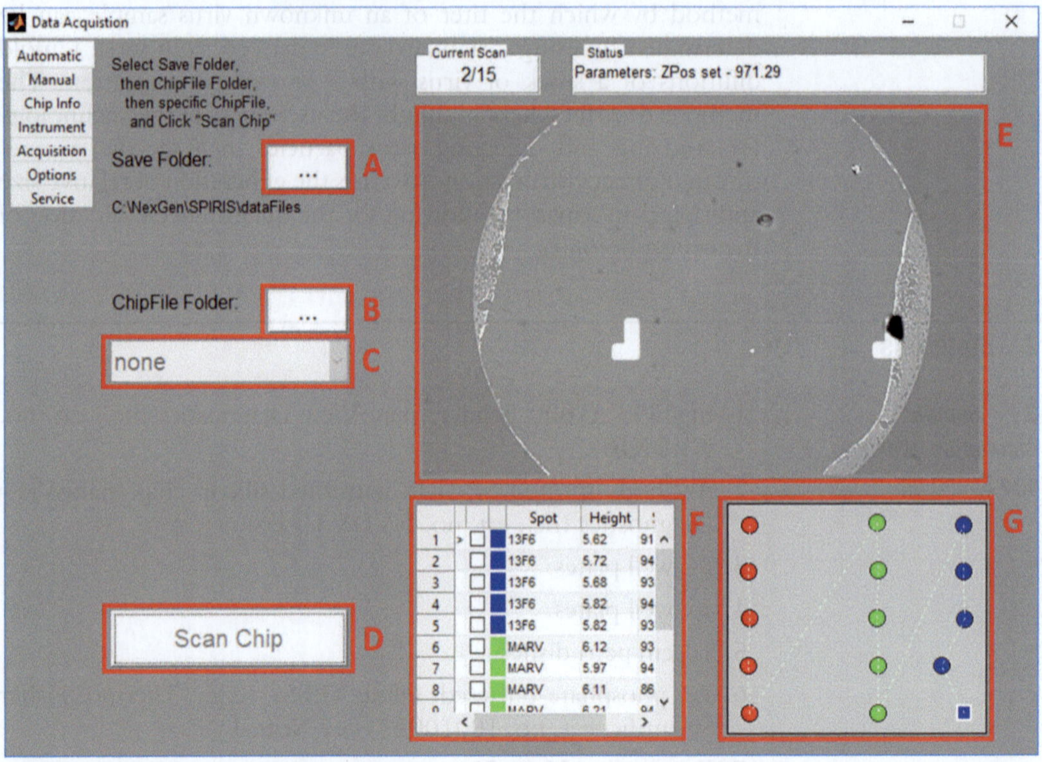

Fig. 2 SP-IRIS Acquisition 2015.05 (V20) window. To collect data, (**a**) the "Save Folder" button allows the user to select the folder in which they would like chip data to be saved. (**b**) Select a chip file folder. (**c**) Select desired chip file. (**d**) The "Scan Chip" button starts the acquisition of data from the SP-IRIS chip. (**e**) Scan viewing window allows the user to view the antibody spots as they are being scanned. (**f**) In the spot legend, the color of spot indicates the antigen specificity of the antibody used and can be labeled by antibody name, antibody target, or some other name of choice. (**g**) Spot array indicates the scan status of each antibody spot (*see* **Note 3**)

Fig. 3 Image of an SP-IRIS chip. The area within the *red shaded box* is where the capture antibody array is situated and where virus particles will bind. It should be facing up during incubation. Avoid touching this area or allowing it to dry between washes until the final ultrapure water wash. When loading the chip into the SP-IRIS instrument, the corner indicated by the *white star* should line up with the back left corner of the stage (*see* **Note 4**)

3. Select the folder into which you would like chip data to be saved (Fig. 2a).
4. To begin scanning a chip, select the folder (Fig. 2b) that contains the chip files of interest, and then select the chip file (Fig. 2c) that corresponds with the chip loaded onto the machine. The chip file specifies the x- and y-coordinates of each antibody spot on the chip, allowing the SP-IRIS Acquisition software to automatically find the spots during the array scanning process. Chip files are created during the spot height analysis of the antibody array with a custom-written MATLAB graphical user interface.
5. Select "Scan Chip" (Fig. 2d). Pre-scans of every chip should be made prior to incubation with virus. Pre-scan particle counts will be subtracted from the post-scan counts to calculate the net number of particles bound during the sample incubation step.

3.2 Virus Binding Assay in a Multi-well Plate Using SP-IRIS

1. After pre-scanning, place IRIS sensor chip facing up in the well of a 24-well plate. Add 1 mL of virus-containing sample to the well with the chip.
2. Place the plate on an orbital shaker. Incubate the chip(s) for 1 h on the shaker at room temperature. To allow adequate mixing, shaking speed should be approximately 150 rpm.
3. After the incubation ends, prepare a 12-well plate with three wells containing 1× PBS (1 mL/well) and one well with 0.1× PBS for each chip.

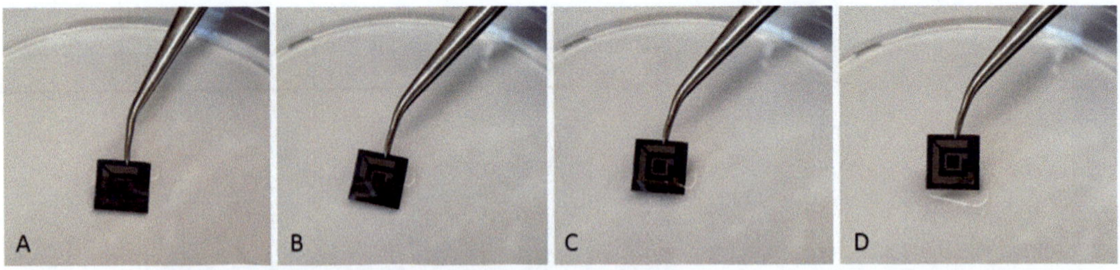

Fig. 4 Proper removal of chip from ultrapure water. (**a**) Grasp the edge of chip from fully submerged position, and lift the edge out of water at a 45° angle. (**b**) Be particularly careful to avoid residual water droplets as virus capture area reaches water front. (**c** and **d**). Proceed to remove the chip fully from the water, avoiding residual water on the top surface of the chip (*see* **Note 5**)

4. Quickly transfer the chips to a well containing 1× PBS (make sure to do so quickly in order to minimize the time that the chip is out of buffer). Place on the shaker for 3 min. Repeat this wash three more times by moving the chip through the wash solution wells (two more times in 1× PBS and once in 0.1× PBS).
5. Finally, place the chip in ultrapure water in a 10 cm petri dish, and pipet water over the center of the chip several times. Pull the chip out slowly so that the water "slides" off the surface without leaving residual droplets (*see* Fig. 4). This works well by grabbing the chip from one edge and carefully lifting the chip out of the water at a 45° angle. The top face of the chip should be completely dry once it is removed from the water (*see* **Note 4**).
6. Place the chip face up on a Kimwipe to dry. Blot any liquid left on the edges of the chip with the Kimwipe, taking care not to touch the antibody array at the center of the chip.
7. Proceed to scan the chip with SP-IRIS, as described in Subheading 3.1.
8. Analysis is performed automatically as the chip is being scanned by using custom software that processes each spot image (both pre- and post-incubation images). The data for both virus quantification (net virus count and net virus density detected) and virus imaging can be viewed in the NanoAnalysis V0.92 (Beta) software, and a copy of the data will be automatically exported in an excel format for further analysis.

3.3 Low-Volume Virus Binding Assay in a Multi-well Plate Using SP-IRIS

1. Pre-scan the chip following Subheading 3.1, **steps 1–5**.
2. Virus samples may be diluted in 1× PBS if desired. The final volume of the dilution(s) should be at least 30 μL.
3. Place the IRIS sensor chip facing up in the well of a 24-well plate. Add 30 μL of the sample over the central square of the chip as highlighted by the red box in Fig. 3 and as shown in Fig. 5.

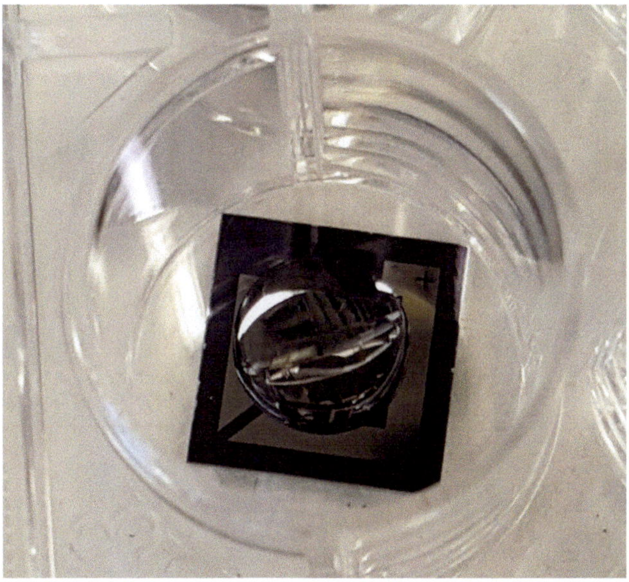

Fig. 5 Incubation of virus samples with the chip during the low-volume virus binding assay. This method allows for virus analysis using only 30 μL of virus-containing sample. As shown, the 30 μL droplet of sample is positioned directly over the virus capture area of chip

4. Leave plate on the benchtop; do not shake to avoid movement of sample from the center of chip. Incubate for 1 h at room temperature.
5. Continue the protocol as described in Subheading 3.2, **steps 3–7**.

3.4 Data Analysis for Virus Quantification Using a Dry Chip Virus Binding Assay

1. Select a "Baseline" data file, usually the pre-scan, and a "Diff." (Differential) data file, usually the post-virus incubation scan (Fig. 6a, b).
2. A "chip file" folder must also be selected which contains the chip alignment data for the chip(s) you wish to analyze (Fig. 6c). A particular chip file may be selected from the drop-down menu to view the data from a desired chip (Fig. 6d).
3. Particle filter parameters can be adjusted as needed. For imaging pseudotyped VSV or filoviruses with dry chip imaging, we recommend 3% minimum contrast, 20% maximum contrast, and 0.75 minimum correlation.
4. Virus particle counts for each antibody will be automatically displayed in the form of a bar graph and can be expressed as the average number of particles on a given spot or the average number of particles/mm^2 (particle density) (Fig. 6g).
5. To further analyze the results, open the data file under which the chips were run (i.e., the file selected in the Browse option

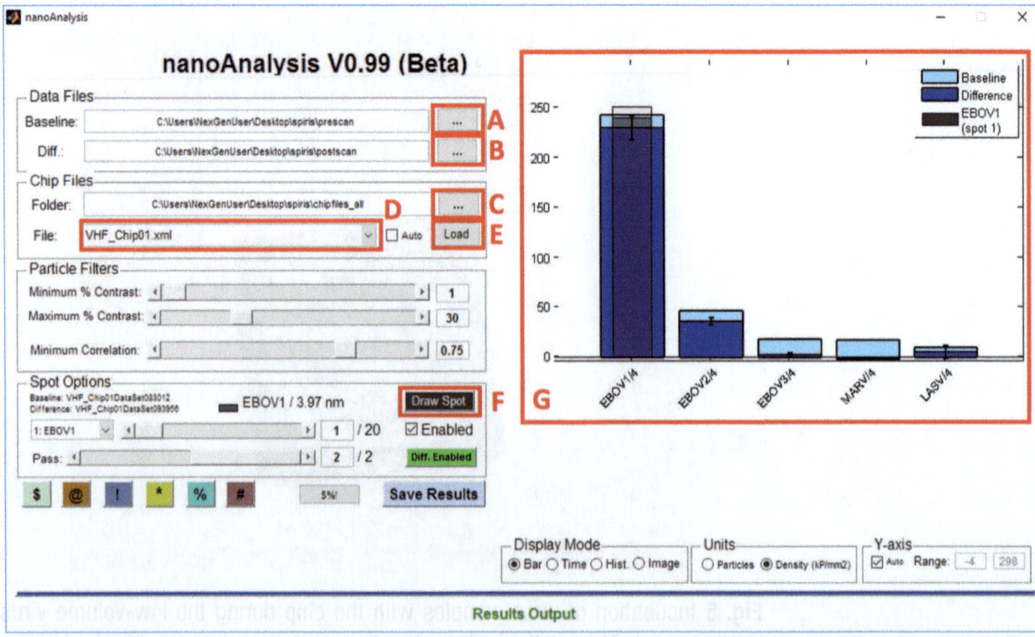

Fig. 6 NanoAnalysis V0.92 (Beta) main page. (**a**) Select a Baseline data file. (**b**) Select a Differential data file. (**c**) Select desired chip file folder. (**d**) Select individual chip file for viewing and analysis. (**e**) Select "Load" to view the selected chip data. (**f**) Select "Draw Spot" for antibody spot visualization. (**g**) Data shown is from a chip incubated with VSV ΔG-EBOV GP (Ebola GP-pseudotyped vesicular stomatitis virus). The *dark blue bars* indicate the average particle densities over five antibodies: the first three are Ebola GP-specific monoclonal antibodies, the fourth is a Marburg GP-specific monoclonal antibody, and the last is a Lassa GP-specific monoclonal antibody. The *lighter blue stacks* on each bar indicate the pre-scan particle counts that were subtracted to give the *dark blue* post-scan bars. The *gray-shaded bar* represents the particle density of an individual spot replicate that was selected at the time (*see* **Note 6**)

found in Fig. 2d), and select the Excel document associated with the chip of interest (chips are typically numbered, and their corresponding Excel files will share the same number). Here, analyzed results, like virus particle counts, are listed for each antibody spot on the chip selected.

3.5 Data Analysis for Virus Imaging Using a Dry Chip Virus Binding Assay

1. After selecting the desired Baseline, Diff., and chip files as described in Subheading 3.4, **steps 1** and **2**, individual spots on a chip can be selected using the Index bar under "Spot Options."
2. Clicking "Draw Spot" (Fig. 6e) will bring up a separate window in which an image of the selected spot can be viewed and interacted with (Fig. 7). Blue circles indicate the particles that were included in the particle count. Precisely which particles are included in the count is determined by the min. and max. contrast settings and the min. correlation (*see* Subheading 3.4, **step 3,** for recommended parameters). The size of the circles correlates to the signal intensity (i.e., particle size). Individual particle morphologies (Fig. 8) can be seen using the zoom function, available through the task bar.

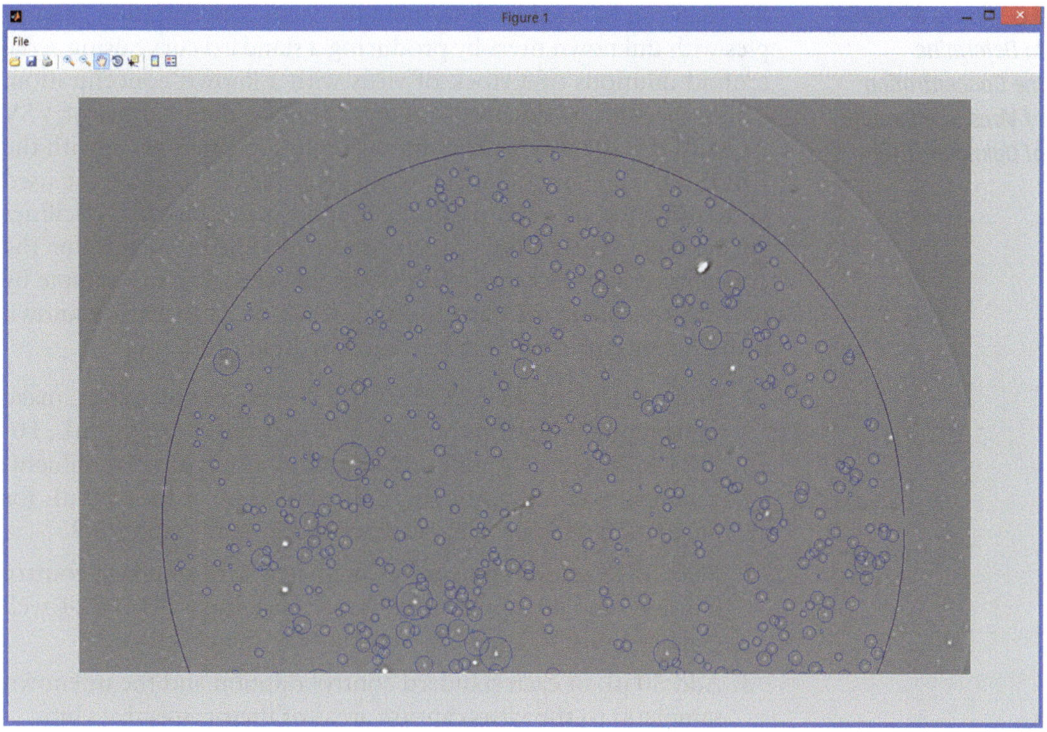

Fig. 7 Visualizing an antibody spot. After selection of "Draw Spot," a secondary window will display the selected antibody spot. The image may be zoomed in/out, panned, cropped, and interacted with in other ways using the tools available through the task bar

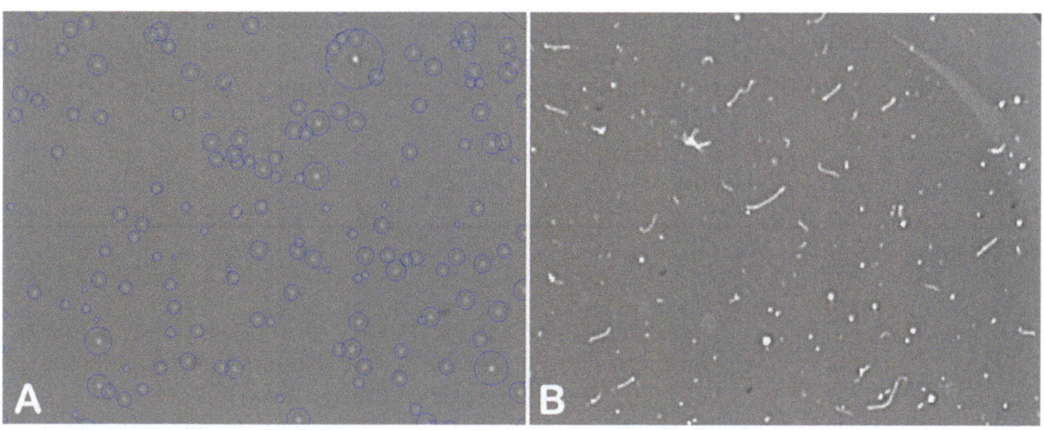

Fig. 8 Visualizing viruses and VLPs. (**a**) Chip incubated with VSV ΔG-EBOV GP. Particles appear as white-rounded dots with little variation in size and signal intensity. (**b**) Chip incubated with Ebola virus-like particles (VLPs) showing filamentous particle morphology. It is typical to *see* VLPs of varying lengths from ~2 to 8 microns and even punctate morphologies

3.6 Using SP-IRIS to Determine the Concentration of Virus in a Sample of Unknown Titer

SP-IRIS can be used to determine the virus concentration of samples with unknown titers by producing a standard curve using serial tenfold dilutions of a stock of virus with a known concentration. This procedure is described using a blinded-titer sample of VSV ΔG-EBOV GP (VSV lacking its native glycoprotein gene, with the EBOV GP gene in its place), but, in principle, it could also be used with authentic filovirus particles, if appropriate biosafety facilities are available. This technique can also be modified to determine the concentration of filovirus VLPs (virus-like particles) in a sample by producing a standard curve using a stock of VLPs with a known protein concentration as determined by Bradford assay.

1. From a stock of VSV ΔG-EBOV GP with a known titer, make serial tenfold dilutions (e.g., 10^7 PFU/mL, 10^6 PFU/mL, 10^5 PFU/mL, etc.), by diluting the virus stock in assay diluent. The final volume of the dilutions should be at least 30 μL for incubation with the chip as described in Subheading 3.3.
2. Place IRIS sensor chips to be incubated with standard control dilutions and unknown samples face up in the wells of a 24-well plate. Be sure to label the wells and/or the cover.
3. Add 30 μL of each standard control dilution and the unknown sample(s) to the virus capture areas of their respective chips.
4. Incubate chips on benchtop for 1 h without shaking.
5. Continue with washing and scanning chips as outlined in Subheading 3.2, **steps 3–7**.
6. From each chip's data file, the average net particle counts from the antibody spots specific for Ebola GP can be gathered and used to create a standard curve based on the titers of the stock dilutions (Fig. 9a). The average net particle counts from the chip incubated with a sample of unknown virus titer can be plotted on the curve to extrapolate the original sample's titer (Fig. 9b).

4 Notes

1. PBS can be made from powder in-house instead of bought commercially but should be sterile-filtered and supplemented with 0.05% sodium azide to prevent bacterial growth.
2. We recommend wide-tipped plastic forceps. They provide better grip and are less prone to damaging the edges of the chips.
3. Definitions of spot shapes: square, to be scanned; "X," currently scanning; circle, successfully scanned; and black star, failed to read.

A.

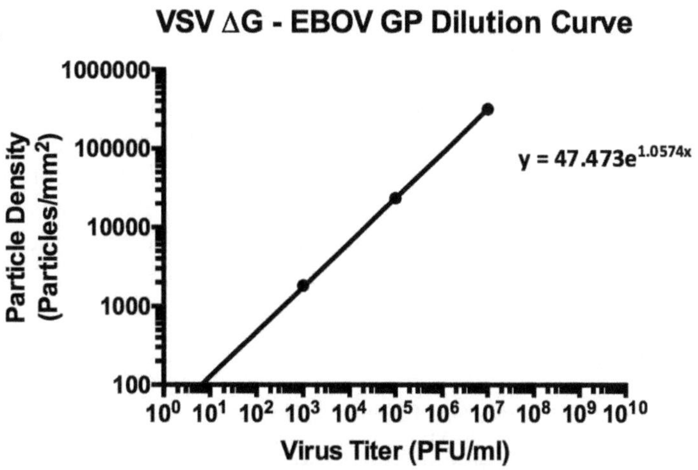

B.

	Virus Particle Density (particles/mm²)	SP-IRIS - Determined Titer (PFU/mL)	Plaque Assay Determined Titer (PFU/mL)
Blinded-Titer Sample	38592.7	1.8×10^6	2.5×10^6
Time to Results		~1.5 hrs.	~30 hrs.

Fig. 9 Determining virus titer using IRIS. (**a**) A standard curve was made using serial tenfold dilutions of a 1×10^8 PFU/mL stock of VSV ΔG-EBOV GP (dilutions were made in assay diluent). (**b**) A blinded-titer sample of VSV ΔG-EBOV GP was prepared, and a chip was incubated as described in Subheading 3.6. The resulting particle density over the anti-Ebola GP antibody spots was used to calculate the titer of the sample against the standard curve

4. The chip shown in Fig. 2 has some damage around the edges due to handling with metal forceps. This will *not* affect the chip's performance. However, be careful not to damage or scratch the virus capture area.
5. This step is important because if small droplets of water are left on the virus capture area to air-dry, it can leave behind salt deposits and other debris, which compromises the quality and accuracy of the imaging.
6. Viewing individual spot particle densities (light blue bar) can be useful to detect samples that vary significantly from the rest of the replicates. The light blue bar can be removed by unselecting "Enabled" below the "Draw Spot" button.

Acknowledgments

We would like to thank Kristen N. Peters, Ph.D.; Michelle Toomey Olsen, Ph.D.; John Ruedas, Ph.D.; and others in the Connor Lab for their help in preparing this manuscript. This work was supported by R01AI1096159 to JHC.

Erik P. Carter and Steven M. Scherr contributed equally to this manuscript.

References

1. Daaboul GG, Lopez CA, Chinnala J, Goldberg BB, Connor JH, Unlu MS (2014) Digital sensing and sizing of vesicular stomatitis virus pseudotypes in complex media: a model for Ebola and Marburg detection. ACS Nano 8(6):6047–6055. doi:10.1021/nn501312q
2. Yurt A, Daaboul GG, Connor JH, Goldberg BB, Unlu MS (2012) Single nanoparticle detectors for biological applications. Nanoscale 4(3):715–726. doi:10.1039/c2nr11562j
3. Lopez CA, Daaboul GG, Vedula RS, Ozkumur E, Bergstein DA, Geisbert TW, Fawcett HE, Goldberg BB, Connor JH, Unlu MS (2011) Label-free multiplexed virus detection using spectral reflectance imaging. Biosens Bioelectron 26(8):3432–3437. doi:10.1016/j.bios.2011.01.019. S0956-5663(11)00043-1 [pii]
4. Ozkumur E, Needham JW, Bergstein DA, Gonzalez R, Cabodi M, Gershoni JM, Goldberg BB, Unlu MS (2008) Label-free and dynamic detection of biomolecular interactions for high-throughput microarray applications. Proc Natl Acad Sci U S A 105(23):7988–7992. doi:10.1073/pnas.0711421105. 0711421105 [pii]
5. Reddington AP, Trueb JT, Freedman DS, Tuysuzoglu A, Daaboul GG, Lopez CA, Karl WC, Connor JH, Fawcett H, Unlu MS (2013) An interferometric reflectance imaging sensor for point of care viral diagnostics. IEEE Trans Biomed Eng 60(12):3276–3283. doi:10.1109/TBME.2013.2272666
6. Scherr SM, Daaboul GG, Trueb JT, Sevenler D, Fawcett H, Goldberg B, Connor JH, Unlu MS (2016) Real-time capture and visualization of individual viruses in complex media. ACS Nano 10(2):2827–2833
7. Ahn S, Freedman DS, Zhang X, Unlu MS (2013) High-throughput label-free detection of DNA hybridization and mismatch discrimination using interferometric reflectance imaging sensor. Methods Mol Biol 1039:181–200. doi:10.1007/978-1-62703-535-4_16

Part IV

Studying Infectious Ebolaviruses In Vivo

Chapter 22

Assessing Antiviral Countermeasures Using Mouse Models of Ebolavirus Infection

Andrea Kroeker, Bryan D. Griffin, Xiangguo Qiu, and Gary Kobinger

Abstract

Mouse models of Ebola virus (EBOV) have demonstrated their utility as important tools for screening the efficacy of candidate therapeutics and vaccines. In this chapter we explain the various mouse models that utilize either wild-type or mouse-adapted EBOV variants.

Key words Ebola virus, Animal model, Mouse, Monoclonal antibodies, Vaccine, Antiviral, Recombinant vesicular stomatitis virus (rVSV), Drug therapy

1 Introduction

Currently there are no licensed vaccines or therapeutics against Ebola virus disease (EVD) [1]; however, several candidate vaccines [2, 3] and post-exposure therapeutic agents [4–7] have demonstrated clinical potential against Ebola virus (EBOV) infections. All EVD-specific countermeasures that have advanced to testing in animals of higher phylogenetic order and eventually clinical trials were initially shown to protect susceptible mice against a lethal EBOV challenge. Current candidate vaccines that have shown efficacy in mice include a DNA vaccine expressing the EBOV envelope glycoprotein (GP) or nucleocapsid protein (NP) [8], an EBOV-GP immune complex subunit vaccine [9], an EBOV GP nanoparticle vaccine [10], an EBOV-GP and matrix protein (VP40) virus-like particle (eVLP) vaccine [11], vaccines based on human adenovirus (Ad) serotype 5 (Ad5) [12], serotype 26 (Ad-26) [13], chimpanzee adenovirus pan7 (AdC7) [14], and EBOV-GP recombinant vesicular stomatitis virus vaccines (VSVΔG/ZEBOVGP) [15]. Investigational therapeutic approaches initially evaluated in mice include antisense phosphorodiamidate morpholino oligomers (PMO), AVI-7537 [16], the antiviral agent favipiravir (T-705; 6-fluoro-3-hydroxy-2-pyrazinecarboxamide) [17],

Thomas Hoenen and Allison Groseth (eds.), *Ebolaviruses: Methods and Protocols*, Methods in Molecular Biology, vol. 1628, DOI 10.1007/978-1-4939-7116-9_22,

passive transfer of convalescent plasma [18], individual components of the monoclonal antibody (MAb) cocktail ZMapp [19], and the cross-reactive MAb 6D6 [20].

The advantages and limitations of the various animal models of EVD have been described previously [21–24]. Briefly, despite their limitations mouse models of EVD have several advantages compared to higher order EVD models, including the relatively low cost for purchasing and housing mice, minimal genetic variations, and the availability of molecular biology reagents and research tools, including genetically defined knockout strains, for mechanistic studies and characterization of host immune responses [25]. Recently, mouse models have been reported that also recapitulate hallmarks of EVD observed in infected nonhuman primates (NHPs), including hemorrhage [26, 27].

To address the limited susceptibility of many mouse models to infection with wild-type EBOV, a mouse-adapted EBOV (MA-EBOV) was generated by serially passaging EBOV in progressively older suckling mice. This produced a virus isolate that results in death when administered to immunocompetent adult BALB/c, C57BL/6, and CD-1 mice by the i.p. route, but not the s.c. route [28]. Exposure of IFNα/βR$^{-/-}$ and STAT1$^{-/-}$ mice to MA-EBOV by the i.p. route also results in uniform lethality [29, 30]. Additionally, SCID, STAT1, and interferon (IFN)-γ knockout mice are susceptible to lethal infection with MA-EBOV via the aerosol route; however BALB/c mice are resistant [31]. Recently, infection of the Collaborative Cross (CC) CC-RIX mice with MA-EBOV resulted in the identification of several susceptible mouse lines that display pathogenic phenotypes that closely mirror the signs of EVD observed in infected NHPs [26]. The mutations in the MA-EBOV genome that are responsible for its high pathogenicity in immunocompetent mice have been identified, and can be introduced into an EBOV reverse genetics system to result in the rescue of EBOV variants that are lethal to mice [32].

This chapter describes the treatment and challenge of mice as a basis for establishing the efficacy of various countermeasures in the mouse model.

2 Materials

2.1 Model Selection and Animal Care

1. Mice (*Mus musculus*), species selected based on project goals (*see* **Note 1**).
2. Materials for marking animals, e.g., multicolored nontoxic markers, an ear hole punch or an electronic transponder chip (Bio-Medic Data Systems).

2.2 Vaccination or Administration of Therapeutic (See Note 2)

1. 26 gauge needles.
2. 1 mL syringes.
3. Vaccine or therapeutic for testing.
4. Appropriate solvent and/or diluent.

2.3 Anesthesia/Euthanasia

1. Isoflurane.
2. Isoflurane chamber.

2.4 Determination of Virus Stock LD_{50} and EBOV Challenge

1. 26 gauge needles.
2. 1 mL syringes with manual stepper with adjustable dispensing volume.
3. Virus stock of wild-type or mouse-adapted EBOV diluted in Dulbecco's Modified Eagle's Medium (DMEM).
4. Clinical scoring chart (*see* **Note 3**).
5. Animal weighing scale and plastic bucket.
6. Implantable programmable temperature transponder and handheld reader.
7. Portable anesthetic machine with oxygen, isoflurane vaporizer, chamber, and rodent breathing circuit.
8. Isoflurane.

2.5 Evaluation of Disease Progression

1. Scale (minimal precision 0.1 g).
2. Plastic beaker.
3. Forceps.
4. Clinical scoring chart.

2.6 Blood, Swab, and Organ Collection

1. 26 gauge needles.
2. 1 mL syringes.
3. Serum-separation tubes.
4. Whole blood tubes spray coated with K_2EDTA.
5. 70% ethanol.
6. 2 sets of sterile forceps.
7. Sterile scissors.
8. 2 mL cryotubes.
9. RNA stabilization reagent such as RNAlater (Qiagen).

2.7 Hematological Analysis and Serum Biochemistry

1. VetScan HM5 (Abaxis).
2. VetScan HM5 Reagent Pack (Abaxis #770-9000).
3. VetScan VS2 (Abaxis).
4. Comprehensive Diagnostic Profile discs (Abaxis #500-0038).

3 Methods

3.1 Model Selection and Animal Care

1. Select a mouse strain for use in experiments based on the desired features of the infection model as outlined in Tables 1 and 2. BALB/c, C57BL/6 and CD1 mice remain the most commonly used strains.
2. Select an animal age consistent with the project goals. It is common to begin experiments when mice are 6–8 weeks of age when

Table 1
Lethality of mouse-adapted ebolavirus in some commonly used mouse strains

		Lethality				
Mouse strain	**Virus**	**I.C.**	**I.P.**	**I.N.**	**S.C.**	**References**
Adult immunocompetent mice	MA-Ebola virus		Full		None	[30]
Adult IFNα/βR$^{-/-}$ mice	MA-Ebola virus		Full		Full	[30]
Adult STAT1$^{-/-}$ mice	MA-Ebola virus		Full	Full	Full	[29, 30]

Table 2
Lethality of wild-type ebolavirus in some commonly used mouse strains

		Lethality				
Mouse strain	**Virus**	**I.C.**	**I.P.**	**I.N.**	**S.C.**	**References**
Adult immunocompetent mice	Ebola virus	None	None	None	None	[30]
Adult IFNα/βR$^{-/-}$ mice	Ebola virus, variant Mayinga (1976)		Full	Full	Full –	[30, 33–35]
	Ebola virus, variant Kikwit (1995)		None, partial		–	[30, 33–35]
	Sudan virus		Full or partial	None	–	[30, 34, 35]
	Bundibugyo virus		None			[30, 34]
	Taï forest virus		None			[30, 34]
	Reston virus		None	Weight loss	–	[30, 34, 35]
Adult STAT1$^{-/-}$ mice	Ebola virus		Full	Full	–	[29–31]
	Sudan virus		Full	Full		[29, 31]
	Reston virus		Mild disease	Mild disease		[29, 36]
Adult SCID mice	Ebola virus		Full	Full	–	[30, 34, 35]
Adult humanized (hu-BLT) mice	Ebola virus, variant Mayinga (1976) and Makona (2014)		Full			[37]

their immune systems are strong and fully developed. Using a common age also allows data generated from different labs to be compared to one another more easily. However, studies that seek to investigate immune responses at very young or old ages should consider altering this timeframe.

3. Select sex of study animals. The majority of ebolavirus mouse studies utilize only female mice due to ease of handling and housing. The few mouse studies that have compared the different sexes have not found any evidence of differences in the course of disease, vaccination or treatment between them [38, 39]. However, when testing novel vaccines or drug therapies it would be prudent to evaluate whether they have the potential to impact hormonal or sex-related pathways and processes and to consider testing in sex-matched groups of both male and female mice.
4. Animals should be housed in an approved facility that is supervised by a veterinarian and equipped with an isolated air system, 12 h light and dark cycles and environmental enrichment, with food and water given ad libitum.
5. Animals should be allowed to acclimate to the animal facility for a period of at least 1 week before the onset of experiments.
6. Prior to the onset of experiments, each mouse in a given cage should be marked, e.g., color-coded with a nontoxic marker, an ear hole punch, or electronic chip.

3.2 Vaccination or Administration of Therapeutic

1. Anesthetize mice lightly (*see* Subheading 3.3) and administer the vaccine. As volumes should be considered 300 μL for i.p. and s.c., and 50 μL for i.n. per nostril and i.m. per site (*see* **Notes 4** and **5**).
2. Dose: The dose will need to be determined empirically for each drug and will depend on its efficacy and route of administration. If possible, it is highly recommended to collect pharmacokinetic information prior to animal experiments such as the EC_{50}, toxicity and half-life of the compound. The treatment window can be extended in either direction from the EC_{50} to determine the optimal treatment regimens.
3. Timing: The dosing regimens can vary considerably for drug compounds and should follow the rationale of the study being performed as well as what is known about the metabolism and mechanism of the drug. Many drugs are initially given once every 24–48 h (*see* **Note 6**).
4. Route of administration: The route of administration can substantially impact the absorption, effect and dose of a drug. Many mouse studies initially screen drugs via the i.p. route, but the most efficacious route should be determined empirically.

3.3 Anesthesia/Euthanasia

1. Anesthesia. Place mice into the isoflurane chamber and given isoflurane (5%) until activity ceases and breathing has slowed (but not stopped). Using smooth ended forceps, the tips of the toes can be pinched to confirm unconsciousness.
2. Euthanasia. Place mice into the isoflurane chamber and give an overdose of isoflurane (5%) until signs of respiration are not observed for more than 2 min, followed by cervical dislocation (*see* **Note 7**).

3.4 Determination of Virus Stock LD_{50} and EBOV Challenge

1. For determination of virus stock LD_{50}, prepare a tenfold serial dilution series of an aliquot of the virus stock to be administered to mice. Monitor daily for 28 days for survival, weight, and clinical signs. For challenge, dilute virus for a challenge dose of 1000× LD_{50} of wild-type or MA-EBOV, a dose of virus that results in uniform lethality.
2. Mice are anesthetized (*see* Subheading 3.3) and administered up to 300 μL of virus via i.p. injection. For determination of LD_{50}, a power calculation using $\alpha = 0.05$ and $\beta = 0.8$ indicated that ten mice should be used for each dilution in order to achieve statistically significant survival data (*see* **Notes 8** and **9**).
3. The LD50 can then be calculated using the Reed–Muench method [40].

3.5 Evaluation of Disease Progression

1. Survival: Monitor mice for survival for 28 days after infection.
2. Clinical scoring: Each day post challenge, monitor mice visually for signs of disease in their cage using the clinical assessment chart (*see* **Note 3**).
3. Weight: Tare the scale with an empty plastic bucket. Grip each mouse by their tail using mouse handling forceps, and transfer it to the plastic bucket and proceed to weigh the mice. Weigh mice before the experiment to establish a baseline weight. During experiments, weigh mice daily for the duration of the study. Mice that lose more than 25% of their initial body weight should be euthanized (regardless of other clinical scores, or following the guidelines of animal care committee as it may change depending on different institutions).

3.6 Blood, Swab, and Organ Collection

1. Collection and storage of whole blood: At sample collection points during the experiment collect 50–150 μL blood. This is done by gently restraining the mouse and extending a leg, shaving away the fur from a small area, applying Vaseline to it, and then pricking a large vein with a needle. For repeated bleeds at short intervals, the sampling volume must be reduced to allow sufficient recovery time between sampling points. Following euthanasia (*see* Subheading 3.3) collect the entire

blood volume (up to 1 mL per mouse) through cardiac puncture or retro-orbital bleed (*see* **Note 10**). Collect the blood in 2 mL tubes and store at −20 to −80 °C. All work is performed in a BSC.

2. Collection and storage of serum: Aliquot 100–500 μL whole blood into serum-separation collection tubes and spin at 1000 ×*g* for 10 min at 4 °C to separate serum. Aliquot serum into a fresh, sterile 0.5 mL tube and store at −80 °C. All work is performed in a BSC.
3. Internal organ collection: After euthanization, surface-decontaminate the mouse by submerging it briefly in 70% ethanol, and harvest the desired internal organs inside a BSC. Place the organs in a tube containing the appropriate solution for downstream processing (*see* **Note 11**).

3.7 Hematological Analysis and Serum Biochemistry

1. For hematological analysis, complete blood counts can be performed with the VetScan HM5 (Abaxis Veterinary Diagnostics) per manufacturer's instruction.
2. For serum biochemistry, add 100 μL of serum to a VetScan VS2 disc (one disc per sample), and then insert disc into the Abaxis VetScan VS2 system and read as per manufacturer's instruction.

4 Notes

1. Mice from the various treatment groups should be matched for age and sex.
2. These items are listed for the most common injectable routes for drug administration routes. If other routes are desired such as oral gavage, additional or alternative materials may be required.
3. Clinical assessment charts are created differently by each institution and should be adapted to each animal model. Typical parameters that are monitored include coat condition, posture, weight loss, mobility (slow motion, paralysis), and response to stimuli (delayed, unresponsive). Determination of the end-point should be determined in consultation with an Animal Care Committee and Veterinarian following the guidelines of the Canadian Council on Animal Care or a similar animal ethics body. For example scoring: 0—no symptoms; 1—ruffled fur, slowing activity, loss of body condition; 2—labored breathing, hunched posture, bleeding; 3—death.
4. For new vectors, the vaccine schedule (single-dose versus prime-boost and timing) should be determined empirically as the efficacy of the vaccine will depend on the mouse strain,

route of administration, the vector used, and the vaccination schedule. The goal is usually to challenge mice with virus after the vaccine has induced the maximum adaptive immune response, e.g., by measuring T-cell activation (cell-mediated immunity) and IgG levels (humoral immunity) which typically peak at 2–4 weeks.

5. Replication competent vaccines should be administered inside a class II biosafety cabinet.
6. Note that some drugs require administration prior to infection. For example, this is particularly true if a drug must deplete a host factor, which may necessitate extended pretreatment for several days to weeks. In contrast, some drugs can be administered at 24 h to 30 min prior to infection. Other drugs are effective when given post infection. It is advisable to start at 24 h post infection and to use those results to adjust the timeframe in subsequent experiments.
7. The Canadian Council on Animal Care dictates that mice should be euthanized at the earliest endpoint that is compatible with the objectives of the study.
8. Lethality can vary by isolate and mouse strain.
9. The LD_{50} for MA-EBOV in adult BALB/c mice challenged via the i.p. route is approximately 10 focus-forming units (FFU) [32].
10. Blood quantities required for different downstream assays include 50 μL blood for $TCID_{50}$, 140 μL blood for viral RNA extraction, 150 μL blood for total IgG (50 μL serum), 150 μL blood for neutralizing IgG (50 μL serum), and 250 μL blood for biochemistry (100 μL serum).
11. For detection of infectious virus, place tissue in 2 mL cryotubes and snap-freeze with dry ice–methanol and store at −80 °C. For RNA extraction place ~30 mg of tissue in 600 μL of RNAlater in a 2 mL tube. Fully submerge the tissue in RNAlater and store the samples at 4 °C at least overnight and then transfer to −80 °C for storage up to 1 month.

References

1. Wong G, Qiu X (2015) Development of experimental and early investigational drugs for the treatment of Ebola virus infections. Expert Opin Investig Drugs 24(8):999–1011. doi:10.1517/13543784.2015.1052403
2. Richardson JS, Dekker JD, Croyle MA, Kobinger GP (2010) Recent advances in Ebolavirus vaccine development. Hum Vaccin 6(6):439–449
3. Bradfute SB, Dye JJM, Bavari S (2014) Filovirius vaccines. Hum Vaccin 7(6):701–711. doi:10.4161/hv.7.6.15398
4. Mendoza EJ, Qiu X, Kobinger GP (2016) Progression of Ebola therapeutics during the 2014-2015 outbreak. Trends Mol Med 22(2):164–173. doi:10.1016/j.molmed.2015.12.005
5. Wong G, Qiu X, Olinger GG, Kobinger GP (2014) Post-exposure therapy of filovirus

infections. Trends Microbiol 22(8):456–463. doi:10.1016/j.tim.2014.04.002

6. Zeitlin L, Whaley KJ, Olinger GG, Jacobs M, Gopal R, Qiu X, Kobinger GP (2016) Antibody therapeutics for Ebola virus disease. Curr Opin Virol 17:45–49. doi:10.1016/j.coviro.2016.01.006
7. Picazo E, Giordanetto F (2015) Small molecule inhibitors of ebola virus infection. Drug Discov Today 20(2):277–286. doi:10.1016/j.drudis.2014.12.010
8. Vanderzanden L, Bray M, Fuller D, Roberts T, Custer D, Spik K, Jahrling P, Huggins J, Schmaljohn A, Schmaljohn C (1998) DNA vaccines expressing either the GP or NP genes of Ebola virus protect mice from lethal challenge. Virology 246(1):134–144. doi:10.1006/viro.1998.9176
9. Phoolcharoen W, Dye JM, Kilbourne J, Piensook K, Pratt WD, Arntzen CJ, Chen Q, Mason HS, Herbst-Kralovetz MM (2011) A nonreplicating subunit vaccine protects mice against lethal Ebola virus challenge. Proc Natl Acad Sci U S A 108(51):20695–20700. doi:10.1073/pnas.1117715108
10. Bengtsson KL, Song H, Stertman L, Liu Y, Flyer DC, Massare MJ, Xu RH, Zhou B, Lu H, Kwilas SA, Hahn TJ, Kpamegan E, Hooper J, Carrion R Jr, Glenn G, Smith G (2016) Matrix-M adjuvant enhances antibody, cellular and protective immune responses of a Zaire Ebola/Makona virus glycoprotein (GP) nanoparticle vaccine in mice. Vaccine 34(16):1927–1935. doi:10.1016/j.vaccine.2016.02.033
11. Warfield KL, Bosio CM, Welcher BC, Deal EM, Mohamadzadeh M, Schmaljohn A, Aman MJ, Bavari S (2003) Ebola virus-like particles protect from lethal Ebola virus infection. Proc Natl Acad Sci U S A 100(26):15889–15894. doi:10.1073/pnas.2237038100
12. Patel A, Zhang Y, Croyle M, Tran K, Gray M, Strong J, Feldmann H, Wilson JM, Kobinger GP (2007) Mucosal delivery of adenovirus-based vaccine protects against Ebola virus infection in mice. J Infect Dis 196(Suppl 2):S413–S420. doi:10.1086/520603
13. Zahn R, Gillisen G, Roos A, Koning M, van der Helm E, Spek D, Weijtens M, Grazia Pau M, Radosevic K, Weverling GJ, Custers J, Vellinga J, Schuitemaker H, Goudsmit J, Rodriguez A (2012) Ad35 and ad26 vaccine vectors induce potent and cross-reactive antibody and T-cell responses to multiple filovirus species. PLoS One 7(12):e44115. doi:10.1371/journal.pone.0044115
14. Kobinger GP, Feldmann H, Zhi Y, Schumer G, Gao G, Feldmann F, Jones S, Wilson JM (2006) Chimpanzee adenovirus vaccine protects against Zaire Ebola virus. Virology 346(2):394–401. doi:10.1016/j.virol.2005.10.042
15. Jones SM, Stroher U, Fernando L, Qiu X, Alimonti J, Melito P, Bray M, Klenk HD, Feldmann H (2007) Assessment of a vesicular stomatitis virus-based vaccine by use of the mouse model of Ebola virus hemorrhagic fever. J Infect Dis 196(Suppl 2):S404–S412. doi:10.1086/520591
16. Iversen PL, Warren TK, Wells JB, Garza NL, Mourich DV, Welch LS, Panchal RG, Bavari S (2012) Discovery and early development of AVI-7537 and AVI-7288 for the treatment of Ebola virus and Marburg virus infections. Virus 4(11):2806–2830. doi:10.3390/v4112806
17. Oestereich L, Ludtke A, Wurr S, Rieger T, Munoz-Fontela C, Gunther S (2014) Successful treatment of advanced Ebola virus infection with T-705 (favipiravir) in a small animal model. Antiviral Res 105:17–21. doi:10.1016/j.antiviral.2014.02.014
18. Gupta M, Mahanty S, Bray M, Ahmed R, Rollin PE (2001) Passive transfer of antibodies protects immunocompetent and imunodeficient mice against lethal Ebola virus infection without complete inhibition of viral replication. J Virol 75(10):4649–4654. doi:10.1128/JVI.75.10.4649-4654.2001
19. Qiu X, Fernando L, Melito PL, Audet J, Feldmann H, Kobinger G, Alimonti JB, Jones SM (2012) Ebola GP-specific monoclonal antibodies protect mice and guinea pigs from lethal Ebola virus infection. PLoS Negl Trop Dis 6(3):e1575. doi:10.1371/journal.pntd.0001575
20. Furuyama W, Marzi A, Nanbo A, Haddock E, Maruyama J, Miyamoto H, Igarashi M, Yoshida R, Noyori O, Feldmann H, Takada A (2016) Discovery of an antibody for pan-ebolavirus therapy. Sci Rep 6:20514. doi:10.1038/srep20514
21. Bente D, Gren J, Strong JE, Feldmann H (2009) Disease modeling for Ebola and Marburg viruses. Dis Model Mech 2(1–2):12–17. doi:10.1242/dmm.000471
22. Bradfute SB, Warfield KL, Bray M (2012) Mouse models for filovirus infections. Virus 4(9):1477–1508. doi:10.3390/v4091477
23. Nakayama E, Saijo M (2013) Animal models for Ebola and Marburg virus infections. Front Microbiol 4:267. doi:10.3389/fmicb.2013.00267
24. Shurtleff AC, Bavari S (2015) Animal models for ebolavirus countermeasures discovery: what defines a useful model? Expert Opin Drug Discovery 10(7):685–702. doi:10.1517/17460441.2015.1035252

25. Eppig JT, Motenko H, Richardson JE, Richards-Smith B, Smith CL (2015) The International Mouse Strain Resource (IMSR): cataloging worldwide mouse and ES cell line resources. Mamm Genome 26(9–10):448–455. doi:10.1007/s00335-015-9600-0
26. Rasmussen AL, Okumura A, Ferris MT, Green R, Feldmann F, Kelly SM, Scott DP, Safronetz D, Haddock E, LaCasse R, Thomas MJ, Sova P, Carter VS, Weiss JM, Miller DR, Shaw GD, Korth MJ, Heise MT, Baric RS, de Villena FP, Feldmann H, Katze MG (2014) Host genetic diversity enables Ebola hemorrhagic fever pathogenesis and resistance. Science 346(6212):987–991. doi:10.1126/science.1259595
27. Bird BH, Spengler JR, Chakrabarti AK, Khristova ML, Sealy TK, Coleman-McCray JD, Martin BE, Dodd KA, Goldsmith CS, Sanders J, Zaki SR, Nichol ST, Spiropoulou CF (2015) Humanized mouse model of Ebola virus disease mimics the immune responses in human disease. J Infect Dis. doi:10.1093/infdis/jiv538
28. Bray M, Davis K, Geisbert T, Schmaljohn C, Huggins J (1998) A mouse model for evaluation of prophylaxis and therapy of Ebola hemorrhagic fever. J Infect Dis 178(3):651–661
29. Raymond J, Bradfute S, Bray M (2011) Filovirus infection of STAT-1 knockout mice. J Infect Dis 204(Suppl 3):S986–S990. doi:10.1093/infdis/jir335
30. Bray M (2001) The role of the type I interferon response in the resistance of mice to filovirus infection. J Gen Virol 82(Pt 6):1365–1373
31. Zumbrun EE, Abdeltawab NF, Bloomfield HA, Chance TB, Nichols DK, Harrison PE, Kotb M, Nalca A (2012) Development of a murine model for aerosolized ebolavirus infection using a panel of recombinant inbred mice. Virus 4(12):3468–3493. doi:10.3390/v4123468
32. Ebihara H, Takada A, Kobasa D, Jones S, Neumann G, Theriault S, Bray M, Feldmann H, Kawaoka Y (2006) Molecular determinants of Ebola virus virulence in mice. PLoS Pathog 2(7):e73. doi:10.1371/journal.ppat.0020073
33. Bray M (1998) A mouse model for evaluation of prophylaxis and therapy of Ebola hemorrhagic fever. J Infect Dis 178(3):651–661
34. Brannan JM, Froude JW, Prugar LI, Bakken RR, Zak SE, Daye SP, Wilhelmsen CE, Dye JM (2015) Interferon alpha/beta receptor-deficient mice as a model for Ebola virus disease. J Infect Dis 212(Suppl 2):S282–S294. doi:10.1093/infdis/jiv215
35. Lever MS, Piercy TJ, Steward JA, Eastaugh L, Smither SJ, Taylor C, Salguero FJ, Phillpotts RJ (2012) Lethality and pathogenesis of airborne infection with filoviruses in A129 alpha/beta −/− interferon receptor-deficient mice. J Med Microbiol 61(Pt 1):8–15. doi:10.1099/jmm.0.036210-0
36. de Wit E, Munster VJ, Metwally SA, Feldmann H (2011) Assessment of rodents as animal models for Reston ebolavirus. J Infect Dis 204(Suppl 3):S968–S972. doi:10.1093/infdis/jir330
37. Prescott J, Feldmann H (2016) Humanized mice: a neoteric animal disease model for Ebola virus? J Infect Dis 213(5):691–693
38. Chen G, Koellhoffer JF, Zak SE, Frei JC, Liu N, Long H, Ye W, Nagar K, Pan G, Chandran K, Dye JM, Sidhu SS, Lai JR (2014) Synthetic antibodies with a human framework that protect mice from lethal Sudan Ebolavirus challenge. ACS Chem Biol 9(10):2263–2273
39. Frei JC, Nyakatura EK, Zak SE, Bakken RR, Chandran K, Dye JM, Lai JR (2016) Bispecific antibody affords complete post-exposure protection of mice from both Ebola (Zaire) and Sudan viruses. Sci Rep 6:19193
40. Reed LJM, Muench H (1938) A simple method of estimating fifty percent endpoints. Am J Hyg 27:493–497

Chapter 23

Evaluation of Ebola Virus Countermeasures in Guinea Pigs

Andrea Marzi

Abstract

Ebola virus (EBOV) pathology in humans remains incompletely understood; therefore, a number of rodent and nonhuman primate (NHP) models have been established to study the disease caused by this virus. While the macaque model most accurately recapitulates human disease, rodent models, which display only certain aspects of human disease but are more cost-effective, are widely used for initial screens during EBOV countermeasure development. In particular, mice and guinea pigs were among the first species used for the efficacy testing of EBOV vaccines and therapeutics. While mice have low predictive value, guinea pigs have proven to be a more reliable predictor for the evaluation of countermeasures in NHPs. In addition, guinea pigs are larger in size compared to mice, allowing for more frequent collection of blood samples at larger volumes. However, guinea pigs have the disadvantage that there is only a limited pool of immunological tools available to characterize host responses to vaccination, treatment and infection. In this chapter, the efficacy testing of an EBOV vaccine and a therapeutic in the guinea pig model are described.

Key words Ebola virus, Guinea pigs, In vivo experiment, Animal model, Countermeasures, Vaccines, Therapeutics

1 Introduction

Despite decades of research into the development of vaccines and treatment options against Ebola virus (EBOV), there is still no approved countermeasure available. In order to understand this disease and facilitate countermeasure development, several animal models for EBOV have been developed, ranging from rodents to nonhuman primates (NHPs). Macaques are considered the gold standard model in EBOV research since wildtype (wt) EBOV isolates readily cause disease in NHPs that is very similar to that observed in humans (reviewed in [1]). However, working with NHPs has its disadvantages, including the very high costs associated with these studies, the ethical considerations, and the infrastructural requirements. As a result only a few facilities worldwide work with NHPs in a biosafety level 4 (BSL4) laboratory. Accordingly, rodent models, such as the mouse, hamster and guinea pig, have been established to overcome the disadvantages

Thomas Hoenen and Allison Groseth (eds.), *Ebolaviruses: Methods and Protocols*, Methods in Molecular Biology, vol. 1628,
DOI 10.1007/978-1-4939-7116-9_23,

presented by NHP work. Unlike NHPs, rodents are relatively cheap and easy to handle, and guinea pigs have the added advantage of being large enough to permit sequential sampling. Notably, however, wtEBOV is incapable of causing severe disease in rodents, with guinea pigs developing only a transient febrile illness and antibody response after infection with wtEBOV [2, 3]. To address this issue, wtEBOV Mayinga 1976 was serially passaged in guinea pigs until a uniformly lethal, guinea pig-adapted EBOV (GPA-EBOV) was obtained [4, 5]. During the adaptation process, the GPA-EBOV acquired nucleotide substitutions resulting in amino acid changes in the NP, GP, VP24, and L genes (reviewed in [6]).

Guinea pigs are routinely used in testing countermeasures for many viral hemorrhagic fever pathogens, including filoviruses (reviewed in [1]). In our hands, infection of Hartley guinea pigs with 10 plaque-forming-units (PFU) of GPA-EBOV [4], which equals 1000 LD_{50}, causes lethal disease within 7–9 days [3]. In this chapter, I describe the procedures for efficacy testing of a vaccine or therapeutic against EBOV using guinea pigs in a BSL4 laboratory.

2 Materials

1. Hartley guinea pigs (*Cavia porcellus*).
2. Vaccine or therapeutic for efficacy testing.
3. Guinea pig-adapted (GPA) EBOV.
4. Dulbecco's Modified Eagle's Medium (DMEM) for inoculum preparation.
5. Isoflurane.
6. 1 mL tuberculin syringe with 25 gauge × 5/8″ needle for infection and blood collection.
7. Micro-Chem Plus (National Chemical Laboratories) or another disinfectant approved for use in the BSL4 laboratory.
8. 1.3 mL collection tube with 1.6 mg EDTA/mL blood for blood collection (Sarstedt).
9. 1.1 mL Z-Gel collection tube for serum collection (Sarstedt).
10. BD Vacutainer Eclipse Blood collection needles, 21 gauge × 1 ¼″.
11. BD Vacutainer one use holders.
12. BD Vacutainer blood collection tubes, 10 mL (serum or EDTA).
13. Backdraft table.
14. Absorbent bench pads.
15. Carcass bags.

3 Methods

The following procedures must be performed in the animal space of a BSL4 laboratory, and they require a high level of expertise in handling animals and sharps in this environment. The procedures described here were approved by the Institutional Biosafety Committee to be carried out at the BSL4 laboratory at the Integrated Research Facility, Division of Intramural Research, National Institute for Allergy and Infectious Diseases, National Institutes of Health, Hamilton, Montana, USA. Several parameters, including the disinfectant used and protocols for disposal of carcasses may vary between BSL4 facilities.

3.1 Infection with GPA-EBOV

1. Acclimate the guinea pigs to the BSL4 environment for 5–7 days before infection with GPA-EBOV (*see* **Note 1**). Provide food and water *ad libitum*.
2. On the day of infection, obtain a baseline weight for each of the animal prior to infection (*see* **Note 2**).
3. Perform the infection on a backdraft table in the animal procedure space of the BSL4 laboratory. Cover the table section where the animal will be infected with an absorbent pad. Place the anesthesia chamber on the table and connect it to an anesthetic vaporizer such that inhalational isoflurane is administered at low doses (≤5%) (*see* **Note 3**).
4. Place one guinea pig at a time in the anesthesia chamber and observe until breathing is slow and deep, indicating sufficient anesthesia of the guinea pig.
5. Even though the animal is anesthetized, two people are required for the procedure, one person handling the syringe/needle, another person properly restraining the animal. Place the anesthetized guinea pig on its back, and while the first person restrains the upper body of the animal with long forceps, the second person straightens and holds the animal's hind feet together and administers 1 mL GPA-EBOV (*see* **Note 4**) diluted in DMEM intraperitoneally (IP) using a 25 gauge × 5/8″ needle into two sites in the lower abdomen (0.5 mL each site) (*see* **Note 5**).
6. Following injection, the second person immediately disinfects the syringe and needle with 5% Micro-Chem Plus before disposal into a sharps collection container (*see* **Note 6**). The first person lays the animal on its side in the cage and monitors breathing until the guinea pig is awake.
7. After infection, the guinea pigs are monitored and weighed daily for signs of disease. Around day 7 post-infection, the animals will develop endstage signs of disease, including dehydration

and weight loss greater than 20%. During this critical phase of the study, the animals are monitored twice daily.

8. Once euthanasia criteria are reached (*see* **Note 7**), animals are humanely euthanized, however, given the rapid disease progression animals will also sometimes be found dead. For euthanasia, the animal is first deeply anesthetized with inhalational isoflurane and subsequently euthanized by exsanguination (i.e., severing the jugular blood vessels with scissors) and placed in a carcass bag (*see* **Note 8**). Carcasses are stored at −80°C until removal from the BSL4 laboratory (via autoclave) and proper disposal.
9. After GPA-EBOV infection (usually 4–6 weeks), surviving animals are deeply anesthetized using inhalational isoflurane and a large volume (greater than 10 mL) terminal blood sample is collected via cardiac puncture using vacutainer tubes. To accomplish this, assemble the 21 gauge × 1 ¼ ″ vacutainer needle with the tube holder, but leave the needle sheath on. Place the anesthetized guinea pig on its back and continue anesthesia using a nose cone. Place a blood collection tube loosely in the tube holder, unsheathe the needle and insert it directly under the sternum of the animal at a 30–45° upward angle. When the needle is 0.5–1 cm inserted, push the vacutainer tube into place in the holder. Push the needle further into the guinea pig until the needle punctures the heart and the tube begins to fill with blood. Remove the full vacutainer tube from the holder before withdrawing the needle from the animal. Dispose of the needle/holder immediately in a sharps container. Place blood collection tubes on a tube roller (EDTA) or in a tube rack (serum) until sample processing.
10. The animal is euthanized by exsanguination (i.e., severing the jugular blood vessels with scissors) and placed in a carcass bag (*see* **Note 8**). Carcasses are stored at −80°C until removal from the BSL4 laboratory (via autoclave) and proper disposal.

3.2 Treatment Administration

In guinea pigs, treatment testing is usually performed using either intramuscular (IM) or IP administration of the drug while the animal is under anesthesia (*see* **Note 9**). Similar to the GPA-EBOV infection, the procedure must be carried out on a back draft table in the animal procedure space of the BSL4 laboratory. IP treatment should be at least 6–8 h before or after IP GPA-EBOV infection to allow for proper virus uptake. Single treatments like monoclonal antibody therapy are generally first tested on day 1 after infection. Depending on the success rate, a later treatment start may be assessed afterwards.

1. Cover the table section where the animal will be handled with an absorbent pad. Place the anesthesia chamber on the table and connect it to an anesthetic vaporizer such that inhalational isoflurane is administered at low doses (≤5%) (*see* **Note 3**).
2. Place one guinea pig at a time in the anesthesia chamber and observe until breathing is slow and deep, indicating sufficient anesthesia of the guinea pig.
3. Even though the animal is anesthetized, two people are required for the procedure, one person handling the syringe/needle, another person properly restraining the animal.

   For IP administration, place the anesthetized guinea pig on its back, and while the first person restrains the upper body of the animal with long forceps, the second person straightens and holds the animal's hind feet together and administers up to 1 mL of drug IP using a 25 gauge × 5/8″ needle into one site in the lower abdomen.

   For IM administration, place the anesthetized guinea pig on its abdomen, and while the first person restrains the upper body of the animal with long forceps, the second person restrains the animal's hind feet and administers up to 0.2 mL of drug IM into the thigh muscle using a 25 gauge × 5/8″ needle. Alternate the thigh muscle into which treatments are administered if giving daily treatments.
4. Following injection, the person holding the syringe/needle immediately disinfects the syringe/needle with 5% Micro-Chem Plus before disposal into a sharps collection container (*see* **Note 6**). The other person lays the animal on its side back in the cage and monitors breathing until the guinea pig is awake.

3.3 Sequential Blood Collection from the anterior vena cava

Some animal experiments, mainly treatment studies, may require serial blood collection in order to monitor viremia or drug levels in blood/serum over time. Small blood volumes can be drawn from guinea pigs via the marginal ear vein, lateral saphenous vein, or cephalic and tarsal veins. Larger volumes (up to 1 mL) can be collected from the jugular vein, *anterior vena cava*, or femoral vein. Collecting blood from the *anterior vena cava* is my preferred method, as the animal does not need to be very deeply anesthetized, and the risk of asphyxiation is therefore minimized.

1. Cover the table section where the animal will be handled with an absorbent pad. Place the anesthesia chamber on the table and connect it to an anesthetic vaporizer such that inhalational isoflurane is administered at low doses (≤5%) (*see* **Note 3**).
2. Place one guinea pig at a time in the anesthesia chamber and observe until breathing is slow and deep, indicating sufficient anesthesia of the guinea pig.

3. Even though the animal is anesthetized, two people are required for this procedure. The first person restrains the lower abdomen/hind feet of the animal by straightening the legs and holding the animal's hind feet together while the second person is collecting the blood.
4. Lay the animal on its back, nose towards the person performing the procedure (Fig. 1). A nose cone providing continuous inhalational isoflurane is recommended. Use a 25 gauge × 5/8″ needle on a 1-mL tuberculin syringe to draw the blood. The needle is inserted at about a 45° angle to the sternum into the thoracic region, 30–45° angle upward (Fig. 1). As soon as the needle is inserted apply sucktion to draw blood. Once the needle is in the vein, pull the syringe plunger back slowly to achieve a slow and continuous blood draw of up to 1 mL.

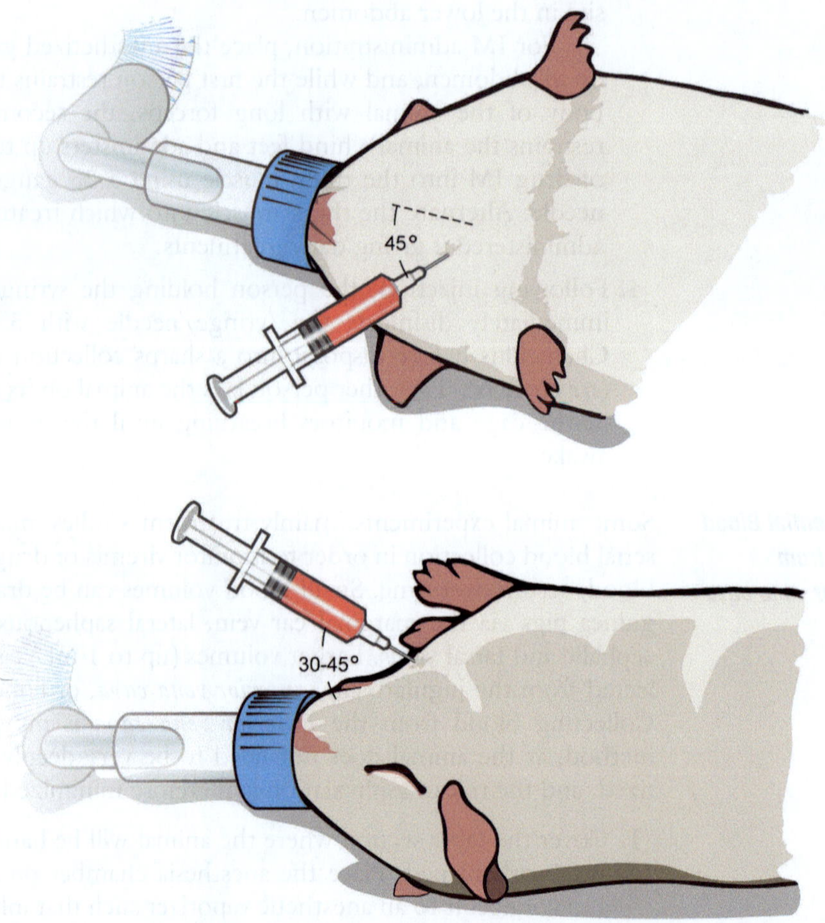

Fig. 1 Blood sample collection from the *anterior vena cava*. The animal is laid on its back under isolfurane anesthesia via a nose cone. The needle is not fully inserted at about a 45° angle to the sternum (top panel, top view) and a 30–45° angle to the guinea pig (bottom panel, side view) for a slow blood draw from the *anterior vena cava* (up to 1 mL)

5. After the blood draw, the wound is immedialtely covered with gauze by the person restraining the animal and pressure is applied to stop the bleeding. Once the bleeding is stopped, the animal is laid on its side in the cage and breathing is monitored until the guinea pig is awake.
6. The blood is immediately transferred into a serum or whole blood (EDTA) collection tube which is placed on a tube roller (EDTA) or in a tube rack (serum) until sample processing. The syringe and needle are disinfected using the 5% Micro-Chem Plus solution before disposal into a sharps collection container (*see* **Note 6**).

3.4 Terminal Sample Collection

Some studies require the determination of disease progression and viremia. If taking a serial blood sample as outlined above is insufficient and the experiment requires determination of tissue virus load and pathological changes, it is recommended to assign a small number of animals per study group for euthanasia and tissue sample collection, e.g., liver and spleen, two main target organs during GPA-EBOV infection, between days 5 and 7 after infection.

1. Cover the table section where the animal will be handled with an absorbent pad. Place the anesthesia chamber on the table and connect it to an anesthetic vaporizer such that inhalational isoflurane is administered at low doses (≤5%) (*see* **Note 3**).
2. Place one guinea pig at a time in the anesthesia chamber and observe until breathing is slow and deep, indicating sufficient anesthesia of the guinea pig.
3. First, a large volume (greater than 10 mL) terminal blood sample is collected via cardiac puncture using vacutainer tubes as described in Subheading 3.1, **step 9**.
4. The animal is euthanized by exsanguination as described in Subheading 3.1, **step 10**.
5. Spray the abdomen of the animal with 70% ethanol to wet the fur for surface disinfection and easier cutting. Use one set of scissors/forceps to cut the abdomen of the animal open. Use a new set of scissors/forceps to remove samples from target organs such as the liver and spleen. Place the samples immediately in proper containers. After the procedure is completed on one animal, place the used instruments in a bucket with 5% Micro-Chem Plus solution for disinfection (*see* **Note 10**).
6. Store the samples properly for the desired downstream application (e.g., titration, histopathology).
7. Carcasses are stored at −80 °C until removal from the BSL4 laboratory (via autoclave) and proper disposal.

4 Notes

1. Guinea pig caging systems often vary between biosafety level 2 laboratory and BSL4 conditions where HEPA-filtered isolator cages are required. Ideally, the animals should be housed in pairs if the BSL4 caging system allows for it. One animal should have an ear tag or some other kind of identifier as individual body weight will be followed over time.
2. Following challenge, weight loss and dehydration are obvious signs of GPA-EBOV infection in guinea pigs. Therefore, the individual guinea pig body weight is measured and recorded every day until at least day 14 after GPA-EBOV infection.
3. Guinea pigs are very sensitive to anesthesia independent of the anesthetic used (inhalational isoflurane for short anesthesia; 2–4 mg Xylazine plus 80–100 mg Ketamine injected IM for longer anesthesia). Animals should be without food for at least 2 h before anesthesia to minimize the risk of asphyxiation after waking up.
4. For this GPA-EBOV strain the LD_{50} in Hartley guinea pigs is 0.01 PFU. Generally challenge doses of 100 or 1000 LD_{50} per animal are used to ensure uniform lethality in the control group. I prefer 1000 LD_{50} (10 PFU) per animal in 1 mL.
5. The guinea pig, mouse, and hamster models for EBOV were developed using 2 IP injection sites in order to obtain consistent and reproducible results. If the infection is performed at only one site, variability of results increases between groups in one study and between different studies.
6. Prepare a 1 L bucket half full with 5% Micro-Chem Plus or any other effective and approved disinfectant. Place the syringe with attached needle in the bucket and suck up 1 mL. Leave the syringe/needle like this until the end of the procedure, when all the animals are infected; minimum contact time is 10 min. Eject Micro-Chem Plus from the syringe/needle and dispose syringe/needle in a sharps container.
7. Euthanasia criteria as approved by the Institutional Animal Care and Use committee at the Rocky Mountain Laboratories, DIR, NIAID, NIH: Any animal with weight loss >20%, ataxia, extreme lethargy (animal is unresponsive to touch), bloody discharge from nose, mouth, rectum or urogenital area, tachypnea, dyspnea or paralysis of the limbs will be immediately euthanized.
8. For this purpose medium size autoclave bags are used. We place 8–12 guinea pigs in one bag depending on their size. In addition, these animals can have sharp claws, and therefore we double-bag the carcasses before putting them into the freezer.

9. Unlike hamsters or mice, guinea pigs are sensitive to anesthesia. They are a great rodent model for vaccine or single-dose treatment efficacy testing, but not my model of choice for daily treatment studies. The risk of asphyxiation is very high and large animal groups are needed in order to compensate for losses while still obtaining statistically significant results. Therefore, I prefer to test drugs requiring daily administration in the hamster model.
10. Instruments are soaked in the 5% Micro-Chem Plus or any other effective and approved disinfectant solution for about 1 h. After a thorough rinse with water and scrubbing if needed, the instruments are placed in a sonicator. Following sonication for 60 min at 60 °C, the instrruments are removed, thoroughly rinsed with water and air-dried. A thin coat of lubricant (Miltex Spray Lube) is applied to the scissors before the instruments are sorted into dishes and autoclaved to sterilize them for the next use.

Acknowledgments

The author is grateful to Logan Banadyga (NIAID) for critically reading the manuscript, and to Austin Athman and Ryan Kissinger (NIAID) for assistance with figure production.

References

1. Safronetz D, Geisbert TW, Feldmann H (2013) Animal models for highly pathogenic emerging viruses. Curr Opin Virol 3(2):205–209. doi:10.1016/j.coviro.2013.01.001
2. Bowen ET, Lloyd G, Harris WJ, Platt GS, Baskerville A, Vella EE (1977) Viral haemorrhagic fever in southern Sudan and northern Zaire. Preliminary studies on the aetiological agent. Lancet 1(8011):571–573
3. Marzi A, Ebihara H, Callison J, Groseth A, Williams KJ, Geisbert TW, Feldmann H (2011) Vesicular stomatitis virus-based Ebola vaccines with improved cross-protective efficacy. J Infect Dis 204(Suppl 3):S1066–S1074. doi:10.1093/infdis/jir348
4. Connolly BM, Steele KE, Davis KJ, Geisbert TW, Kell WM, Jaax NK, Jahrling PB (1999) Pathogenesis of experimental Ebola virus infection in guinea pigs. J Infect Dis 179(Suppl 1):S203–S217. doi:10.1086/514305
5. Cross RW, Fenton KA, Geisbert JB, Mire CE, Geisbert TW (2015) Modeling the disease course of Zaire ebolavirus infection in the outbred Guinea pig. J Infect Dis 212(Suppl 2):S305–S315. doi:10.1093/infdis/jiv237
6. Banadyga L, Dolan MA, Ebihara H (2016) Rodent-adapted Filoviruses and the molecular basis of pathogenesis. J Mol Biol. doi:10.1016/j.jmb.2016.05.008

Chapter 24

Evaluation of Medical Countermeasures Against Ebolaviruses in Nonhuman Primate Models

Chad E. Mire and Thomas W. Geisbert

Abstract

Several ebolavirus species, with varying lethality rates, have caused sporadic outbreaks in Africa resulting in human disease. Ebolaviruses also have the potential for use as biological weapons. Currently, there are no licensed vaccines or therapeutics to respond to outbreaks or deliberate misuse of ebolaviruses. Vaccine or therapeutic efficacy testing of medical countermeasures against ebolaviruses requires an animal model of disease; in vitro testing in cell culture cannot reproduce the complicated balance between host-pathogen interactions required for the ultimate licensure of a countermeasure. Depending on the target of the countermeasure, demonstration of efficacy in the nonhuman primate ebolavirus disease models will most likely be required before licensure. Here, we describe the selection and use of nonhuman primates for vaccine and therapeutic studies against ebolaviruses.

Key words Ebolavirus, Ebola, Zaire, Sudan, Bundibugyo, Vaccine, Therapeutic, Nonhuman primate model

1 Introduction

Ebolaviruses are the causative agents of Ebola hemorrhagic fever (EHF) and are characterized as filoviruses within the family *Filoviridae*. Within the family *Filoviridae*, the *Ebolavirus* genus is composed of five distinct species: (1) *Zaire ebolavirus* (EBOV), (2) *Sudan ebolavirus* (SUDV), (3) *Reston ebolavirus* (RESTV), (4) *Bundibugyo ebolavirus* (BDBV), and (5) *Tai Forest ebolavirus* (TAFV) (also known as *Ivory Coast ebolavirus*) [1]. The genomes of these viruses encode seven gene products: the nucleoprotein (NP), virion protein (VP)35, VP40, glycoprotein (GP), VP30, VP24, and polymerase (L). In addition, these viruses express two additional nonstructural proteins from the GP gene referred to as soluble (s) GP [2] and small soluble (ss)GP [3]. The GP, with potential contribution from the NP, appears to be the key immunogenic antigens in vaccine protection, with most vaccines using GP as the antigen of choice. Therapeutic targets tested to date have been the genes/

Thomas Hoenen and Allison Groseth (eds.), *Ebolaviruses: Methods and Protocols*, Methods in Molecular Biology, vol. 1628, DOI 10.1007/978-1-4939-7116-9_24, © Springer Science+Business Media LLC 2017

mRNAs of NP, VP35, VP24, or L, with GP being the sole protein target for antibody therapy.

An understanding of the pathogenesis of ebolaviruses in relevant animal models is important for the evaluation of the efficacy vaccine and therapeutic candidates and is also essential in light of the "Animal Rule" enacted by the US Food and Drug Administration (FDA) in 2002 [4], which established guidance for the data needed to demonstrate effectiveness of new drugs and biological products when human efficacy studies are not ethical or feasible, such as with ebolaviruses. This rule establishes that a product can be licensed through effectiveness shown in well-characterized animal models along with the typical demonstration of biological activity and safety in humans. Currently, rodents, ferrets [5], and nonhuman primates (NHPs) are the animal models used for ebolavirus studies. Initial assessment of vaccines and therapeutics against ebolaviruses typically occurs in mice, guinea pigs, and hamsters and uses viruses adapted to cause ebolavirus-mediated disease in those models (i.e., mouse- or guinea pig-adapted EBOV strains) [6–12]. While rodent models are useful for initial screens, ebolavirus isolates from humans or NHPs do not cause severe disease in immune-competent rodents after first exposure to the virus; though serial adaptation has produced uniformly lethal infection for some ebolaviruses. While there is adaptation, the disease pathogenesis seen in immune-competent rodent models does not as fully recapitulate the human conditions or the disease as observed in NHPs [7, 10]. As such, data from vaccine and therapeutic studies using solely rodents may not be suitable for supporting applications for licensure of filovirus vaccines and treatments, requiring studies performed in NHPs after initial screening in rodents. Additionally, the validation of NHPs as accurate and reliable models of EHF has been and will be critical to the final evaluation and testing of candidate vaccines and therapeutics. Ultimately, no vaccine or therapeutic against ebolavirus infection will be approved for human use until it can protect NHPs from viremia and clinical illness (for vaccines) or show protection when treatment begins at the onset of signs of clinical illness and viremia (for therapeutics).

Over the last 10 years, much progress has been made toward developing medical countermeasures against ebolaviruses with the most recent efforts focused on non-replicating and replication-competent vaccine vectors and a wide array of antivirals. This progress has mainly been measured through success in the NHP models of ebolaviruses. There are a number of species that represent NHP models of ebolavirus infection: hamadryad baboons (*Papio hamadryas*), marmosets (*Callithrix jacchus*), African green monkeys (*Chlorocebus aethiops*), rhesus macaques (*Macaca mulatta*), and cynomolgus macaques (*Macaca fascicularis*) [1, 13–16]. Out of the available NHP models, the rhesus and cynomolgus macaque models are the most frequently used for therapeutic and vaccine studies, respectively. Here, we describe which model to consider for therapeutic or vaccine studies and how to conduct these ebolavirus countermeasure evaluation studies based on a parenteral route of exposure.

2 Requirements and Materials

2.1 Requirements to Work with Ebolaviruses in NHPs

1. US Centers for Disease Control and Prevention (CDC), or equivalent, approved animal biosafety level 4 (ABSL4) laboratory for animal work.
2. US CDC, or equivalent, approval to work with ebolaviruses.
3. Institutional/university, or equivalent, approval to work with ebolaviruses in NHPs at ABSL4.
4. Institutional Animal Care and Use Committee (IACUC), or equivalent, approval for protocols to vaccinate, infect, treat, and handle NHPs in the ABSL4.
5. Skilled staff that has been specially trained to work with ebolavirus-infected NHPs at ABSL4.

2.2 General

1. Well-characterized seed stocks of EBOV, SUDV, or BDBV: at a minimum, these should be deep sequenced to assess the GP editing site [17, 18] and be confirmed negative for endotoxin and mycoplasma in addition to confirmation of seed stock lethality in NHP pilot studies.
2. NHPs: rhesus or cynomolgus macaques, male or female, 2–10 kg.
3. Animal medical record (AMR) to record procedures for each individual NHP under study.
4. Cage with squeeze mechanism to house NHPs with dimensions recommended in the National Research Council's *Guide for the Care and Use of Laboratory Animals* eighth edition [19] or equivalent.
5. NHP chow/biscuits.
6. Fruit/vegetables.
7. Environmental enrichment such as toys.
8. Procedure table of adequate size to comfortably rest an anesthetized NHP while allowing room to place documentation required for procedures.
9. Versi-Dry Lab soaker, or equivalent absorbent bench pads.
10. Digital rectal thermometer.
11. Scale and weighing tub/platform.
12. Gauze.
13. Rubbing alcohol, or equivalent disinfectant.

Table 1
Clinical observations sheet

IACUC protocol #	NHP type	Virus
NHP ID#: Gender:		
Day	Day 0 [actual date]	Day 1 [actual date] Add more columns as needed
Weight		
Rectal temperature		
Anorexia		
Depression		
Weakness		
Hunched posture		
Unresponsive		
Dyspnea		
Nasal exudate		
Edema		
Petechial rash		
Ecchymotic rash		
Bleeding		
Diarrhea		
Dehydration		

Back of the page used for specific comments about observations

14. Anesthetic, such as ketamine (5–20 mg/kg) or Telazol (2–6 mg/kg).
15. Sharps containers.
16. Shallow metal pan to secure sharps while moving about room, e.g., to a cage from the procedure table.
17. Clinical observation sheet (Table 1).
18. Clinical score sheet (Table 2).

2.3 Exposure of NHP to Ebolaviruses

1. Hanks' balanced salt solution (1×) supplemented up to 2% certified heat-inactivated fetal bovine serum (2% FBS-HBSS).
2. 15 and 50 mL conical tubes.
3. Disposable 1 mL "Luer lock" syringes.
4. Disposable 25-gauge needles.

Table 2
Clinical score sheet

Parameter	Degree of parameter	Possible	Score
Respiration	Normal	0	
	Increased or decreased rate	#	
	Mild dyspnea; increase abdominal breathing	#	
	Moderate dyspnea; mainly abdominal breathing	#	
	Severe dyspnea; agonal breathing*	#	
Appetite	Normal	0	
	No biscuits eaten but enrichment eaten	#	
	No biscuits or enrichment eaten	#	
Activity/appearance	Normal	0	
	Hunched but active most of the time	#	
	Hunched; requires toy or treat stimulation to respond	#	
	Hunched with head between knees; no interest in treats or toys	#	
	Lies down occasionally; gets up without approach	#	
	Lies down; gets up when approached	#	
	Lies down; gets up with some prodding but not when approached*	#	
Bleeding/ hemorrhage	No signs	0	
	Petechiation (less than 20%) and/or ecchymosis (less than 5%)	#	
	Petechiation (more than 20%) and/or ecchymosis (5–10%)	#	
	Ecchymosis (more than 10%)	#	
	Observable bleeding; controlled by clotting (not menses)	#	
	Uncontrolled steady bleeding*	#	
Total		#	
Observer			
Time			

Bold* No matter the total score, an NHP showing these clinical signs should be humanely euthanized
The maximum score for each category is up to the investigator and/or defined by their institutional policies
Date, time, and animal ID# should be on the score sheets as well

2.4 Vaccination, Treatment, Sampling of NHPs

1. Therapeutic or vaccine; preparation dependent on the type being examined.
2. 1 mL up to 20 mL "Luer lock" syringes.
3. Disposable 25-gauge needles and 21-gauge intravenous needles.
4. Vacutainer one-use blood collection needle holders.
5. 21- or 22-gauge blood collection needles.
6. Vacutainer blood collection tubes appropriate for the desired analyses:

   K_2-EDTA—for viral load/immunoglobulin (Ig) analysis/hematology.

Serum separator—for viral load/Ig analysis/blood chemistries.

Sodium citrate—for coagulopathy.

Sodium heparin—for large peripheral blood mononuclear cell (PBMC) preparation/Ig analysis/plasma analysis.

7. Laryngoscope.
8. Gavage tube.
9. Disposable "breakaway" cotton-tipped swabs.

2.5 Euthanasia of NHPs at Humane and Scientific Endpoints

1. Disposable 1 and 5 mL "Luer lock" syringes.
2. 20-gauge intravenous needles.
3. American Veterinary Medical Association (AVMA), or equivalent, approved euthanasia solution; typically a pentobarbital sodium solution.

3 Methods

3.1 Selection of NHP for Studies

While cynomolgus macaques are the most widely used species for evaluating vaccines against ebolaviruses, rhesus monkeys are the most widely used species for evaluating treatments against ebolaviruses [20]. Therefore, the "gold standard" would be selection of cynomolgus macaques for vaccine studies or rhesus macaques for therapeutic studies. There are, however, examples of other NHP species being used in medical countermeasure studies [21, 22]. Cynomolgus macaques typically have a slightly more acute disease course and, therefore, represent a more stringent test for vaccines to protect against ebolavirus. The use of rhesus macaques stems from historical use of this species for the generation of pharmacokinetic data and rhesus having a slightly more protracted disease course, as do humans, allowing for the evaluation of postexposure treatments.

3.2 Route of Exposure and Dose

Ebolaviruses are lethal in NHPs through aerosol, oral, and conjunctival exposure routes; however, the typical exposure route in medical countermeasure studies is intramuscular (i.m.) as this represents a worst case scenario of an accident with healthcare worker managing an infected patient or a lab accident with an infected animal. The typical dose chosen is 1000 plaque-forming units (PFU) or tissue culture infectious dose 50 ($TCID_{50}$) which mimics the potential exposure from an i.m. accident with a needle containing or contaminated with material that has a ~1×10^7 PFU/mL titer. This dose represents a stringent benchmark to overcome, but protection against a worst case exposure will aid confidence in the medical countermeasure being evaluated, which in turn will make it more likely to proceed toward advanced development.

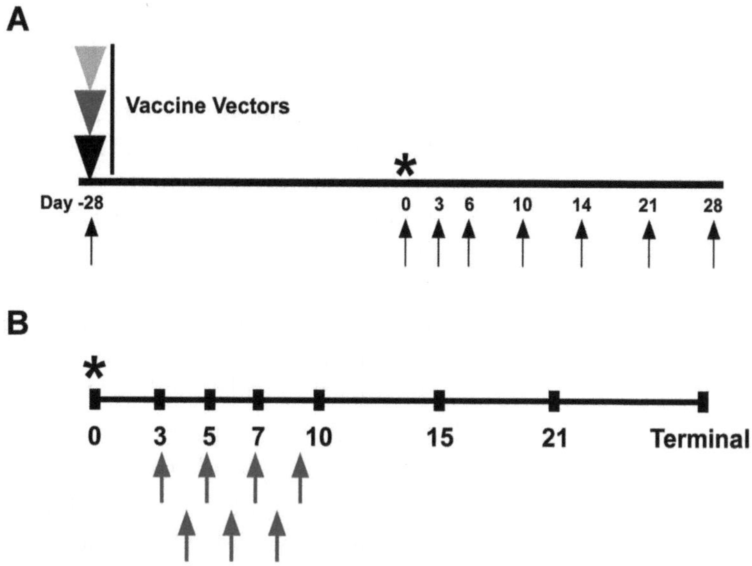

Fig. 1 Ebolavirus medical countermeasure study design. (**a**) Flow chart showing the days of vaccination (*triangles*), days of sampling (*black arrows*), day of ebolavirus exposure (*). (**b**) Diagram of therapeutic treatment regimen days post-ebolavirus challenge in NHPs. *Grey arrows* depict days of treatment. * depicts the day of ebolavirus exposure

3.3 Vaccination

Figure 1a depicts a replication-competent vaccine vector with a single-injection vaccination regimen, where cynomolgus macaques are vaccinated and sampled for replication of the vaccine vector, detection of cell-mediated immune response, and detection of humoral immune response. This phase will typically last 21–35 days postvaccination before exposure of the NHPs to ebolaviruses.

1. Assess the NHPs cage-side for general health and whether or not they are fit for anesthesia and procedures.
2. Set up the procedure table and scale/tub with absorbent bench pads to place the NHPs on.
3. Using the squeeze mechanism of the cage, pull the first NHP to the front of the cage, and maneuver and immobilize the NHP so that the triceps or thigh is available for i.m. injection of anesthetic (e.g., ketamine (5–20 mg/kg) or Telazol (2–6 mg/kg)) using a 25-gauge needle and syringe.
4. Once the NHP is verified as sedated, remove them from the cage, weigh them, and place them on the procedure table. Assess their overall health by observing the eyes, mucus membranes, fur, skin, and hydration status as well as measuring their rectal temperature.
5. Disinfect the inguinal region for phlebotomy. Depending on the size of the NHP, use a 22-gauge (<4 kg) or 21-gauge

(>4 kg) needle for blood collection. Attach the needle to the needle holder, and penetrate the skin toward the femoral vein. Once under the skin, place the desired blood tube type onto the blood collection needle, and insert the needle into the femoral vein. After collecting the desired amount of blood in each of the different tubes needed for downstream analysis, remove the needle and place pressure on the venipuncture site with the thumb and a piece of gauze to assist in clotting. Swab the mucus membranes if desired.

6. Inject the vaccine to be tested using a 25-gauge needle and syringe into as many sites as desired; typically one injection site in either the caudal thigh or deltoid is used.
7. Assess the venipuncture site again, and place NHP back into the cage on their side facing the front of the cage.
8. Fill out the AMR with the health of the animal, weight, temperature, type and amount of anesthetic used, details of the vaccination given, and time of recovery from the anesthetic. Remain in the room until all NHPs have recovered from the anesthetic.
9. If multiple vaccination doses are to be given, repeat **steps 1–8** on subsequent vaccination days.
10. Repeat **steps 1–8**, except for the vaccination step (**step 6**), on desired sampling days; see examples in Fig. 1a.
11. At the desired time point postvaccination, proceed with exposure to ebolavirus, as described in Subheading 3.4.

3.4 Ebolavirus Exposure

1. Thaw the desired ebolavirus seed stock, and dilute the virus in 2% FBS-HBSS starting with at least 0.5 mL of virus stock (*see* **Note 1**). Dilute preparation to 2000 PFU/mL and place on ice to keep cool until use.
2. Assess the NHPs cage-side for general health and whether or not they are fit for anesthesia and procedures.
3. Set up the procedure table and scale/tub with absorbent bench pads to place NHPs on.
4. Using the squeeze mechanism of the cage, pull the first NHP to the front of the cage, and maneuver and immobilize the NHP so that the triceps or thigh is available for i.m. injection of anesthetic [e.g., ketamine (5–20 mg/kg) or Telazol (2–6 mg/kg)] using a 25-gauge needle and syringe (*see* **Note 2**).
5. Once the NHP is sedated, remove them from the cage, weigh them, and place them on the procedure table. Assess their overall health by observing the eyes, mucus membranes, fur, skin, and hydration status as well as measuring their rectal temperature.
6. Disinfect the inguinal region for phlebotomy. Depending on the size of the NHP, use a 22-gauge (<4 kg) or 21-gauge

(>4 kg) needle for blood collection. Attach the needle to the needle holder, and penetrate the skin toward the femoral vein. Once under the skin, place the desired blood tube type onto the blood collection needle, and insert the needle into the femoral vein. After collecting the desired amount of blood in each of the different tubes needed for downstream analysis, remove the needle and place pressure on the venipuncture site with the thumb and a piece of gauze to assist in clotting. Swab the mucus membranes if desired.

7. Inject 0.5 mL (1000 PFU) of the diluted ebolavirus preparation using a 25-gauge needle and syringe into one site on the caudal thigh.
8. Assess the venipuncture site again, and place the NHP back into the cage on their side facing the front of the cage (*see* **Note 3**).
9. Fill out the AMR with the health of the animal, weight, temperature, type and amount of anesthetic used, and time of recovery from the anesthetic. Remain in the room until all NHPs have recovered from the anesthetic.
10. Repeat steps, minus the virus exposure step, on desired sampling days; *see* examples in Fig. 1.
11. If conducting an evaluation of postexposure treatment, continue with treatments as described in Subheading 3.5 starting at the appropriate time point post-challenge.

3.5 Postexposure Treatment

Figure 1b depicts a therapeutic treatment schedule that begins when most NHPs should have detectable levels of viral RNA by RT-qPCR (i.e., around 3 days postexposure). Depending on the candidate therapeutic used, the number of treatments will vary with some examples being 3 [23] or 7 [24] doses. Considering the recent advances in therapeutics for post-ebolavirus exposure treatment, the benchmark for beginning treatment would seem to be 3 days postexposure at a minimum, although treatments that work only when administered at times before 3 days postexposure would certainly not be excluded from consideration for human use.

1. Assess the NHPs cage-side for general health and whether or not they are fit for anesthesia and procedures.
2. Set up the procedure table and scale/tub with absorbent bench pads to place NHPs on.
3. Using the squeeze mechanism of the cage, pull the first NHP to the front of the cage, and maneuver and immobilize the NHP so that the triceps or thigh is available for i.m. injection of anesthetic [e.g., ketamine (5–20 mg/kg) or Telazol (2–6 mg/kg)] using a 25-gauge needle and syringe (*see* **Note 2**).

4. Once the NHP is sedated, remove them from the cage, weigh them, and place them on the procedure table. Assess their overall health by observing the eyes, mucus membranes, fur, skin, and hydration status as well as measuring their rectal temperature.
5. If bleeding on a treatment day, disinfect the inguinal region for phlebotomy. Depending on the size of the NHP, use a 22-gauge (<4 kg) or 21-gauge (>4 kg) needle for blood collection. Attach the needle to the needle holder, and penetrate the skin toward the femoral vein. Once under the skin, place the desired blood tube type onto the blood collection needle, and insert the needle into the femoral vein. After collecting the desired amount of blood in each of the different tubes needed for downstream analysis, remove the needle and place pressure on the venipuncture site with the thumb and a piece of gauze to assist in clotting. Swab the mucus membranes if desired.
6. For intravenous (i.v.) treatment: disinfect the inguinal region. Use a 21-gauge intravenous needle and syringe to deliver the treatment into the femoral or saphenous vein. Once under the skin, insert the needle into the femoral vein (*see* **Note 4**), and pull back on the plunger until blood is drawn into syringe; at this point, slowly depress the plunger releasing the treatment into the femoral vein. Remove the needle and place pressure on the venipuncture site with the thumb and a piece of gauze to assist in clotting. For oral treatment: sit the NHP upright and have a second person keep the mouth open using gauze hooked around canines. Use a laryngoscope to locate the opening of the esophagus, and insert the gavage tube attached to a syringe filled with the treatment, and then depress syringe plunger to release the treatment.
7. If i.v. treatment is used, assess the venipuncture site again, and place the NHP back into the cage on their side facing the front of the cage (*see* **Note 3**).
8. Fill out the AMR with the health of the animal, weight, temperature, type and amount of anesthetic used, and time of recovery from the anesthetic. Remain in the room until all NHPs have recovered from the anesthetic.
9. Repeat all steps on additional treatment days. Repeat all steps, minus the treatment step (**step 6**), on desired sampling days; see examples in Fig. 1b.

3.6 Clinical Samples, Clinical Observations, and Humane Endpoint Scoring

Infection of NHPs with ebolaviruses leads to disease characterized by signs of fever, malaise, loss of appetite, coagulopathy, petechial rash, thrombocytopenia, lymphopenia, neutrophilia, and liver damage, with varying degrees of severity among the different ebolaviruses. Each of these characteristics/signs can be used to assess the efficacy of medical countermeasures as the countermeasures should be able to prevent or diminish signs of disease. Below is a description of the analysis that can be performed on NHPs postexposure to ebolaviruses.

1. On bleed days, EDTA-plasma tubes, serum tubes, and sodium-citrate tubes can be collected for hematology, serum biochemistry, and coagulopathy, respectively. Remove 100 μL of whole blood from EDTA-plasma tube, and run it on a hematology machine; the total number of lymphocytes, granulocytes, and platelets is of the most interest. Remove 100 μL of serum from serum tube, and run it in in a biochemistry panel disc; alanine aminotransferase (ALT), aspartate aminotransferase (AST), albumin, total protein (TP), blood urea nitrogen (BUN), and C-reactive protein (CRP) are the parameters that are of the most interest. Remove 100 μL of sodium-citrate plasma, and assess coagulation parameters (i.e., prothrombin time (PT), activated partial thromboplastin time (APTT), thrombin time, and fibrinogen) on a coagulation instrument; D-dimers can also be measured by ELISA if desired (*see* **Note 5**).
2. On the day of and all days after exposure of NHPs to ebolaviruses (up to the study endpoint), clinical observation sheets (Table 1) are filled out cage-side or while on the procedure table. The time (i.e., day) postexposure and observed manifestations of depression, anorexia, fever, petechial rash or other hemorrhagic observations, recumbency, diarrhea, and the identity of the person doing the observation are critical for documentation in order to compare data between a control cohort and medical countermeasure cohorts.
3. When animals are left untreated or not vaccinated, EBOV and SUDV cause 100% lethality in cynomolgus macaques and nearly 100% in rhesus macaques, while BDBV causes ~67–75% lethality in cynomolgus macaques [25, 26], and the rhesus macaque model may be closer to 40% lethality (C.E. Mire and T.W. Geisbert personal observations), although the number of control rhesus macaques infected with BDBV is still too low to make a definitive statement about BDBV lethality. Since these ebolaviruses cause lethality, it is necessary to ensure that NHPs in ebolavirus studies do not suffer unduly without also diluting the NHP scientific purpose as proper controls to test medical countermeasures. To this end, euthanasia criteria should be set for each study based on respiration, appetite, activity, and hemorrhagic manifestations as shown in Table 2. The scoring and euthanasia criteria (*see* **Note 6**) can be investigator specific; however, the ultimate decision on what is acceptable resides with the IACUC of the institution involved in NHP ebolavirus studies. NHPs are scored daily at a minimum

beginning day 1 postexposure through the acute phase of the disease with certain scores/categories triggering additional monitoring in order to be prepared to intervene with euthanasia (*see* **Note 7**).

3.7 Euthanasia

1. At the study endpoint or when a clinical score triggers euthanasia, set up the procedure table and scale/tub with Versi-Dry® absorbent bench pads to place the NHPs on.
2. Using the squeeze mechanism of the cage, pull the NHP to the front of the cage, and maneuver and immobilize the NHP so that the triceps or thigh is available for i.m. injection of Telazol (6 mg/kg) using a 25-gauge needle and syringe (*see* **Note 2**).
3. Once the NHP is sedated, remove them from the cage, weigh them, and place them on the procedure table. Assess their overall health by observing the eyes, mucus membranes, fur, skin, and hydration status as well as measuring their rectal temperature.
4. Disinfect the inguinal region for phlebotomy. Depending on the size of the NHP, use a 22-gauge (<4 kg) or 21-gauge (>4 kg) needle for blood collection. Attach the needle to the needle holder, and penetrate the skin toward the femoral vein. Once under the skin, place the desired blood tube type onto the blood collection needle, and insert the needle into the femoral vein. After collecting the desired amount of blood in each of the different tubes needed for downstream analysis, remove the needle and place pressure on the venipuncture site with the thumb and a piece of gauze to assist in clotting. Swab the mucus membranes if desired.
5. Deliver pentobarbital sodium solution (<4 kg use 1.5 mL; >4 kg 2.0 mL) via intracardiac injection using a 20-gauge needle and 5 ml syringe. Insert the needle in between the ribs near the left nipple and pull back on the plunger until blood is drawn into the syringe; at this point, depress the plunger injecting the pentobarbital sodium solution into the heart and circulatory system (*see* **Note 8**).
6. When euthanasia is required or the study endpoint is reached, NHPs are typically necropsied with gross pathology observed and noted with further analysis through hematoxylin and eosin (H&E) staining for cytopathology and immunohistochemistry (IHC) staining for virus antigen. Pathology and virus load in tissues can be used to evaluate the effectiveness of medical countermeasures with lymph nodes, the liver, and adrenal glands being the major tissues to observe closely for virus replication (*see* **Note 9**).

4 Notes

1. Most ebolavirus stocks will need to be diluted around 5000-fold to reach 2000 PFU/mL, so an initial tenfold dilution is made by placing 0.5 mL of virus stock into 4.5 mL of 2% FBS-HBSS. Pulse vortex the tube of virus stock three times so that all liquid goes up the sides of the tube, and then remove the 0.5 mL. This assures the user pulls in a more homogenous mixture of virus than pulling a smaller volume (e.g., 0.001 mL and placing it into 4.999 mL for 5000-fold dilution). Pulse vortex at each dilution step using at least 2 mL for tenfold dilution.
2. As NHPs become ill, they experience hypovolemia, through leaky blood vessels, which reduces the heart's ability to maintain normal blood circulation. Without proper blood circulation, the effectiveness of the anesthetic given i.m. is diminished as the anesthetic will not circulate quickly or at all. When this is encountered, it is best to administer the anesthetic in the triceps as this is closer to being put into circulation by the heart, as compared to the legs. Typically this is experienced from day 5 onward with Ebola-infected control cynomolgus macaques as an example.
3. As coagulopathy develops following ebolavirus exposure, it becomes more important to assess the venipuncture site, and it may take longer than usual for clotting to occur. Time the anesthetizing of NHPs accordingly to account for this possibility.
4. If treating numerous times via the femoral vein, scar tissue will build up (also consider coagulopathy), so it is necessary to alternate veins for bleeding and treating on the same day and potentially on successive days of treatment. The saphenous vein can also be used as an alternative.
5. It is imperative that all values are compared to the day of challenge which represents the baseline for each individual animal. Ranges for hematology and blood chemistry that represent advanced disease: greater than 30–35% change in hematology total cell/platelet numbers and greater than 10–20-fold changes in liver enzymes, 30% reduction in albumin and TP, and changes in CRP levels over 20-fold.
6. Table 2 is a representation of the signs typically observed that can be scored. The investigators can adjudicate the scores as they deem suitable as long as there is an established score that triggers humane intervention and while still ensuring that the control animals have contributed to the proper assessment of the medical countermeasure being investigated.
7. While the clinical scores are meant to provide a humane endpoint for the NHPs on study, the disease course of ebolaviruses

can be slightly variable as the NHPs are an outbred animal model. Additionally, due to the coagulopathy experienced by the NHPs, blood clots can become dislodged resulting in quick expiration of an NHP even when not scoring near a typical euthanasia score, resulting in an unintended death even when the intent was to intervene with euthanasia.

8. Euthanasia can be confirmed by shining a light at the eyes to observe a lack of iris reflex, observing cessation of chest movement (no longer breathing), and/or by "squeezing" the ribs on left and right side between the thumb and fingers to feel for the lack of a heartbeat. Death is assured by opening the chest cavity at the diaphragm.
9. With delayed treatment studies, the eyes, reproductive organs, and central nervous system have become tissues of interest as these represent tissues found to harbor virus in survivors even after convalescence [27, 28]. It is recommended to collect these tissues for virological and pathological analysis if an NHP shows signs of disease past day 10–15 postexposure but still becomes convalescent.

References

1. Feldmann H, Sanchez A, Geisbert TW (2013) Filoviridae: Ebola and Marburg viruses. In: Knipe DM, Howley P (eds) Fields virology, 6th edn. Lippincott Williams and Wilkins, Philadelphia, PA
2. Volchkova VA, Feldmann H, Klenk HD, Volchkov VE (1998) The nonstructural small glycoprotein sGP of Ebola virus is secreted as an antiparallel-orientated homodimer. Virology 250(2):408–414. doi:10.1006/viro.1998.9389. [pii] S0042-6822(98)99389-8
3. Mehedi M, Falzarano D, Seebach J, Hu X, Carpenter MS, Schnittler HJ, Feldmann H (2011) A new Ebola virus nonstructural glycoprotein expressed through RNA editing. J Virol 85(11):5406–5414. doi:10.1128/JVI.02190-10. [pii] JVI.02190-10
4. UDoHaHS (DHHS) (2014) Guidance for industry: product development under the animal rule. UDoHaHS, Silver Spring, MD
5. Cross RW, Mire CE, Borisevich V, Geisbert JB, Fenton KA, Geisbert TW (2016) The domestic ferret (mustela putorius furo) as a lethal infection model for 3 species of ebolavirus. J Infect Dis. doi:10.1093/infdis/jiw209
6. Bray M, Davis K, Geisbert T, Schmaljohn C, Huggins J (1998) A mouse model for evaluation of prophylaxis and therapy of Ebola hemorrhagic fever. J Infect Dis 178(3):651–661
7. Bray M, Hatfill S, Hensley L, Huggins JW (2001) Haematological, biochemical and coagulation changes in mice, guinea-pigs and monkeys infected with a mouse-adapted variant of Ebola Zaire virus. J Comp Pathol 125(4): 243–253. doi:10.1053/jcpa.2001.0503. [pii] S0021-9975(01)90503-1
8. Connolly BM, Steele KE, Davis KJ, Geisbert TW, Kell WM, Jaax NK, Jahrling PB (1999) Pathogenesis of experimental Ebola virus infection in guinea pigs. J Infect Dis 179(Suppl 1):S203–S217. doi:10.1086/514305
9. Ebihara H, Zivcec M, Gardner D, Falzarano D, LaCasse R, Rosenke R, Long D, Haddock E, Fischer E, Kawaoka Y, Feldmann H (2013) A Syrian golden hamster model recapitulating ebola hemorrhagic fever. J Infect Dis 207(2):306–318. doi:10.1093/infdis/jis626. [pii] jis626
10. Geisbert TW, Pushko P, Anderson K, Smith J, Davis KJ, Jahrling PB (2002) Evaluation in nonhuman primates of vaccines against Ebola virus. Emerg Infect Dis 8(5):503–507
11. Ryabchikova E, Kolesnikova L, Smolina M, Tkachev V, Pereboeva L, Baranova S, Grazhdantseva A, Rassadkin Y (1996) Ebola virus infection in guinea pigs: presumable role of granulomatous inflammation in pathogenesis. Arch Virol 141(5):909–921
12. Warfield KL, Bradfute SB, Wells J, Lofts L, Cooper MT, Alves DA, Reed DK, VanTongeren SA, Mech CA, Bavari S (2009) Development and characterization of a mouse model for

Marburg hemorrhagic fever. J Virol 83(13): 6404–6415. doi:10.1128/JVI.00126-09. [pii] JVI.00126-09

13. Carrion R Jr, Ro Y, Hoosien K, Ticer A, Brasky K, de la Garza M, Mansfield K, Patterson JL (2011) A small nonhuman primate model for filovirus-induced disease. Virology 420(2):117–124. doi:10.1016/j.virol.2011.08.022
14. Davis KJ, Anderson AO, Geisbert TW, Steele KE, Geisbert JB, Vogel P, Connolly BM, Huggins JW, Jahrling PB, Jaax NK (1997) Pathology of experimental Ebola virus infection in African green monkeys. Involvement of fibroblastic reticular cells. Arch Pathol Lab Med 121(8):805–819
15. Geisbert TW, Jahrling PB, Larsen T, Davis KJ, Hensley LE (2004) Filovirus pathogenesis in nonhuman primates. In: Klenk HD, Feldmann H (eds) Ebola and Marburg viruses: molecular and cellular biology. Horizon Bioscience, Norfolk, pp 203–238
16. Reed DS, Lackemeyer MG, Garza NL, Sullivan LJ, Nichols DK (2011) Aerosol exposure to Zaire ebolavirus in three nonhuman primate species: differences in disease course and clinical pathology. Microbes Infect 13(11):930–936. doi:10.1016/j.micinf.2011.05.002
17. Trefry JC, Wollen SE, Nasar F, Shamblin JD, Kern SJ, Bearss JJ, Jefferson MA, Chance TB, Kugelman JR, Ladner JT, Honko AN, Kobs DJ, Wending MQ, Sabourin CL, Pratt WD, Palacios GF, Pitt ML (2015) Ebola virus infections in nonhuman primates are temporally influenced by glycoprotein poly-u editing site populations in the exposure material. Viruses 7(12):6739–6754. doi:10.3390/v7122969
18. Volchkova VA, Dolnik O, Martinez MJ, Reynard O, Volchkov VE (2011) Genomic RNA editing and its impact on Ebola virus adaptation during serial passages in cell culture and infection of guinea pigs. J Infect Dis 204(Suppl 3):S941–S946. doi:10.1093/infdis/jir321. [pii] jir321
19. National Research Council (2011) Guide for the care and use of laboratory animals, 8th edn. The National Academies Press, Washington, DC
20. Geisbert TW, Strong JE, Feldmann H (2015) Considerations in the use of nonhuman primate models of ebola virus and marburg virus infection. J Infect Dis 212(Suppl 2):S91–S97. doi:10.1093/infdis/jiv284
21. Meyer M, Garron T, Lubaki NM, Mire CE, Fenton KA, Klages C, Olinger GG, Geisbert TW, Collins PL, Bukreyev A (2015) Aerosolized Ebola vaccine protects primates and elicits lung-resident T cell responses. J Clin Invest 125(8):3241–3255. doi:10.1172/JCI81532
22. Qiu X, Audet J, Wong G, Pillet S, Bello A, Cabral T, Strong JE, Plummer F, Corbett CR, Alimonti JB, Kobinger GP (2012) Successful treatment of ebola virus-infected cynomolgus macaques with monoclonal antibodies. Science Transl Med 4(138):138ra181. doi:10.1126/scitranslmed.3003876
23. Qiu X, Wong G, Audet J, Bello A, Fernando L, Alimonti JB, Fausther-Bovendo H, Wei H, Aviles J, Hiatt E, Johnson A, Morton J, Swope K, Bohorov O, Bohorova N, Goodman C, Kim D, Pauly MH, Velasco J, Pettitt J, Olinger GG, Whaley K, Xu B, Strong JE, Zeitlin L, Kobinger GP (2014) Reversion of advanced Ebola virus disease in nonhuman primates with ZMapp. Nature 514(7520):47–53. doi:10.1038/nature13777
24. Thi EP, Mire CE, Lee AC, Geisbert JB, Zhou JZ, Agans KN, Snead NM, Deer DJ, Barnard TR, Fenton KA, MacLachlan I, Geisbert TW (2015) Lipid nanoparticle siRNA treatment of Ebola-virus-Makona-infected nonhuman primates. Nature 521(7552):362–365. doi:10.1038/nature14442
25. Falzarano D, Feldmann F, Grolla A, Leung A, Ebihara H, Strong JE, Marzi A, Takada A, Jones S, Gren J, Geisbert J, Jones SM, Geisbert TW, Feldmann H (2011) Single immunization with a monovalent vesicular stomatitis virus-based vaccine protects nonhuman primates against heterologous challenge with Bundibugyo ebolavirus. J Infect Dis 204(Suppl 3):S1082–S1089. doi:10.1093/infdis/jir350
26. Mire CE, Geisbert JB, Marzi A, Agans KN, Feldmann H, Geisbert TW (2013) Vesicular stomatitis virus-based vaccines protect nonhuman primates against Bundibugyo ebolavirus. PLoS Negl Trop Dis 7(12):e2600. doi:10.1371/journal.pntd.0002600
27. Uyeki TM, Erickson BR, Brown S, McElroy AK, Cannon D, Gibbons A, Sealy T, Kainulainen MH, Schuh AJ, Kraft CS, Mehta AK, Lyon GM 3rd, Varkey JB, Ribner BS, Ellison RT 3rd, Carmody E, Nau GJ, Spiropoulou C, Nichol ST, Stroher U (2016) Ebola virus persistence in semen of male survivors. Clin Infect Dis 62(12):1552–1555. doi:10.1093/cid/ciw202
28. Varkey JB, Shantha JG, Crozier I, Kraft CS, Lyon GM, Mehta AK, Kumar G, Smith JR, Kainulainen MH, Whitmer S, Stroher U, Uyeki TM, Ribner BS, Yeh S (2015) Persistence of ebola virus in ocular fluid during convalescence. N Engl J Med 372(25):2423–2427. doi:10.1056/NEJMoa1500306

Chapter 25

Quantification of Filovirus Glycoprotein-Specific Antibodies

Wakako Furuyama, Hiroko Miyamoto, Reiko Yoshida, and Ayato Takada

Abstract

Serological methods such as the enzyme-linked immunosorbent assay (ELISA) and virus neutralization test are fundamental tools used in diagnosis, seroepidemiological studies of filovirus transmission/prevalence, and the evaluation of vaccine immunogenicity and potential therapeutic antibodies. Filoviruses have a single transmembrane glycoprotein (GP), which is the only known target of neutralizing antibodies. Here we describe serological methods to quantify filovirus GP-specific antibodies.

Key words Filovirus, *Ebolavirus*, *Marburgvirus*, Antibody, Glycoprotein, ELISA, Neutralization, Antibody-dependent enhancement, ADE

1 Introduction

Filoviruses are divided into three genera, *Marburgvirus*, *Ebolavirus*, and *Cuevavirus*. The genus *Marburgvirus* has one species, *Marburg marburgvirus*, with two viruses, Marburg virus (MARV) and Ravn virus. The genus *Cuevavirus* also has a single species, *Lloviu cuevavirus*, with one virus, Lloviu virus (LLOV). In contrast, five distinct species are known in the genus *Ebolavirus*: *Zaire ebolavirus*, *Sudan ebolavirus*, *Taï Forest ebolavirus*, *Bundibugyo ebolavirus*, and *Reston ebolavirus*, represented by Ebola virus (EBOV), Sudan virus (SUDV), Taï Forest virus (TAFV), Bundibugyo virus (BDBV), and Reston virus (RESTV), respectively.

Filoviruses express a single transmembrane glycoprotein (GP) that is responsible for both receptor binding and membrane fusion [1, 2] and thus the only known target of neutralizing antibodies. On the other hand, some GP-specific antibodies are also reported to induce antibody-dependent enhancement (ADE) of infection in vitro through interaction with the cellular Fc receptor (FcR) or complement component C1q and its ligands, which likely promotes viral entry into cells [3–6]. FcRs are expressed exclusively on particular immune cells such as monocytes/macrophages, neutrophils, B-cells, and granulocytes, whereas C1q ligands are found on most

Thomas Hoenen and Allison Groseth (eds.), *Ebolaviruses: Methods and Protocols*, Methods in Molecular Biology, vol. 1628, DOI 10.1007/978-1-4939-7116-9_25,

mammalian cells, suggesting a ubiquitous mechanism for ADE of filovirus infection. Importantly, standard neutralization tests using Vero E6 cells fail to detect ADE activity since the presence of cellular FcR or supplemented C1q and its ligands on the target cells are required to detect filovirus ADE in serological assays in vitro (Fig. 1).

It should also be noted that there is a large antigenic difference among the filovirus GPs [7, 8]. Even within the genus *Ebolavirus*, the majority of antibodies induced against one virus species are not cross-reactive to the other species. Therefore, both neutralizing and ADE activities observed in immunized or infected animal sera are generally filovirus species specific [5, 7, 8]. In this chapter, we describe serological assays that can be used for serological experiments such as detection of GP-specific antibodies in animal sera and screening of the biological activities of GP-specific monoclonal antibodies.

2 Materials

2.1 Preparation of Soluble Filovirus GPs with Histidine Tag (GP-His)

1. Human embryonic kidney (HEK) 293T cells (ATCC: CRL-3216).
2. Complete growth DMEM (DMEM high glucose + 10% fetal calf serum [FCS] + 2 mM L-glutamine + 100 U/mL penicillin + 0.1 mg/mL streptomycin).
3. Opti-MEM.
4. TransIT-LT1 (Mirus).
5. Square cell culture dish (500 cm^2).
6. GP-expressing plasmids (e.g., pCAGGS encoding GP-His) [7, 8] (Fig. 2).
7. Bottle-top filter (0.45 μm).
8. Ni-NTA Agarose (Invitrogen).
9. 5× native purification buffer (250 mM NaH_2PO_4, pH 8.0, 2.5 M NaCl), binding buffer, wash buffer, and elution buffer (*see* **Note 1**).
10. Purification column (e.g., Bio-Rad Econo-Column).
11. Chromatography pump (e.g., Atto Perista pump), if available.
12. Amicon Ultra-4 spin columns (30,000 molecular weight cut-off; Millipore).

2.2 ELISA

1. Phosphate-buffered saline (PBS; Wako 048-29805).
2. PBST (0.05% Tween 20 in PBS).
3. Purified filovirus GP antigens (GP-His).
4. FCS (cell culture)-derived control antigens (*see* **Note 2**).
5. BSA_{10} (10 mg/mL in PBS).

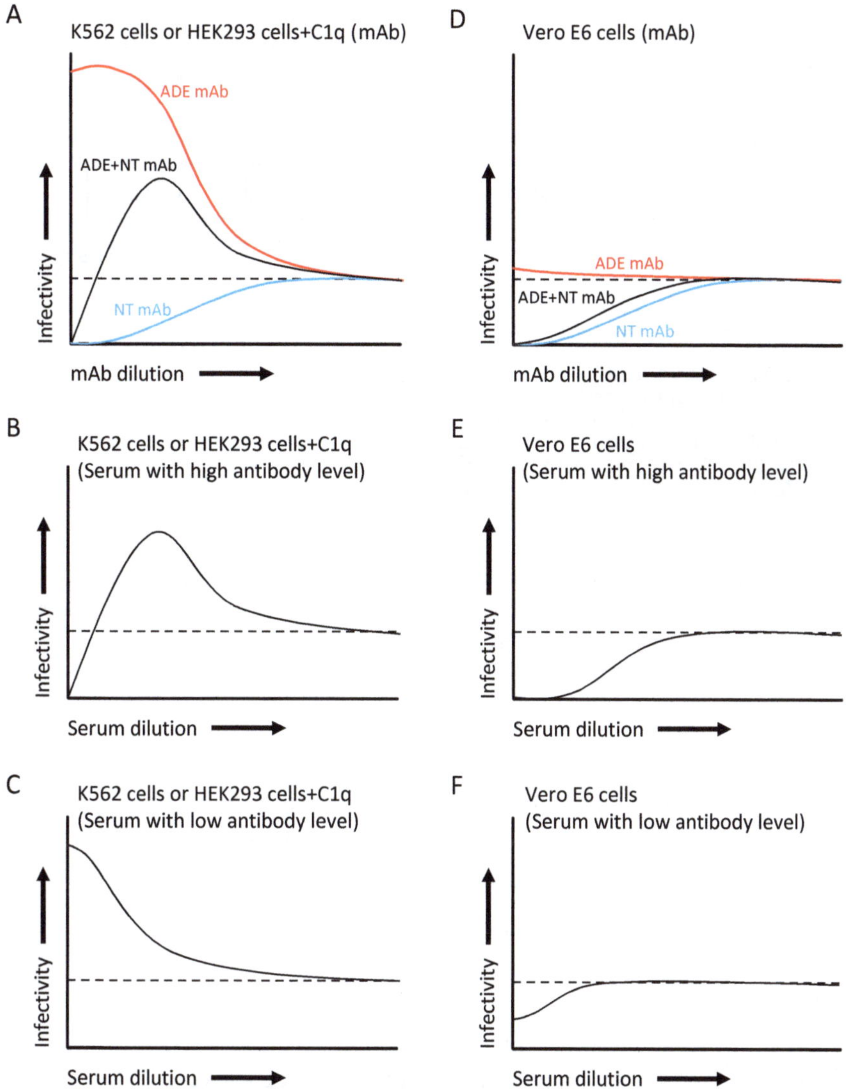

Fig. 1 Antibody-dependent enhancement (ADE) and neutralizing (NT) activity models of monoclonal antibodies (mAbs) and polyclonal antisera against filoviruses. Schematic representation of virus infectivity in K562 or HEK293 cells with C1q (**a–c**) and Vero E6 cells (**d–f**) is shown. In K562 or HEK293 cells with C1q, ADE and NT mAbs clearly show unique activities, respectively, in a dose-dependent manner, but the mixture of these mAbs shows neutralizing activity only at high concentrations, and ADE activity is dominantly seen (**a**). A polyclonal antiserum (e.g., hyperimmune serum to filoviruses) potentially shows a similar curve to the ADE and NT mAb mixture (**b**), but only ADE activity can be detected if overall antibody levels are low (**c**). In contrast, the ADE mAb is not expected to show activity in Vero E6 cells, and the ADE and NT mAb mixture and serum antibodies only show NT activity (**d–f**). Thus, ADE activity of serum samples may sometimes be detectable in K562 cells or HEK293 cells with C1q even if NT activity is not significant in Vero E6 cells (**c**, **f**). *Dotted lines* represent baseline virus infectivity without antibodies

6. BSA_5T (5 mg/mL in PBST).
7. BSA_5T-FCS (5 mg/mL BSA and 2% FCS in PBST).
8. Primary antibodies (e.g., serum, plasma, and monoclonal antibody).

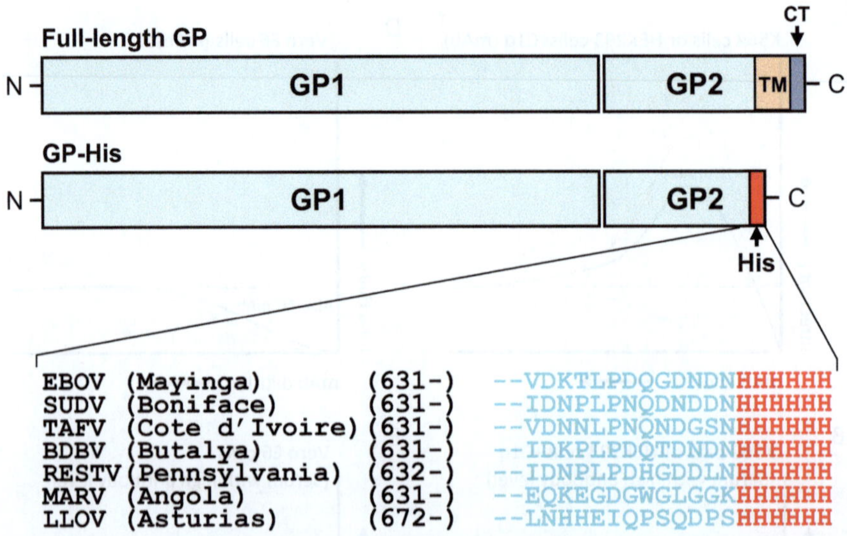

Fig. 2 Structure of filovirus GPs. Full-length GP has a transmembrane domain (TM) and a cytoplasmic tail (CT) at the C-terminus. Each GP-His has a polyhistidine-tag at the indicated position in place of the TM and CT regions. Amino acid sequences of the CT of representative isolates of each filovirus species are shown

9. Secondary antibodies (horseradish peroxidase-conjugated antibodies).
10. Substrate solution (e.g., TMB: 3,3′,5,5′-Tetramethylbenzidine Liquid Substrate System for ELISA; Sigma T0440-1L).
11. 1 M phosphoric acid.
12. ELISA plate.
13. Plate reader.

2.3 Preparation of Vesicular Stomatitis Virus (VSV)Pseudotyped with Filovirus GPs

1. HEK293T cells.
2. Complete growth DMEM (see above).
3. Opti-MEM.
4. TransIT-LT1 (Mirus).
5. 10 cm culture dish.
6. GP-expressing plasmids (e.g., pCAGGS encoding full-length GP).
7. Replication-incompetent pseudotyped VSVs containing green fluorescent protein (GFP) instead of the VSV G gene (VSVΔG-VSVG) [1].

2.4 Neutralization Assay

1. African green monkey kidney Vero E6 cells (ATCC: CRL-1586).
2. Complete growth DMEM (see above).
3. 96-well tissue culture plates.
4. VSVs pseudotyped with filovirus GP or authentic filoviruses.

2.5 ADE Assay

1. Human leukemia K562 cells (ATCC: CCL-243).
2. HEK 293 cells (*see* **Note 3**).
3. Complete growth DMEM (see above).
4. Complete growth RPMI (RPMI-1640 + 10% FCS + 2 mM L-glutamine + 100 U/mL penicillin + 0.1 mg/mL streptomycin).
5. 96-well tissue culture plates.
6. VSVs pseudotyped with filovirus GP or authentic filoviruses.

3 Methods

3.1 Expression and Purification of GP-His

1. Seed HEK293T cells in a 500 cm^2 dish (3–3.5 × 10^5 cells/mL) in 100 mL of complete growth DMEM.
2. Incubate cells for 16–20 h at 37 °C in a 5% CO_2 incubator. Cells should be ~70% confluent at the time of transfection.
3. Prior to transfection, warm the TransIT-LT1 and Opti-MEM to room temperature.
4. For 1 dish, add 100 μg of GP-His-expressing plasmid to 5 mL of OPTI-MEM, and mix well.
5. Add 200 μL of TransIT-LT1 into the plasmid/Opti-MEM solution and mix well.
6. Incubate for 30 min at room temperature.
7. Add TransIT-LT1/plasmid mixture dropwise to different areas of the dish.
8. Gently rock the dish for even distribution of complexes.
9. Incubate cells for 48–72 h at 37 °C in a 5% CO_2 incubator.
10. Harvest the supernatant from the cells.
11. Centrifuge harvested supernatant at 4000 × *g* for 20 min at 4 °C.
12. Collect the supernatant and filter through the bottle-top filter.
13. Distribute filtered supernatant into 50 mL centrifuge tubes, and keep on ice or at 4 °C.
14. To prepare the Ni-NTA Agarose, resuspend Ni-NTA Agarose in its bottle by inverting and gently tapping the bottle repeatedly (see manufacturer's instructions for further details).
15. Pipet or pour 0.5 mL resin per 50 mL supernatant into 15 mL centrifuge tubes, and fill the tube with sterile, distilled water (*see* **Note 4**).
16. Centrifuge at 1000 × *g* for 2 min and gently aspirate the supernatant (*see* **Note 5**).
17. Add sterile, distilled water (4× resin volume), and resuspend the resin by inverting and gently tapping the tube (*see* **Note 6**).

18. Centrifuge at 1000 × *g* for 2 min and gently aspirate the supernatant (*see* **Note 5**).
19. Add binding buffer (4× resin volume), and resuspend the resin by inverting and gently tapping the tube (*see* **Note 6**).
20. Centrifuge at 1000 × *g* for 2 min and gently aspirate the supernatant (*see* **Note 5**).
21. Add binding buffer (3× resin volume) and resuspend the resin by gently tapping the tube.
22. Pipet 2 mL (i.e., 0.5 mL of resin) of the prepared Ni-NTA Agarose into 50 mL of the clarified supernatant prepared in **step 13**.
23. Rotate the tube slowly (<1 rpm) overnight at 4 °C.
24. Centrifuge at 1000 × *g* for 2 min and gently aspirate the supernatant (*see* **Note 5**).
25. Add binding buffer (3× resin volume) and resuspend the resin by gently tapping the tube.
26. Pipet the Ni-NTA Agarose into a purification column of an appropriate size (*see* **Note 7**).
27. Add additional binding buffer (3× resin volume) to the tube, resuspend the residual resin by gently tapping the tube, and transfer it to the column (*see* **Note 8**).
28. Allow the resin in the column to settle completely by gravity (5–10 min).
29. Wash the column with wash buffer (25× resin volume) at a flow speed of 1 mL/min at 4 °C. If a chromatography pump is not available, wash the column by gravity flow.
30. Carefully remove the supernatant and gently load a small amount of elution buffer onto the resin.
31. Elute the protein with elution buffer (5× resin volume) at a flow speed of 1 mL/min at 4 °C. Collect 0.5–1 mL fractions for analysis by SDS-PAGE and immunoblotting. If a chromatography pump is not available, elute the protein with elution buffer by gravity flow.
32. Pool the fractions containing GP-His.
33. Replace the elution buffer with PBS (e.g., using Amicon Ultra-4 spin columns according to the manufacturer's directions or by dialysis).
34. Concentrate the eluted protein if needed to achieve a concentration of at least 2–4 μg GP/mL (e.g., using Amicon Ultra-4 spin columns).

3.2 Detection of Anti-GP Antibodies in ELISA

1. Coat (50 μL/well) 96-well ELISA plates with purified GP-His (2–4 μg GP/mL in PBS), and incubate them at room temperature for 2 h or overnight at 4 °C. Plates coated with control

antigens (0.4–0.6 μg protein/mL in PBS) should also be prepared when serum/plasma samples are used (*see* **Note 2**).

2. Discard antigens.
3. Add BSA10 (100–150 μL/well), and incubate for 2 h at room temperature or overnight at 4 °C.
4. During this incubation time, prepare primary antibody dilutions (e.g., 1:100, 1:1000, or serial dilutions). Use BSA5T-FCS for dilution (*see* **Note 9**).
5. Discard BSA10 and wash the plates with PBST once.
6. Add sample (primary antibody) dilutions (50 μL/well), and incubate overnight at 4 °C.
7. Discard sample dilutions and wash the plates with PBST four times.
8. Add appropriately diluted secondary antibody (50 μL/well), and incubate for 1 h at room temperature (*see* **Note 10**).
9. During this incubation time, turn on and set up an ELISA reader.
10. Discard secondary antibody and wash the plates with PBST five times.
11. Add TMB solution (50 μL/well) and incubate for 15–30 min at room temperature (*see* **Note 11**).
12. Add 1 M phosphoric acid (50 μL/well) to stop the reaction.
13. Measure absorbance at 450 nm.
14. When serum/plasma samples are used, calculate GP-His-specific OD values by subtracting nonspecific binding (i.e., [OD values given by GP-His] – [OD values given by control antigens] for each sample) (*see* **Note 2**).

3.3 Preparation of VSV Pseudotyped with Filovirus GPs

1. Seed HEK293T cells in a 10 cm dish (2.5–3.5 × 10^5 cells/mL) in 10 mL of DMEM containing 10% FCS, L-glutamine, and antibiotics (*see* **Note 12**).
2. Incubate cells for 12–20 h at 37 °C in a 5% CO_2 incubator. Cells should be ~80% confluent at the time of transfection.
3. Prior to transfection of HEK293T cells with GP-expressing plasmids, warm OPTI-MEM and TransIT-LT1 to room temperature.
4. For 1 dish, add 10 μg of GP-expressing plasmid to 500 μL of OPTI-MEM and mix.
5. Add 20 μL of TransIT-LT1 and mix well.
6. Incubate for 30 min at room temperature.
7. Add TransIT-LT1/plasmid mixture dropwise to different areas of the dish.

8. Gently rock the dish for even distribution of complexes.
9. Incubate cells for 12–20 h at 37 °C in a 5% CO_2 incubator (*see* **Note 13**).
10. For infection, prepare an appropriate dilution of VSVΔG-VSVG (*see* **Note 14**).
11. Carefully remove the medium of transfected HEK293T cells, and gently inoculate VSVΔG-VSVG.
12. Incubate for 1 h at 37 °C in a 5% CO_2 incubator, gently rocking the dish every 10–15 min.
13. Remove the inoculum and gently wash the dish with FCS-free DMEM three times.
14. Add 5 mL of DMEM containing 10% FCS, L-glutamine, and antibiotics.
15. Incubate cells for 16–20 h at 37 °C in a 5% CO_2 incubator.
16. Harvest infected HEK293T cell culture supernatant.
17. Centrifuge at 2000 × *g* for 10 min at 4 °C.
18. Aliquot the supernatant (0.1–1.0 mL) and store at −80 °C.

3.4 Neutralization Assay Using Vero E6 Cells

1. Seed Vero E6 cells in a 96-well plate (2.5–4 × 10^4 cells/100 μL/well) in DMEM containing 10% FCS, L-glutamine, and antibiotics.
2. Incubate cells for 12–20 h at 37 °C in a 5% CO_2 incubator. Cells should be 100% confluent at the time of infection.
3. To prepare the virus-antibody mixture, first prepare the required amounts of virus solution (e.g., 55 μL/well) appropriately diluted with FCS-free DMEM (*see* **Note 15**).
4. Prepare antibody dilutions (e.g., 55 μL/tube or well of 1:100, 1:1000, or serial dilutions) in 0.5–1.5 mL tubes or 96-well plates with FCS-free DMEM for dilution.
5. Add equal amounts of the virus solution (e.g., 55 μL/tube) to all antibody dilutions and mix well.
6. Incubate 30–60 min at room temperature.
7. To infect Vero E6 cells with the virus-antibody mixture, inoculate 100 μL of virus-antibody mixture into each well.
8. Incubate for 18–24 h at 37 °C in a 5% CO_2 incubator.
9. Count infected (GFP-positive) cells under a fluorescence microscope (*see* **Note 16**).

3.5 Fc Receptor-Mediated ADE Assay Using K562 Cells

1. Seed K562 cells in a 96-well plate (1–2 × 10^5 cells/100 μL/well) in RPMI containing 10% FCS, L-glutamine, and antibiotics.
2. Incubate cells for 10–60 min at 37 °C in a 5% CO_2 incubator. Cells should be 80–100% confluent at the time of infection.

3. To prepare the virus-antibody mixture, first prepare the required amounts of virus solution (e.g., 55 μL/well) appropriately diluted with FCS-free RPMI (*see* **Note 17**).
4. Prepare antibody dilutions (e.g., 55 μL/tube or well of 1:100, 1:1000, or serial dilutions) in 0.5–1.5 mL tubes or 96-well plates with FCS-free RPMI for dilution.
5. Add equal amounts of virus solution (e.g., 55 μL/tube) to all antibody dilutions and mix well.
6. Incubate for 30–60 min at room temperature.
7. To infect K562 cells with the virus-antibody mixture, inoculate 100 μL of virus-antibody mixture into each well (*see* **Note 18**).
8. Incubate for 18–24 h at 37 °C in a 5% CO_2 incubator.
9. Count infected (GFP-positive) cells under a fluorescence microscope or by flow cytometry (*see* **Notes 16** and **19**).

3.6 C1q-Mediated ADE Assay Using HEK293 Cells

1. Seed HEK293 cells in a 96-well plate (2–4 × 10^4 cells/100 μL/well) in DMEM containing 10% FCS, L-glutamine, and antibiotics.
2. Incubate cells for 12–20 h at 37 °C in a 5% CO_2 incubator. Cells should be 100% confluent at the time of infection.
3. To prepare the virus-antibody mixture, first prepare the required amounts of virus solution (e.g., 55 μL/well) appropriately diluted with FCS-free DMEM (*see* **Note 20**).
4. Prepare antibody dilutions (e.g., 55 μL/tube or well of 1:100, 1:1000, or serial dilutions) in 0.5–1.5 mL tubes or 96-well plates with FCS-free DMEM containing C1q (20–100 μg/mL) (*see* **Notes 21** and **22**).
5. Add equal amounts of virus solution (e.g., 55 μL/tube) to all antibody-C1q dilutions and mix well.
6. Incubate for 30–60 min at room temperature.
7. To infect HEK293 cells with the virus-antibody mixture, inoculate 100 μL of virus-antibody-C1q mixture into each well.
8. Incubate for 18–24 h at 37 °C in a 5% CO_2 incubator.
9. Count infected (GFP-positive) cells under a fluorescence microscope (*see* **Note 16**).

4 Notes

1. Compositions of these buffers are indicated in the protocol supplied by the company (Invitrogen). However, imidazole concentrations in the wash buffer should be optimized (10–20 mM) depending on the individual conditions (e.g., GP concentrations in the supernatant). A reduced concentration (e.g., 15 mM) may sometimes give better results.

2. When serum/plasma samples are used, it is important to prepare control antigens (FCS-derived proteins nonspecifically bound to the Ni-NTA resin). Use the supernatant of mock-transfected cells to prepare control antigens using the Ni-NTA column under the same conditions (10–20 mM wash buffer may be used). Since the GP-His preparation contains small amounts of FCS-derived impurities and some serum/plasma samples often contain nonspecific antibodies highly reactive to the impurities, it is quite important to compare the OD values of the His-GP and control antigens.
3. Note that HEK293T is derived from HEK293, but these are different cell lines. Use HEK 293 cells for ADE assays.
4. This resin volume should be a little more than actually needed since it will decrease during the repeated washing and centrifuge steps.
5. Use a gentle brake setting if the centrifuge has a function to control braking.
6. If the resin volume is small, gently mix only by tapping.
7. The column size depends on the resin volume. Also consider the volume of binding buffer (6× resin volume in total) added together into the column.
8. This step is optional to collect the residual resin in the tube and thus can be skipped if larger amounts of the resin are used.
9. As mentioned in **Note 2**, some serum samples contain nonspecific antibodies highly reactive to FCS-derived impurities. To absorb such antibodies, this diluent contains 2% FCS.
10. For example, anti-human IgG (Jackson ImmunoResearch; 109-036-088), anti-monkey IgG (ROCKLAND; 617-103-012), anti-pig IgG (BETHYL; A100-105P), anti-bat IgG (BETHYL; A140-118P), and anti-mouse IgG (Jackson ImmunoResearch; 115-035-062) can be used.
11. Keep this value constant between plates and experiments once this incubation time is established. In addition, the temperature of the substrate solution is important. The TMB solution should be warmed to room temperature before use for each experiment.
12. Since HEK293T cells do not strongly attach to plastic dishes or plates, Poly-L-Lysine coating is recommended. Use 0.01% Poly-L-Lysine, and rinse the dish or plate at least twice with sterile water, PBS, or cell culture medium before seeding the cells.
13. The expression of filovirus GPs causes cell rounding.
14. It is important to infect cells at a multiplicity of infection of more than 1.0.

15. When replication-incompetent GFP-expressing pseudotyped VSVs are used, use 1000–1500 infectious units (IU)/well (i.e., virus dilutions that give 1000–1500 GFP-positive cells/well). Anti-VSV G protein neutralizing mAbs are sometimes used to reduce the background infectivity conferred by the residual VSVΔG-VSVG in GP-pseudotyped virus stock solutions.
16. Modify conditions and methods (plates, incubation time, fluorescent staining, etc.) as needed when authentic filoviruses are used. In these cases, assays for determining focus-forming units or plaque-forming units may be used to quantify virus titers.
17. Note that K562 cells are much less susceptible to infection than Vero E6 cells. Determine the titer of each virus stock in K562 cells before ADE assays are performed. When replication-incompetent GFP-expressing pseudotyped VSVs are used, dilute the virus to give 50–150 IU/well in K562 cells.
18. Gentle pipetting is recommended to agitate the cells.
19. If the cells are not dispersed in one layer (i.e., there are too many cells), a flow cytometer should be used.
20. Note that HEK293 cells are less susceptible to infection than Vero E6 cells. Determine the titer of each virus stock in HEK293 cells before ADE assays are performed. When replication-incompetent GFP-expressing pseudotyped VSVs are used, dilute the virus to give 100–200 IU/well in HEK293 cells.
21. Since C1q activity differs depending on the production lot, it may be necessary to optimize its concentration.
22. When intact (i.e., heat-untreated) serum samples are used at low dilutions (e.g., 1:10), 0.05 M ethylene glycol tetraacetic acid (EGTA) treatment, instead of the addition of C1q, often increases the C1q-mediated ADE [3, 9].

Acknowledgments

This work was supported by the Agency for Medical Research and Development (AMED) and Japan International Cooperation Agency (JICA) within the framework of the Science and Technology Research Partnership for Sustainable Development (SATREPS), the Japan Initiative for Global Research Network on Infectious Diseases (J-GRID), and a Grant-in-Aid from the Ministry of Education, Culture, Sports, Science and Technology (MEXT).

References

1. Takada A, Robison C, Goto H, Sanchez A, Murti KG, Whitt MA, Kawaoka Y (1997) A system for functional analysis of Ebola virus glycoprotein. Proc Natl Acad Sci U S A 94:14764–14769
2. Wool-Lewis RJ, Bates P (1998) Characterization of Ebola virus entry by using pseudotyped viruses: identification of receptor-deficient cell lines. J Virol 72:3155–3160
3. Takada A, Feldmann H, Ksiazek TG, Kawaoka Y (2003) Antibody-dependent enhancement of Ebola virus infection. J Virol 77:7539–7544
4. Takada A, Kawaoka Y (2003) Antibody-dependent enhancement of viral infection: molecular mechanisms and in vivo implications. Rev Med Virol 13:387–398
5. Takada A, Ebihara H, Feldmann H, Geisbert TW, Kawaoka Y (2007) Epitopes required for antibody-dependent enhancement of Ebola virus infection. J Infect Dis 196(Suppl 2):S347–S356
6. Nakayama E, Tomabechi D, Matsuno K, Kishida N, Yoshida R, Feldmann H, Takada A (2011) Antibody-dependent enhancement of Marburg virus infection. J Infect Dis 204(Suppl 3):S978–S985
7. Nakayama E, Yokoyama A, Miyamoto H, Igarashi M, Kishida N, Matsuno K, Marzi A, Feldmann H, Ito K, Saijo M, Takada A (2010) Enzyme-linked immunosorbent assay for detection of filovirus species-specific antibodies. Clin Vaccine Immunol 17:1723–1728
8. Maruyama J, Miyamoto H, Kajihara M, Ogawa H, Maeda K, Sakoda Y, Yoshida R, Takada A (2014) Characterization of the envelope glycoprotein of a novel filovirus, Lloviu virus. J Virol 88:99–109
9. Takada A, Watanabe S, Okazaki K, Kida H, Kawaoka Y (2001) Infectivity-enhancing antibodies to Ebola virus glycoprotein. J Virol 75:2324–2330

Chapter 26

Monitoring Innate Immune Gene Responses in the Hamster Model of Ebola Virus Disease by RT-PCR

Marko Zivcec

Abstract

Ebola virus (EBOV) disease is a severe, acute human syndrome associated with high case fatality rates. Immune responses to EBOV are thought to be at least partially responsible for disease pathogenesis and must therefore be investigated to get a better understanding of underlying mechanisms of pathogenesis. Syrian hamsters are susceptible to EBOV infection and develop a disease more consistent with human EBOV disease than other rodent disease models. Quantitative RT-PCR (qRT-PCR) is ideal for monitoring immune responses during EBOV infection in low- to medium-throughput applications. A relatively straightforward protocol for monitoring immune responses, based on information gleaned from experimental EBOV infection of hamsters, is presented.

Key words Syrian hamster, Animal model, qRT-PCR, Gene response

1 Introduction

Ebolaviruses are zoonotic pathogens that cause severe, life-threatening disease in humans. The recent Ebola virus (EBOV) outbreak in West Africa underscores the need to continue research on this important pathogen. We recently described a Syrian hamster (*Mesocricetus auratus*) model of EBOV disease (EVD) which mimics hallmarks of human EVD and here present reverse transcriptase-polymerase chain reaction (RT-PCR) protocols that offer a sensitive and specific method that may be used to quantify immune responses to EBOV infection in Syrian hamsters [1, 2]. As Syrian hamsters develop EVD when infected with mouse-adapted EBOV, but not wild-type EBOV, researchers may also use this model to compare differences in protective versus non-protective innate immune responses to EBOV infection [2]. The Syrian hamster model, despite its advantages compared to other rodent EVD models, suffers from a lack of commercially available reagents and, until recently, lacked genome and transcriptome sequence information [3]. RT-PCR is a sensitive technique for quantifying RNA

Thomas Hoenen and Allison Groseth (eds.), *Ebolaviruses: Methods and Protocols*, Methods in Molecular Biology, vol. 1628, DOI 10.1007/978-1-4939-7116-9_26, © Springer Science+Business Media LLC 2017

in fluids and tissues. Due to its simplicity, robustness, speed, and target sequence specificity, quantitative RT-PCR (qRT-PCR) is still the method of choice for quantifying genes rapidly and for detecting partial sequences and genes of low relative abundance. Furthermore, qRT-PCR is ideal for low- to medium-throughput applications and requires fairly low investment in terms of equipment, number of trained staff, computing power and storage, lab space, and maintenance/upkeep compared to high-throughput systems. Furthermore, this method may be adapted to other less frequently used animal disease models that lack full-length genome or transcriptome sequence. A relatively uncomplicated protocol for monitoring innate immune gene transcription by qRT-PCR in response to EBOV infection in hamsters is presented.

2 Materials

2.1 Sample Collection

1. Hamsters, or other susceptible animal species, infected with EBOV (*see* **Note 1**).
2. Buffer RLT (for tissues) or AVL (for fluids) (Qiagen) (*see* **Notes 2** and **3**).
3. Optional: RNAlater for storing tissues (Qiagen).
4. Metal beads (5–7 mm diameter, depending on the diameter of the tubes used) and Tissue Lyser II (Qiagen).

2.2 RNA Extraction

1. QIAamp Viral RNA Mini Kit (for body fluids) or RNeasy Plus Mini Kit (for tissues) (Qiagen) (*see* **Note 2**).
2. UV-Vis spectrophotometer (e.g., NanoDrop 8000).

2.3 RT-PCR

1. Superscript III reverse transcriptase.
2. Random hexamers.
3. Target gene primers (Table 1).
4. *Taq* PCR Master Mix Kit (Qiagen).

2.4 qRT-PCR

1. Target gene primers and probes (Table 1) [1].
2. Internal control (housekeeping) gene primers and probes (Table 1) [2] (*see* **Notes 4** and **5**).
3. Rotor-Gene Probe RT-PCR kit (Qiagen).
4. Rotor-Gene Q 5-plex HRM or 2-plex HRM instrument (Qiagen).
5. Computer with Rotor-Gene Q software (Qiagen) and Excel software (Microsoft).

Table 1
Sequence information of the primers and probes

Gene (accession number)	Oligo name	Oligo sequence	Size (bp)
β-2-Microglobulin (X17002)	**B2M F**	**GGCTCACAGGGAGTTTGTAC**	76
	B2M R	**TGGGCTCCTTCAGAGTTATG**	
	B2M TM	**YAK-CTGCGACTGATAAATACGCCTGCA-BBQ**	
β-actin (AJ312092)	**bactin F**	**ACTGCCGCATCCTCTTCCT**	73
	bactin R	**TCGTTGCCAATGGTGATGAC**	
	bactin TM	**YAK-CCTGGAGAAGAGCTATGAGCTGCCTGATG-BBQ**	
B-cell lymphoma 2 protein (bcl-2) (AJ582074)	Bcl2 F	CTTCGCAGAGATGTCCAGTC	165
	Bcl2 R	CATCTCCCTGTTGACGCTC	
	Bcl2 TM	6FAM-TGACGCCCTTCACCGCGA-BBQ	
Bcl-2-associated protein (AJ582075)	Bax F	GGCAACTTCAACTGGGG	163
	Bax R	CCACCCTGGTCTTGGATC	
	Bax TM	6FAM-CCAGCCCATGATGGTTCTGATTAGC-BBQ	
CD83 protein (DQ094177)	CD83 F	AACCTGGTACGGAACAAGCT	200
	CD83 R	CAAAGGAAGGTTGCCGTC	
	CD83 TM	6FAM-TCCAGGCAGCATTCAGGTACACTGA-BBQ	
Chemokine (C-X-C motif) ligand 10 (IP-10) (AY007988)	IP-10F	GCCATTCATCCACAGTTGACA	66
	IP-10 R	CATGGTGCTGACAGTGGAGTCT	
	IP-10 TM	6FAM-CGTCCCGAGCCAGCCAACGA-BBQ	
Chemokine CCL20/MIP-3 α (AY924377)	CCL20 F	AGTCAGTCAGAAGCAAGCAACT	201
	CCL20 R	TGAAGCGGTGCATGATCC	
	CCL20 TM	6FAM-CACAAGGAGCACTATCCCACCCAGA-BBQ	
Chemokine ligand 17 (FJ664143)	CL17 F	CGAGTGCTGCCTGGAGATC	163
	CL17 R	TGATGGCCTTCTTCACATGC	
	CL17 TM	6FAM-TGGACCTGCCCTGGACAGTCACA-BBQ	

(continued)

Table 1
(continued)

Gene (accession number)	Oligo name	Oligo sequence	Size (bp)
Chemokine ligand 22 (FJ664144)	CL22 F	CGCGTAGTGAAGGAGTTCTTC	143
	CL22 R	TCTTCACCAGGCCAGCTTA	
	CL22 TM	6FAM-ACCTCAAAGTCCTGCCGCAAGCC-BBQ	
Claudulin-1 (EU856105)	ham cld1 F2	GCCACAGCATGGTATGGAA	146
	ham cld1 R1	GCAAGAAAGTAGGGCACCTC	
	ham cld1 TM	6FAM-CCCGTCAATGCCAGGTATGAATT-BBQ	
Complement C3 (complement C3d region) (AB024425)	CC3d F	GGAGCCTTACCTCAGCAAGT	188
	CC3d R	TAGCCGCCTCCGTAGTATCT	
	CC3d TM	6FAM-CAGAAGCTCTACAATGTGGAGGCCA-BBQ	
Complement component 5 (DQ369042)	CC5 F	GTAGTTCCCGATGCTGAAGTG	118
	CC5 R	TGATTAACTCCATTGACCAACG	
	CC5 TM	6FAM-TGTGACTTGCATCGCTTTCGGC-BBQ	
Complement protein C1qBP (DQ367730)	CP1qBP F	CAGAGGATGAGGTTGGACAA	155
	CP1qBP R	CCATTAGGTGGTCATACAAGGC	
	CP1qBP TM	6FAM-TCCATTCAGAGTCACCAGTGGTCTGGA-BBQ	
E-cadherin (DQ237892)	Ecad F	GTTAAGGTTCTGGAGATGAGATTGG	76
	Ecad R	CATCTTTCCCCTCCGAGACA	
	Ecad TM	6FAM-TTATGTAGATGACCATGACTTTAATGACAA-BBQ	
Epithelial mucin (Muc1) (L41545)	Muc1 F	CGGAAGAACTATGGGCAGCT	181
	Muc 1 R	GCCACTACTGGGTTGGTGTAAG	
	Muc 1 TM	6FAM-TGCCTGCCGAGACCTCCTCGTA-BBQ	
Fibrinogen A α-chain (D43757)	FAAC F	GCACAAGCACGACACGT	144
	FAAC R	TGGGTCATGCCTAAGTCTCC	
	FAAC TM	6FAM-CGATGGTCACCGAGAAGTGGTCA-BBQ	
Forkhead box P3 (FJ664148)	FbP3 F	AAGCAGATCACCTCCTGGAT	81
	FbP3 R	AGCTGCTGCTCCAGAGAC	
	FbP3 TM	6FAM-CACCACTTCTCTCTGGAGGAGGCAC-BBQ	

Hypoxanthine phosphoribosyltransferase (AF047041)	**HPRT F**	**TGCGGATGATATCTCAACTTTAACTG**	**93**
	HPRT R	**AAAGGAAAGCAAAGTTTGTATTGTCA**	
	HPRT TM	**YAK-AAAGAATGTCTTGATTGTTGAAGGTAAAACTGA CATTGG-BBQ**	
Intracellular adhesion molecule-1 (DQ093373)	ICAM1 F	TGCAGCCGGAGAACAGATG	64
	ICAM1 R	ATCTCCCGTGTGACAGTCTTCA	
	ICAM1 TM	6FAM-AGCCCTGCTGCCCATCGGG-BBQ	
Interferon-α inducible protein (p27-h) (AF212039)	p27 F	TCGTTGCTGCTCCCGTAGTC	66
	p27 R	ATGGATCCCGCTGCAATTC	
	p27 TM	6FAM-TGGGTGCTGTGGGCTTCACTGG-BBQ	
Interferon-γ (AF034482)	IFNg F	GGCCATCCAGAGGAGCATAG	72
	IFNg R	TTTCTCCATGCTGCTGTTGAA	
	IFNg TM	6FAM-CACCATCAAGGCAGACCTGTTTGCTAACTT-BBQ	
Interferon regulatory factor-1 (DQ092344)	IRF1 F	GGCATACAACATGTCTTCACG	229
	IRF1 R	GCTATGCTTTGCCATGTCAA	
	IRF1 TM	6FAM-CACAATGACGCCAGACCTTGCTCA-BBQ	
Interferon regulatory factor-2 (AY714581)	IRF2 F	AATGCCTTCAGAGTGTACCG	213
	IRF2 R	TGTTCACCGTACTATCCACTTCAT	
	IRF2 TM	6FAM-CTGAAGTCAGGACCGCATACTCAGGA-BBQ	
Interleukin-1β (AB028497)	IL-1b F	GGCTGATGCTCCCATTCG	63
	IL-1b R	CACGAGGCATTTCTGTTGTTCA	
	IL-1b TM	6FAM-CAGCTGCACTGCAGGCTCCGAG-BBQ	
Interleukin-2 (EU729351)	IL-2F	GTGCACCCACTTCAAGCTCTAA	73
	IL-2 R	AAGCTCCTGTAAGTCCAGCAGTAAC	
	IL-2 TM	6FAM-AGGAAACCCAGCAGCACCTCGAGC-BBQ	
Interleukin-4 (AF046213)	IL-4F	CCACGGAGAAAGACCTCATCTG	71
	IL-4 R	GGGTCACCTCATGTTGGAAATAAA	
	IL-4 TM	6FAM-CAGGGCTTCCCAGGTGCTTCGCAAGT-BBQ	

(continued)

Table 1
(continued)

Gene (accession number)	Oligo name	Oligo sequence	Size (bp)
Interleukin-6 (AB028635)	IL-6F IL-6 R IL-6 TM	CCTGAAAGCACTTGAAGAATTCC GGTATGCTAAGGCACAGCACACT 6FAM-AGAAGTCACCATGAGGTCTACTCGGCAAAA-BBQ	78
Interleukin-10 (AF046210)	IL-10F IL-10 R IL-10 TM	GTTGCCAAACCTTATCAGAAATGA TTCTGGCCCGTGGTTCTCT 6FAM-CAGTTTTACCTGGTAGAAGTGATGCCCCAGG-BBQ	76
Interleukin-12 p35 subunit (AB085791)	IL-12p35 F IL-12p35 R IL-12p35 TM	GGCCTTCCCTGGCAGAA ATGCTGAAAGCCTGCAGTAGAAT 6FAM-CGGATCCCTACAAAGTGAAAATGAAGCTCTG-BBQ	72
Interleukin-12 p40 subunit (AB085792)	IL12p40 F IL12p40 R IL12p40 TM	TGGTTACCTCCTTAGCAGTCC TCAGCCTGATGATGAACCTGA 6FAM-TCCAGAGTGCCATAATAGCCACACAAA-BBQ	175
Interleukin-21 (FJ664142)	IL21 F IL21 R IL21 TM	TCAACTGATGTGAAAGGAGC ATCTTGTGGAGCTGGCAG 6FAM-TCAGGGTCCTAGCCAAAAGAGAATC-BBQ	94
Interleukin-2 receptor-α (DQ093372)	IL2Ra F IL2Ra R IL2Ra TM	AAAGCAAGCTACACCTAACCC GCCTTGTATCCTTGAATGCG 6FAM-CAGAAATCAGCACAGTCTGTGCACCA-BBQ	200
Junction adhesion molecule (EU856104)	ham jam F1 ham jam R1 ham jam TM	CGTCCAAGTTCCCGAGAGTA CGTGATCTGGCTGTTATAGCA 6FAM-TAGTGCCACCCTGGACGAACTTC-BBQ	129
Matrix metalloproteinase-2 (AF260254)	MM2 F MM2 R MM2 TM	GATGCTGCCTTTAACTGGAGT GAGCTTAGGGAAACCAGGAT 6FAM-CATACATCTTCGCTGGAGACAAGTTC-BBQ	110
MHC class II antigen alpha chain (DQ092501)	MHCAAC F1 MHCAAC R MHCAAC TM	CAGGGAGGACTGCAAGCTATA TGTCCACGAAGCAGATGAG 6FAM-TGCAGCAAAGCAGAACTTGGACATC-BBQ	162

Myxovirus resistance protein-2 (EU616539)	Mx2 F Mx2 R Mx2 TM	CCAGTAATGTGGACATTGCC CATCAACGACCTTGTCTTCAGTA 6FAM-TGTCCACCAGATCAGGCTTGGTCA-BBQ	134
Nitric oxide synthase-2 (DQ355357)	NOS2 F NOS2 R NOS2 TM	TGCCTTGCATCCTCATTGG GTCGCTGTTGCCAGAAACTG 6FAM-CCTGGCACGGGCATCGCTC-BBQ	65
Occludin (EU856106)	ham occ F1 ham occ R1 ham occ TM	CTATTCTGGGCATCCTGGT TTGCACATGGCATAGATCTG 6FAM-AGTCAACCCAACTGCCCAGGCT-BBQ	111
p75 tumor necrosis factor membrane receptor (AF315291)	p75 F p75 R p75 TM	CCCCAGGCCACAGTCAC GCCGTGGGAGGAATCTGAA 6FAM-CTGCACAGGCCTCCTGAGACCCT-BBQ	80
Platelet endothelial adhesion molecule (AF508040)	PECAMF PECAMR PECAM TM	CAGGATCAGAACTTCAGCAAGAT GCAGCTGATGGTTATAGCATGT 6FAM-TGTACCGCAGGCATCGGCAGA-BBQ	181
Protein kinase R (DQ645944)	Eif2ak2 F Eif2ak2 R Eif2ak2 TM	ACGGACCTAAGAGATGGCAT AGGTAACTAAAGCGGAGTGC 6FAM-CCACGGATCGACCTAGTGCTTCTGA-BBQ	208
Ribosomal protein L 18 (DQ403027)	**RPL18 F** **RPL18 R** **RPL18 TM**	**GTTTATGAGTCGCACTAACCG** **TGTTCTCTCGGCCAGGAA** **YAK-TCTGTCCCTGTCCCGGATGATC-BBQ**	**79**
Signal transducer and activator of transcription-1 (DQ092343)	STAT1 F STAT1 R STAT1 TM	GCCAACGATGATTCCTTTGC GCTATATTGGTCATCCAGCTGAGA 6FAM-ACCATCCGTTTCCATGACCTCC-BBQ	68
Signal transducer and activator of transcription-1 β (AB177397)	STAT1b F STAT1b R STAT1b TM	AGGTCCGTCAGCAGCTTAA GCCGTTCCACCACAAAT 6FAM-TCTGAATGAGCTGCTGGAAGAGGACA-BBQ	149

(continued)

Table 1
(continued)

Gene (accession number)	Oligo name	Oligo sequence	Size (bp)
Tight junction protein 2 (EU856099)	ham tjp2 F1	CTACACTGACAATGAGCTGGA	111
	ham tjp2 R1	CTCTGGGCTGGATTTCCTTA	
	ham tjp2 TM	6FAM-TCATGCTGCACCGGCTCCGA-BBQ	
Tissue inhibitor of matrix metalloproteinase-2 (AF260255)	TIMM2 F	AGAGCCTGAACCACAGGT	235
	TIMM2 R	CGGGTCCTCGATGTCAA	
	TIMM2 TM	6FAM-CGAGTGCAAGATCACACGCTGCC-BBQ	
Transforming growth factor-β1 (AF046214)	TGFb F	TGTGTGCGGCAGCTGTACA	63
	TGFb R	TGGGCTCGTGAATCCACTTC	
	TGFb TM	6FAM-CGACTTTCGCAAGGACCTGGGCT-BBQ	
Transforming growth factor 2 (AY007214)	TGFb2 F	TGCTGCCCTCCTACAGACT	186
	TGFb2 R	GCACAGAAGTTGGCATTATACC	
	TGFb2 TM	6FAM-CACAACAGTCCAATCGGCGGA-BBQ	
Transforming growth factor-β 3 (AF298188)	TGFb3 F	CAAGCTCAGGCTCACCAGT	148
	TGFb3 R	CCGACTCTGTGTTCTCCTGAG	
	TGFb3 TM	6FAM-AGCCATCGGTGATGACCCACGT-BBQ	
Transforming growth factor-β type I receptor (AF298187)	TGFbTIR F	ATCAAACTTGCTCTGTCTACGG	183
	TGFbTIR R	TGTCTGTGGCAGAATCATGC	
	TGFbTIR TM	6FAM-ACAGCCAGTCCCAAGTCTGCAATAC-BBQ	
Tumor necrosis factor-α (AF315292)	TNFa F	GGAGTGGCTGAGCCATCGT	65
	TNFa R	AGCTGGTTGTCTTTGAGAGACATG	
	TNFa TM	6FAM-CCAATGCCCTCCTGGCCAACG-BBQ	
Vascular endothelial growth factor (AF297627)	VEGF F	CAGGAGTACCCCGATGAGATAGA	67
	VEGF R	CCCCCACACCGCATCA	
	VEGF TM	6FAM-TCTTCAAGCCGTCCTGTGTGCCC-BBQ	

Primers and probes were designed from Syrian golden hamster-specific sequences available on GenBank and are displayed next to their expected amplicon size. The melting temperatures of all the primers and probes were ~55 °C and ~65 °C, respectively. 6FAM and Yakima Yellow (YAK) fluorescent dyes and the BlackBerry Quencher (BBQ) from TIB MOLBIOL were used in all experiments. Genes used as internal controls are highlighted in bold. *See* **Note 10**

3 Methods

For an overview of the procedure *see* Fig. 1.

3.1 Harvesting Animal Organs

1. In a biosafety level 4 (BSL-4) laboratory, infect animals as dictated by the study plan and approved by the Institutional Animal Care and Use Committee (IACUC) (*see* **Note 1**).
2. Anesthetize animals as necessary, draw blood, and euthanize according to approved animal study protocol using institutional standard operating procedures (SOPs) approved by the Institutional Biosafety Committee (IBC), e.g., by barbiturate overdose.

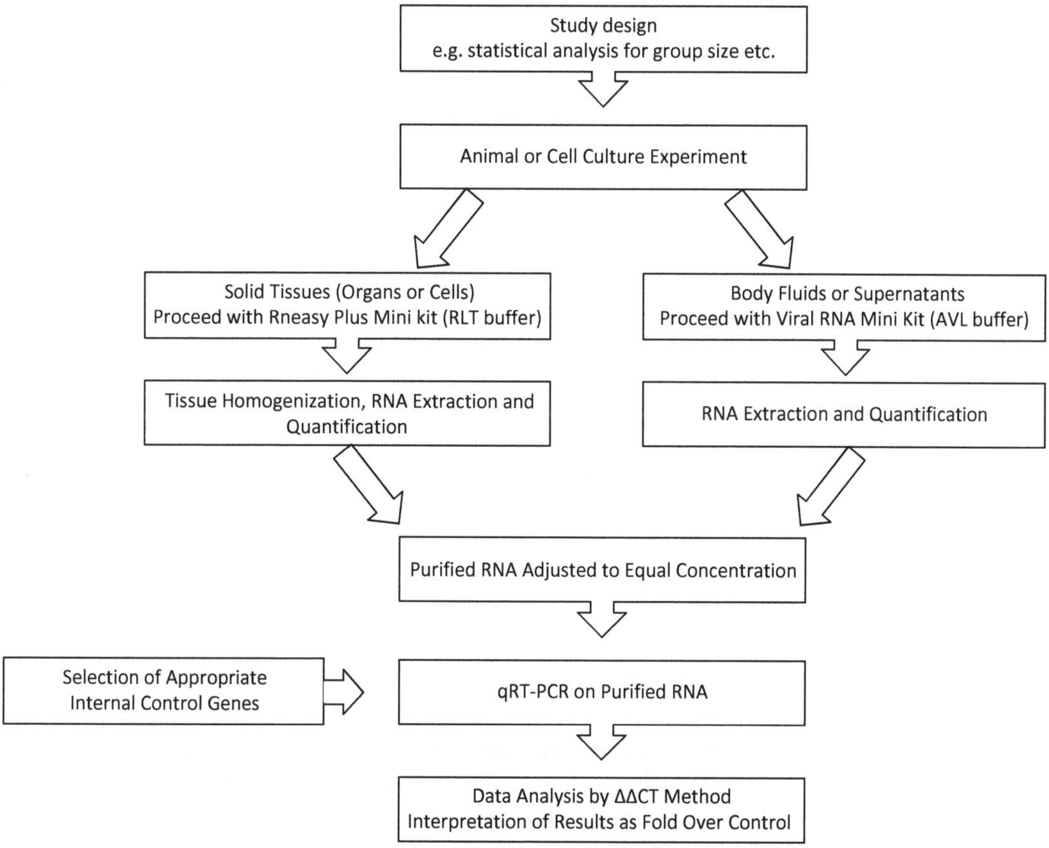

Fig. 1 Flowchart of quantitative RT-PCR (qRT-PCR) for Syrian hamster innate immune genes. Prior to initiating experiments, conduct the appropriate statistical analysis to determine the required group size. Larger group sizes allow for discerning of smaller differences and generate more sound data than smaller groups. Next, conduct the animal experiments according to approved animal study protocol using approved institutional standard operating procedures. As the experiment progresses, harvest tissues and fluids as required and place in RLT or AVL buffer, respectively. Extract RNA according to manufacturer's instructions and quantify the concentration. Adjust the RNA to the same concentration prior to conducting qRT-PCR. Select appropriate internal genes for standardization. In the case of EBOV infection of hamsters, β-actin and ribosomal protein L18 RNA were not altered significantly during the course of the disease. Perform qRT-PCR on RNA with a selection of genes described for hamsters [1]. Analyze the data using the ΔΔCT method

3. Perform necropsy on animals and remove target organs.
4. Place approximately 30 mg of tissue (approximately a 3 mm^3 cube) into cryovials with ~1 mL RNAlater (optional). Flash freeze at ≤−70 °C, or proceed immediately to **step 6**. If using RNAlater for storage, remove the RNAlater prior to weighing vials and continuing the protocol at **step 5**.
5. Weigh amount of tissue harvested using a scale, and add 600 μL of RLT buffer per 30 mg of tissue (*see* **Note 3**).
6. Place a metal bead into each tube, and homogenize the sample in a Tissue Lyser II at 30 Hz for 10 min.
7. Remove ~630 μL of the tissue/RLT homogenate, incubate at room temperature for 10 min, and add 630 μL of 70% ethanol. Do not centrifuge or freeze the homogenates once ethanol has been added to the sample homogenate. Transfer the sample homogenate into a new vial. Prepare for surface decontamination as per approved institutional SOPs.
8. Remove samples from BSL-4 as per institutional SOPs for downstream processing.

3.2 Preparing Samples from Tissue Culture

1. Infect cells as directed by the study plan (*see* **Note 1**).
2. Remove culture medium and lyse cells with 600 μL RLT buffer per 10^7 cells (~50–55 cm^2 of confluent cells) for 10 min at room temperature, with shaking (*see* **Note 3**).
3. Collect cell lysates into vials and add 630 μL of 70% ethanol. Do not centrifuge or freeze the homogenates once ethanol has been added to the sample homogenate.
4. Transfer to a new vial and prepare for surface decontamination as per approved institutional SOPs.
5. Remove samples from BSL-4 as per institutional SOPs.

3.3 Preparing Samples from Body Fluids

1. Add 140 μL of body fluids to 560 μL of AVL buffer (*see* **Note 3**).
2. Incubate for 10 min at room temperature and add 560 μL of 95–100% ethanol. Do not centrifuge or freeze the homogenates once ethanol has been added to the sample.
3. Prepare for surface decontamination as per approved institutional SOPs.
4. Remove samples from BSL-4 as per institutional SOPs.

3.4 RNA Extraction

1. Extract RNA from tissues and animal cells (RNeasy Mini Kit) or body fluids and cell supernatants (QIAamp Viral RNA Mini Kit).

   For RNeasy Mini Kit:

   (a) Add ≤700 μL of RLT/ethanol containing sample homogenate to the RNeasy column and briefly (15–60 s) centrifuge at maximum speed (≥8000 × *g*). Discard flow through and repeat until all sample homogenate has passed through the column.

(b) Add 700 μL of wash buffer RW1 and centrifuge briefly (15–60 s) at maximum speed. Discard flow through.

(c) Add 500 μL of wash buffer RPE and centrifuge briefly (15–60 s) at maximum speed. Discard flow through.

(d) Add 500 μL of wash buffer RPE and centrifuge at maximum speed for 2 min.

(e) Remove the column from the 2 mL collection tube containing the flow through and place into new 2 mL collection tube. Centrifuge at maximum speed for 1 min.

(f) Remove the column from the collection vial and place it into a 1.5 mL collection tube (snap cap tube). Add 50 μL of RNase-free water, or elution buffer, directly onto the RNA column membrane and incubate for approximately 1 min at room temperature.

(g) Centrifuge the sample at 6000–8000 × *g* (to prevent shattering the snap cap tube lids) for 1 min. Collect the purified RNA and proceed to **step 2**.

For QIAamp Viral RNA Mini Kit:

(a) Add ≤700 μL of AVL/ethanol containing sample homogenate to the viral RNA column and centrifuge for 1 min at ≥6000 × *g*. Discard flow through and repeat until all sample homogenate has passed through the column.

(b) Add 500 μL of wash buffer AW1 and centrifuge for 60 s at ≥6000 × *g*. Place column into a new, clean 2 mL collection vial.

(c) Add 500 μL of wash buffer AW2 and centrifuge 3 min at maximum (≥8000 × *g*) speed. Place column into a new, clean 2 mL collection vial.

(d) Remove the column from the collection vial containing the flow through and place into new, clean 2 mL collection vial. Centrifuge at maximum speed for 1 min.

(e) Remove the column from the 2 mL collection vial and place it into a 1.5 mL collection tube (snap cap tube). Add 60 μL of RNase-free water, or buffer AVE, directly onto the RNA column membrane and incubate for approximately 1 min at room temperature.

(f) Centrifuge the sample at 6000 × *g* (to prevent shattering the snap cap tube lids) for 1 min. Collect the purified RNA and proceed to **step 2**.

2. Following extraction measure the total amount of RNA extracted from each sample using a UV spectrophotometer, and dilute the RNA (either with RNase-free water or buffer AVE) to a concentration of 40 ng/μL (*see* **Note 6**).

3.5 Primers and Probes

Sequence information for the primers and probes we used for the various hamster genes [1] is supplied in Table 1. We found that β-actin and ribosomal protein L18 (RPL18) had the most stable expression in the tissues of infected and control hamsters [2], and thus these are used as housekeeping gene targets. However, depending on the host species used, other internal control genes may need to be selected [4] (*see* **Notes 4** and **5** for additional information). All experimental gene probes should be labeled with the same probe color; we used 5′ 6FAM dye (green) quenched by 3′ BlackBerry Quencher (BBQ). All internal control gene probes should be labeled with a secondary probe color; we used Yakima Yellow (YAK) dye (yellow) quenched by BBQ. Prior to use in qRT-PCR assays, all of the primer sets should be tested by conventional RT-PCR to ensure that they amplify the expected target transcripts. Appearance of nonspecific bands and/or smears indicates problems with either primer or RNA quality and needs to be rectified prior to proceeding to qRT-PCR.

1. Generate a cDNA pool by using ~100 ng of extracted RNA and using SuperScript III Reverse Transcriptase and random hexamers.
   (a) In a PCR-grade microcentrifuge tube add:

5 μL	Random hexamers (50 ng/μL)
2.5 μL	Extracted hamster RNA
1 μL	dNTP (10 mM)
4.5 μL	RNase-free water
13 μL	Total volume

   (b) Heat mixture in thermal cycler to 65 °C for 5 min followed by incubation at 0–4 °C for 1–5 min.
   (c) Briefly centrifuge to collect the condensation in the tube and add:

4 μL	5× First-strand buffer
1 μL	0.1 M DTT
1 μL	RNaseOUT™ Recombinant RNase Inhibitor
1 μL	SuperScript™ III RT (200 units/μL)
20 μL	Total volume

   (d) Mix the components gently, return the tube to the thermal cycler, and incubate at 25 °C for 5 min followed by 50 °C for 30–60 min. Stop the reaction by heating to 70 °C for 15 min. Keep the cDNA on ice until ready to proceed with the PCR.

(e) (Optional) Add 1 μL RNase H to the cDNA mixture and heat to 37 °C for 15 min to remove complementary RNA.

2. Use the generated cDNA as a template for conventional PCR using *Taq* or equivalent.

(a) In a PCR-grade microcentrifuge tube, add per reaction:

2 μL	10× PCR Buffer or CoralLoad PCR Buffer
1 μL	dNTP (10 mM)
1 μL	Forward/reverse primer mixture (10 μM each)
0.5 μL	Taq DNA polymerase
2.5 μL	cDNA (from **step 1**)
13 μL	Nuclease-free water
20 μL	Total volume

(b) Primer Tm for qRT-PCR primer pairs is ~55 °C, and the PCR products are approximately 100–400 nucleotides long. Therefore the recommended cycling conditions are 95 °C for 2 min, followed by 25–35 cycles of 95 °C for 20 s, 50 °C for 20 s, and 72 °C for 30 s and a final extension for 5 min at 72 °C.

(c) Do not purify PCR mixtures using columns, as most commercial columns will remove DNA fragments <100 bp in size (the size of many qRT-PCR fragments, *see* Table 1).

3. Resolve PCR fragments on an agarose gel (*see* **Note 7**). Due to the short nucleotide length of the amplified DNA, it is recommended to resolve on a 1.5–2% agarose gel and to use a 100 bp DNA standard.

4. Dilute all primers and probes to appropriate stock concentrations, typically 100 μM.

5. Prepare working stocks of primers and probes at a 2:1 ratio (*see* **Note 8**). Avoid free-thaw cycles to minimize probe degradation and high background signal.

For target genes:

Forward primer	20 μM
Reverse primer	20 μM
Fluorescent probe	10 μM

For internal control genes:

Forward primer	10 μM
Reverse primer	10 μM
Fluorescent probe	5 μM

3.6 qRT-PCR Methods

qRT-PCR experiments that will be compared to each other should be performed on the same type of instrument and analysis software and with the same qRT-PCR reagents, e.g., Rotor-Gene 6000 thermal cyclers and Rotor-Gene Probe RT-PCR kits.

1. In a tube, prepare the master mix. The RT-PCR reagents are relatively stable and do not require storage on ice if they will be dispensed and loaded into a thermal cycler within ~30 min. For longer storage (<6 h) keep on ice or at 4 °C. Keeping the master mix for >6 h is not recommended. Combine the volumes listed below for each reaction (*see* **Notes 5, 6**, and **8**).

12.5 μL	2× Rotor-Gene probe RT-PCR master mix
0.25 μL	Target gene primer/probe stock (from Subheading 3.5, **step 5**)
0.25 μL	Control primer/probe stock (from Subheading 3.5, **step 5**)
0.25 μL	Rotor-Gene RT mix
6.75 μL	Nuclease-free water
20.0 μL	Total volume

2. Dispense 20.0 μL of the qRT-PCR master mix per reaction tube.
3. Add 5.0 μL of extracted RNA to each tube and cap the tubes.
4. Move tubes to a Rotor-Gene thermal cycler.
5. Recommended cycling conditions: 50 °C for 10 min and 95 °C for 5 min, followed by 40 cycles of 95 °C for 5 s and 60 °C for 10 s.
6. Acquire data in the green (510 nm) and yellow (555 nm) channels. As the standards are multiplexed with the target genes and the C_T values are normalized in each run, threshold effects will not have as much of an impact as when comparing C_T values between different qRT-PCR runs.

3.7 Data Analysis

The data should be analyzed using the $\Delta\Delta C_T$ method for relative quantification [5] (*see* **Note 9**). The C_T (cycle threshold) of each test gene is defined as the number of cycles required for the fluorescent signal to cross the threshold (i.e., to exceed background level). The C_T value for an infected hamster is then first normalized to the C_T of the control gene (ΔC_T) and then compared to the same normalized gene in an uninfected (calibrator) hamster to determine the $\Delta\Delta C_T$. Finally, the fold change between the infected and uninfected hamsters is determined by using the formula $2^{-\Delta\Delta CT}$.

In other words: Fold increase = $2^{-(\text{Infected}(\text{CTtarget} - \text{CTRPL18}) - \text{Uninfected}(\text{CTtarget} - \text{CTRPL18}))}$

Sample calculation:

In determining IL-6 gene levels in the liver of an infected hamster during the terminal stage of EBOV infection [2], the sample from the infected hamster yielded an IL-6 C_T of 22.19 and RPL18 yielded a C_T of 19.65, while the average C_T from uninfected control samples were 27.99 (IL-6) and 20.17 (RPL18), respectively. The fold increase in the expression of the IL-6 gene in the infected hamster is calculated as:

$$\text{Fold increase} = 2^{-(\text{Infected}(\text{CTtarget} - \text{CTRPL18}) - \text{Uninfected}(\text{CTtarget} - \text{CTRPL18}))}$$

$$\text{Fold increase} = 2^{-((22.19 - 19.65) - (27.99 - 20.17))}$$

$$\text{Fold increase} = 38.85$$

Thus, the expression of IL-6 in the infected hamster is 38.85-fold higher than in the livers of control hamsters.

4 Notes

1. Study design is essential to the success of any experiment. For qRT-PCR experiments, researchers should ensure that enough biological replicates (cell line infections, animals used) are involved to develop statistically significant results. Technical replicates, such as loading each sample multiple times, only serve to determine the accuracy of pipetting between samples and do not indicate the natural biological variation that is being studied. Ideally, statistical analysis should be conducted prior to undertaking any experiments to determine appropriate group sizes.
2. We recommend the products, instruments, and methods that have worked for us. Other reagents and methods for homogenization of tissues or RNA extraction should yield similar results. Equivalent products or protocols by other manufacturers may also be highly effective.
3. Do not add carrier RNA to buffers AVL or RLT. AVL and RLT will be standardized based on the total amount of RNA isolated, and use of carrier RNA will alter this result.
4. Cellular responses to infection can sometimes alter the transcriptional levels of housekeeping genes, which would alter the baseline to which the levels of the genes of interest are compared. For example, if the levels of β-actin mRNA decrease during infection, an apparent increase in the levels of the gene tested would be detected. Therefore, care should be taken to select reference genes that do not change greatly during infection. In hamsters, we found that RPL18 and β-actin mRNA levels were not significantly altered over the course of EBOV infection, and we used these genes as reference genes [2]. In using a different species of animal or virus, first verify that the levels

of the reference genes do not significantly fluctuate during the course of infection prior to multiplexing with immune gene qRT-PCR.

5. Multiplexing housekeeping genes into each qRT-PCR improves standardization of the results. However, care should be taken when selecting housekeeping genes, as highly expressed housekeeping genes will silence the differences observed in qRT-PCR of genes not expressed at high levels. This is likely because during the PCR, the housekeeping gene consumes the bulk of the active polymerase and nucleotides in the reaction. Ideally, the C_T of the housekeeping gene should be around 25. To achieve this C_T, consider selecting another housekeeping gene or lowering the final concentrations of the housekeeping gene primer and probe.
6. In our experiments, 50 ng/reaction was the lowest amount of RNA required to detect all target genes; however, we generally used 200 ng/reaction [1]. Thus, RNA concentration was adjusted to 40 ng/μL.
7. When resolving RT-PCR fragments generated using TaqMan primers, the size should be between 100 and 400 nucleotides in length, as indicated in Table 1.
8. Whenever possible, make a large stock or master mix of all buffers and reagents, as this will further minimize technical errors during pipetting. Always set up an extra half a reaction, or 5% of total reaction volume (whichever is greater), for any master mix to ensure enough reagent is available. If the master mix will not be dispensed and used immediately (within 30 min of being made), keep on at 4 °C until use. Do keep the master mix for >6 h.
9. Do not compare unstandardized, raw C_T values between samples. qRT-PCR is always a relative value and cannot be compared between samples from different sources without appropriate standardization to a calibrator or, ideally, both to a calibrator and control gene.
10. While hamster immune gene qRT-PCR primers and probes are used as an example here, qRT-PCR can be efficiently developed for other less frequently used animal models of Ebolavirus without the need for complete genome sequence information. When designing primers and probes, ideally there should not be off-target reactivity; the primers should have a high amplification efficiency (≥90%) and a Tm approximately 10 °C lower than the probe; and the amplicons should be relatively short (≤400 bp, in order to be less sensitive to minor RNA degradation) [1].

Acknowledgments

This work was supported by the Intramural Research Program of the National Institute of Allergy and Infectious Diseases (NIAID), National Institutes of Health (NIH). The author would like to thank Tatyana Klimova and Christina Spiropoulou (NCEZID, CDC) and Elaine Haddock (NIAID, NIH) for critical editing of the manuscript. The findings and conclusions in this report are those of the author and do not necessarily represent the official position of the NIAID, NIH.

References

1. Zivcec M, Safronetz D, Haddock E et al (2011) Validation of assays to monitor immune responses in the Syrian golden hamster (Mesocricetus auratus). J Immunol Methods 368:24–35. doi:10.1016/j.jim.2011.02.004
2. Ebihara H, Zivcec M, Gardner D et al (2013) A Syrian golden hamster model recapitulating ebola hemorrhagic fever. J Infect Dis 207:306–318. doi:10.1093/infdis/jis626
3. Tchitchek N, Safronetz D, Rasmussen AL et al (2014) Sequencing, annotation and analysis of the Syrian hamster (Mesocricetus auratus) transcriptome. PLoS One. doi:10.1371/journal.pone.0112617
4. Dheda K, Huggett JF, Bustin SA et al (2004) Validation of housekeeping genes for normalizing RNA expression in real-time PCR. Biotechniques 37:112–119
5. Livak KJ, Schmittgen TD (2001) Analysis of relative gene expression data using real-time quantitative PCR and the 2(-Delta Delta C(T)) Method. Methods 25:402–408. doi:10.1006/meth.2001.1262

Part V

Ebolaviruses in the Field

Chapter 27

Real-Time and End-Point PCR Diagnostics for Ebola Virus

Allen Grolla

Abstract

Reverse transcriptase polymerase chain reaction (RT-PCR)-based techniques allow for highly sensitive and specific detection of RNA viruses. Detection of the amplification products can be achieved using several methods. The following are descriptions of the detection of ebolavirus RNA using end-point RT-PCR (agarose gel visualization of amplification products) and quantitative RT-PCR (Q-RT-PCR), with fluorescent detection using an intercalating dye or detection with the use of 5′ hydrolysis probe assays. All of these techniques can be used to accurately detect the presence of ebolavirus in samples.

Key words Q-RT-PCR, RT-PCR, Inactivation, Diagnostics, Intercalating dye, Agarose gel electrophoresis

1 Introduction

Reverse transcriptase polymerase chain reaction (RT-PCR) techniques are essential tools in ebolavirus diagnostics. Transcription of viral RNA genomes into cDNA, amplification of virus specific sequences by PCR, and detection of these amplification products can be achieved by several approaches. Traditional RT-PCR utilizes sense and antisense primers and targets a region of between 200 and 1000 bp. Detection of amplification is carried out by agarose gel electrophoresis. This approach can be very sensitive and has the advantage of only requiring two primers to be identified in well-conserved regions to establish an assay that can be sensitive as well as broadly reactive [1, 2].

Q-RT-PCR assays are available in several formats, but all allow for the real-time detection of amplification through monitoring for an increase in fluorescence in each reaction site. SYBR Green assays, like traditional RT-PCR, rely on two primers placed in well-conserved regions to define an assay. Fluorescence is detected through the incorporation of the intercalating dye SYBR Green I into double-stranded DNA produced during amplification. Also a very sensitive assay and potentially broadly reactive, SYBR Green I

Thomas Hoenen and Allison Groseth (eds.), *Ebolaviruses: Methods and Protocols*, Methods in Molecular Biology, vol. 1628,
DOI 10.1007/978-1-4939-7116-9_27,

assays require the addition of melting curve analysis at the end of the amplification to verify that amplification products are appropriately sized and not nonspecific products or primer dimers [3]. Melting curve analysis can require subjective interpretation of the results that can be challenging with some samples, and this may limit its adoption in many laboratories.

Q-RT-PCR assays using 5′ hydrolysis probes are currently the most common approach in rapid filovirus diagnostics [2, 4–6]. Design limitations on the primer and probes for these assays can make it difficult to identify target regions that are well conserved in all the known isolates and thus produce an assay that is robust enough to reliably detect new outbreak strains or species [2]. However once designed these assays are very easy to run and interpret and are also conducive to multiplexing to allow multiple assays to be run in a single tube [7].

In practical terms, sensitivity of PCR assays is not normally an issue in diagnosing cases of filovirus infection. Virus titers in Ebola virus infections can reach 10^{10} RNA copies per mL when collected near death and 10^{5} RNA copies per mL at onset of symptoms [4]. What is of importance is the specificity of the assays used (i.e., can the assays allow for potential mismatches that may be present in new strains?). This can be addressed by developing well-designed assays targeting conserved regions, testing for multiple targets, and using multiple assays on initial samples.

Here I describe the techniques used in our facility with specific PCR assay reagents detailed. Alternate suppliers of PCR reagents are widely available, and primer and probe sets have been identified by many groups. Most, if not all, will provide satisfactory results.

2 Materials

2.1 Sample Inactivation and RNA Extraction

1. QIAgen Viral RNA mini kit.
2. 95–100% ethanol.
3. 2 mL screw cap tubes.
4. Disinfectant (quaternary ammonia or hypochlorite solutions: *see* **Note 1**): Prepare working solutions daily from concentrated stocks and fill spray bottles and 1–2 L containers for use as waste disinfection buckets.
5. Exogenous PCR control (MS2 phage or armored RNA) (*see* **Note 2**).
6. Pipetters and aerosol-resistant pipette tips.
7. Microcentrifuge capable of 15,000 × *g*.

2.2 End-Point RT-PCR Analysis

1. Superscript III One-Step RT-PCR System with Platinum Taq DNA Polymerase.

2. Pipetters and aerosol-resistant pipette tips.
3. Control RNA.
4. 1.2% FlashGel cassettes and FlashGel Dock (Lonza).
5. 100–1500 bp DNA ladder and FlashGel Loading Dye (Lonza).
6. Thermocycler.
7. 0.2 mL thin-walled PCR tubes.

2.3 Q-RT-PCR Analysis: Hydrolysis Probe

1. Lightcycler 480 RNA Master Hydrolysis Probe kit.
2. Pipetters and aerosol-resistant pipette tips.
3. Control RNA.
4. Q-PCR thermocycler.
5. Tubes or plates/films compatible with a thermocycler.
6. Centrifuge suitable for tubes or plates.

2.4 Q-RT-PCR Analysis: SYBR Green I

1. QuantiFast SYBR Green RT-PCR kit.
2. Pipetters and aerosol-resistant pipette tips.
3. Control RNA.
4. Q-PCR thermocycler.
5. Tubes or plates/films compatible with a thermocycler.
6. Centrifuge suitable for tubes or plates.

3 Methods

3.1 Sample Inactivation and RNA Extraction

1. Each sample being analyzed will require three 2 mL screw cap tubes prepared: one with 560 μL of buffer AVL (*see* **Note 2**), one with 560 μL of 95–100% ethanol, and an empty tube for long-term storage of the unprocessed sample. Label the tubes in a manner that makes it very easy to identify and distinguish the AVL from the ethanol and from the storage aliquot (*see* **Note 3**).
2. Transfer the samples and the prepared screw cap tubes to the available containment space (*see* **Note 4**). Transfer the sample from its original tube into the long-term sample storage tube and discard the original tube into a disinfectant waste bucket. Transfer 140 μL of sample from the long-term storage tube into a labeled AVL tube. Seal the long-term storage tube and then mix the AVL + sample tube by vigorous inversion for 10 s. Repeat with any additional samples (*see* **Note 5**).
3. Allow for a minimum of 10 min incubation of the sample in AVL and then transfer this entire volume (700 μL) to the screw cap tube containing 95–100% ethanol and discard the now empty AVL tube in the waste bucket. Seal the sample + AVL + ethanol tube and mix by inversion again.

4. Clean up the work area ensuring all surfaces are thoroughly decontaminated using disinfectant spray and waste is sealed and readied for disposal. In a containment lab setting, the long-term storage tube can be placed in an appropriate freezer/refrigeration space, and the sample + AVL + ethanol tube can be brought out of the lab using the appropriate procedures (e.g., sample dunk tank directly to CL2 space or a dunk tank inside the chemical shower). In a field lab, both tubes need to be surface decontaminated for a minimum of 10 min prior to removal from the BSC or glovebox; the long-term storage sample is placed in secure storage on site and the sample + AVL + ethanol tube transferred to the designated sample extraction station.
5. Spin down sample + AVL + ethanol tubes at maximum speed in a microcentrifuge (>15,000 × *g*) for 60 s. Each sample being processed will require one QIAamp mini column and five collection tubes and one 1.5 mL Eppendorf tube.
6. Apply 630 μL of sample + AVL + ethanol to a QIAamp mini column and centrifuge at 15,000 × *g* for 30 s.
7. Transfer column to a new collection tube and add remaining sample onto column and centrifuge as in **step 6**. Avoid adding to the column any pelleted material that might be present in the sample.
8. Transfer column to new collection tube and add 500 μL of buffer AW1 and centrifuge as in **step 6**.
9. Repeat **step 8**. This second wash helps to remove potentially interfering substances that can be present in whole blood samples.
10. Transfer column to new collection tube and add 500 μL of buffer AW2 and centrifuge at >15,000 × *g* for 3 min.
11. Transfer column to new collection tube and centrifuge at >15,000 × *g* for 1 min.
12. Transfer column to a 1.5 mL tube and add 60 μL of buffer AVE and allow to sit for 1 min followed by centrifugation at 6000 × *g* for 1 min. Discard the column, seal the sample tube, and store at −20 °C or below for long-term storage.

3.2 RT-PCR Analysis

1. Collect primer sets (*see* Table 1) and the Superscript III One-Step RT-PCR Kit from the freezer, place in the designated master mix preparation area to thaw, and store at 4 °C until use.
2. Prepare enough master mix solution for the number of reactions needed (i.e., all samples including mock, a no template control [NTC] and at least one positive control) plus one (*see* **Note 6**).

Table 1

Primer and probe sequences for RT-PCR

Ebolaviruses	EVSP (273 bp)	EVSP5	ACATCTTTCTTTCTTTGGGTAAT
		EVSP3	CAGTTCTCAGCCCATTCACCAGCTT
	Filo AB (419 bp)	Filo-A	ATCGGAATTTTTCTTTCTCATT
		Filo-B	ATGTGGTGGGTTATAATAATCACTGACATG
Marburgviruses	MVSP (223 bp)	MVSP5	AAAGTTGCTGATTCCCCTTTGGA
		MVSP3	GCATGAGGGTTTTGACCTTGAAT
	Filo AB (419 bp)	Filo-A	ATCGGAATTTTTCTTTCTCATT
		Filo-B	ATGTGGTGGGTTATAATAATCACTGACATG
MS2	MS2gel (272 bp)	MS2gelF	TGCCTTCGATGTTCTTCGGTG
		MS2gelR	ACTCCGAAGTGCGTATAACGCG

Sequences (5′ to 3′) are shown for sets used in filovirus diagnostics. PCR product sizes are indicated)

Component	Volume for one reaction (μL)	Volume for *N* reactions (μL)
Nuclease-free water	6.2	$(N + 1) \times 6.2$
2× Reaction mix	12.5	$(N + 1) \times 12.5$
20 μM primer mix	0.3	$(N + 1) \times 0.3$
Superscript III RT/Taq mix	1.0	$(N + 1) \times 1.0$
Template RNA	5.0	$(N + 1) \times 5.0$
Total volume	25.0	$(N + 1) \times 25.0$

3. Aliquot the master mix into PCR tubes, add nuclease-free water into the no template control (NTC) tube, and cap all tubes.
4. Transfer tubes to a designated sample addition area and add 5 μL of sample to each tube and firmly cap. Finally add 5 μL of positive control RNA and the cap tube; mix all tubes and briefly spin.
5. Place tubes in the thermocycler and run the following program.

Temperature (°C)	Time (s)	Cycles
50	1200	1
94	120	1
94	15	40
50	30	
68	60	
68	300	1
4	Hold	1

6. Prepare gel for electrophoresis by removing from packaging and then removing the protective strips and adding nuclease-free water to fill each well.
7. Mix 1 μL of FlashGel Loading Dye with 5 μL of PCR product and load onto the gel along with a similarly prepared sample of the DNA ladder and run at 200 V for 8 min (*see* **Note 7**). All samples except the NTC should be positive for the exogenous control, and all patient samples should be positive for RNase P. Record or document the results (*see* **Note 8**).

3.3 Q-RT-PCR Analysis: Hydrolysis Probes

1. Collect primer and probe sets (*see* Table 2) and Lightcycler 480 RNA Master Hydrolysis Probe kit from the freezer, place in the designated master mix preparation area to thaw, and then store at 4 °C until use.
2. Prepare enough master mix solution for the number of reactions needed plus one (*see* **Note 6**).

Component	Volume for one reaction (μL)	Volume for *N* reaction (μL)
Nuclease-free water (vial 3)	7.55	$(N + 1) \times 7.55$
LC 480 master mix (vial 1)	9.25	$(N + 1) \times 9.25$
Activator (vial 2)	1.6	$(N + 1) \times 1.6$
Enhancer (vial 4)	1.0	$(N + 1) \times 1.0$
20 μM primer mix	0.3	$(N + 1) \times 0.3$
10 μM probe mix	0.3	$(N + 1) \times 0.3$
Template RNA	5.0	$(N + 1) \times 5.0$
Total volume	25.0	$(N + 1) \times 25.0$

3. Aliquot the master mix into PCR tubes, add nuclease-free water into the NTC tube, and cap all tubes.
4. Transfer tubes to a designated sample addition area and add 5 μL of sample to each tube and firmly cap. Finally add 5 μL of positive control RNA and cap tube; mix all tubes and briefly spin.
5. Place tubes in the thermocycler and run the following program.

Temperature (°C)	Time (s)	Cycles
61	300	1
95	30	1

Temperature (°C)	Time (s)	Cycles
95	15	45
60	40[a]	
37	30	1

[a]Fluorescent reading taken at the end of this incubation step

6. Cycle threshold (*Ct*) values for reactions should be documented. All samples except the NTC should be positive for the exogenous control, and all patient samples should be positive for RNase P (*see* **Note 9**).

Table 2
Primer and Probe Sequences for Q-RT-PCR with hydrolysis probes

Target	Set	Primer/probe names	Primer/probe sequences
Ebola virus	EBOV L	EBOV LF	CAGCCAGCAATTTCTTCCAT
		EBOV LR	TTTCGGTTGCTGTTTCTGTG
		EBOV LP1	FAM-ATCATTGGCGTACTGGAGGAGCAG
		EBOV LP2	FAM-TCATTGGCGTACTGGAGGAGCAGG
	EBOV NP	EBOV NPF	GACGASGAGGACACTAAGCC
		EBOVNP R	TGGCCCTTTTGACTGTTSTT
		EBOVNP P	FAM-TGCCTAATAGATCRACCAAGGGTGG
Sudan virus	SUDV L	SUDVLF	CAGAAGACAATGCAGCCAGA
		SUDVLR	TTGAGGAATATCCCACAGGC
		SUDV LP	FAM-CTGCTAGCTTGGCCAAAGTCACAAG
	SUDV NP	SUDVNPF	GTGACGAAGATGKTGAGAGC
		SUDVNPR	TTGTAGACTGGTGCTGGTGG
		SUDVNPP	FAM-CAGRGGAGAACAMCCCAACTGTAGC
Bundibugyo virus	BDBV L	BDBVLF1	CCGAGAAAATCCACCAGAAG
		BDBVLF2	TCGGGAAGATCCCCCGGAAG
		BDBVLR	TGTTGRAGTCCCTCAATYCC
		BDBV LP	FAM-YCCAAGCTCTTACCGTGGTCATCTTGG
	BDBV NP	BDBVNPF	RAATGARATCAGCTTCCAGCA
		BDBVNPR	GAGCTTGGCTAGCCTTTCCT
		BDBVNPR2	GTCAATTTGGCCAATCTCTCTT
		BDBVNPP FAM	FAM-ACRACAGCCATGGTCACACTRCGGA

(continued)

Table 2 (continued)

Target	Set	Primer/probe names	Primer/probe sequences
Taï Forest virus	TAFV L	TAFV LF	CGTCATCGCATTGTTGCAA
		TAFV LR	CACTCGACTGTGGGCTTCTG
		TAFV LP	FAM-ATGAGTCCTCCCACGATCATGTTTGTGC
	TAFV NP	TAFV NPF	ACCAGCACCTGTTTATCGGAGTA
		TAFV NPR	ACTCACTTGGTTTGGTTGCTTCT
		TAFV NPP	FAM-AGAAAAGGAGCCCCTCCCCGCAAG
Marburg virus	MARV L	MARVL2F	GCAAAAGCATTCCCTAGTAACATGA
		MARVL2F2	GCGAAGGCATTCCCTAGTAATATGA
		MARVL2R	CACCCCTCACTATRGCGTTYTC
		MARVL2R2	CACCTCTTACTATGGCATTCTC
		MARVL2P	FAM-TGGCACCAYAATTCAGCAAGCATAGG
	MARV GP	MARVGP1F	GTRTGCTCCGGRACYCTCCA
		MARVGP1R	YTGCCCRCTCAGTGTRAATC
		MARVGP1P	FAM-RAARACAGAAGAYGTYCATCTGATGG
MS2	MS2	MS2-TM3-F	GGCTGCTCGCGGATACCC
		MS2-TM3-R	TGAGGGAATGTGGGAACCG
		MS2-TM-P	FAM-ACCTCGGGTTTCCGTCTTGCTCGT
RNase P	RNase P	RNP309F	AGATTTGGACCTGCGAGCG
		RNP353R	GAGCGGCTGTCTCCACAAGT
		RNPpr	FAM-TTCTGACCTGAAGGCTCTGCGCG

Sequences (5′ to 3′) are shown for sets used in filovirus diagnostics. Probes are all shown as dual labeled with 6FAM and BHQ1; alternate probes and quenchers can be used to allow for multiplexing according to the capabilities of the thermocycler being used

3.4 Q-RT-PCR Analysis: SYBR Green I

1. Collect primer sets (*see* Table 3) and the QuantiFast SYBR Green RT-PCR kit from the freezer, place in the designated master mix preparation area to thaw, and then store at 4 °C until use.
2. Prepare enough master mix solution for the number of reactions needed plus one (*see* **Note 6**).

Table 3

Primer and probe sequences for Q-RT-PCR with SYBR Green

Ebolaviruses	EVSP	EVSP5	ACATCTTTCTTTCTTTGGGTAAT
		EVSP3	CAGTTCTCAGCCCATTCACCAGCTT
Marburgviruses	MVSP	MVSP5	AAAGTTGCTGATTCCCCTTTGGA
		MVSP3	GCATGTGGGTTTTGACCTTGAAT
MS2	MS2	MS2-TM3-F	GGCTGCTCGCGGATACCC
		MS2-TM3-R	TGAGGGAATGTGGGAACCG
RNase P	RNase P	RNP309F	AGATTTGGACCTGCGAGCG
		RNP353R	GAGCGGCTGTCTCCACAAGT

Sequences (5′ to 3′) are shown for sets used in filovirus diagnostics

Component	Volume for one reaction (μL)	Volume for *N* reaction (μL)
Nuclease-free water	6.95	$(N+1) \times 6.95$
2× Master mix	12.5	$(N+1) \times 12.5$
20 μM primer mix	0.3	$(N+1) \times 0.3$
QuantiFast RT mix	0.25	$(N+1) \times 0.25$
Template RNA	5.0	$(N+1) \times 5.0$
Total volume	25.0	$(N+1) \times 25.0$

3. Aliquot the master mix into PCR tubes, add nuclease-free water into the NTC tube, and cap all tubes.
4. Transfer tubes to a designated sample addition area and add 5 μL of sample to each tube and firmly cap. Finally add 5 μL of positive control RNA and cap tube; mix all tubes and briefly spin.
5. Place tubes in thermocycler and run the following program.

Temperature (°C)	Time (s)	Cycles
50	600	1
95	300	1
95	10	40
60	40[a]	
Melting curve analysis[b]		

[a]Fluorescent reading taken at the end of this incubation step

[b]Performed according to thermocycler supplier's recommendations

6. Cycle threshold (*Ct*) values and melting curve analysis data for reactions should be documented. All samples except the NTC should be positive for the exogenous control, and all patient samples should be positive for RNase P (*see* **Note 9**).

4 Notes

1. Commonly used disinfectants include 3% Lysol, 5% Micro-Chem Plus, and 0.5% hypochlorite solutions. Other solutions can be used, but their effectiveness in inactivating filoviruses should be fully tested prior to adoption.
2. We utilize the MS2 phage as an exogenous PCR control [8] adding the phage so that 560 μL of AVL contains 500 pfu (~50,000 RNA copies). Make sure that any precipitate present in the AVL is completely dissolved by warming prior to aliquoting or using. We also use detection of the housekeeping gene RNase P mRNA as an endogenous control [9] which is especially useful when the integrity of the sample is in question due to long storage times or high storage temperatures.
3. Buffer AVL has been determined to not fully inactivate filoviruses, and the addition of ethanol in the procedure is necessary to ensure full inactivation of samples [10]. To avoid serious incidents due to misidentification of tubes, we employ different colored caps as an easy way to quickly and accurately distinguish between the different tubes. A red-capped tube is used for the long-term storage sample, red indicating that the sample is still active and caution is needed in handling it; yellow-capped tubes (containing buffer AVL) are used to signify that inactivation may not be complete and caution is still needed; green-capped tubes (containing ethanol) indicate the sample is fully inactivated and is ready for use outside of the containment space once surface decontaminated.
4. The appropriate space to handle the sample will vary on where the sample is being manipulated. In a containment level (CL)4 lab, this is normally in a dedicated BSC or glovebox; in a field setting, a BSC or glovebox would also be used but in areas repurposed for this use. In this situation the personal protective equipment and protocols used are very important as the technical protection offered in fully equipped laboratories is not available. Training of staff and following of protocols are very important to ensure the safe handling and inactivation of patient samples and maintenance of a safe work environment.
5. Be careful to avoid any cross contamination between samples; only one sample tube should be open at any one time. A mock sample consisting of PBS is run through the entire procedure for each batch of samples; if desired, mock samples can be used

prior to handling the test samples and at the end of work with the test samples. Mock samples should be positive for the exogenous control only.

6. A reaction is needed for all test samples, the mock sample(s), a no template control sample, and a positive control. Multiple positive controls can be used to account for the various sized amplicons produced by different species.
7. Opening tubes containing amplified PCR product are the key step that must be controlled to avoid contamination of the work space with amplicon. The analysis area must be separated from the extraction and PCR master mix preparation area, and steps taken to ensure transfer of amplicons from this area to others do not occur. Dedicated lab coats and pipettes for this area are crucial, and tube racks must be cleaned to remove and/or destroy amplified material. Critical to all PCR-based techniques is to ensure all steps of the procedure are separated in space and the equipment used is dedicated for specific purposes. Rooms or, at a minimum, defined areas should be identified for each critical step (sample inactivation, sample extraction, master mix preparation, sample addition, positive control addition, thermocycling, gel electrophoresis). Equipment, supplies, and lab coats should all be for use only in specific areas and not circulate through the labs. For example, materials used in sample extraction should not then be used in preparation of the master mixes. This is especially critical in traditional RT-PCR where tubes containing amplified products must be opened and then run on a gel and thus potentially can contaminate the work area. Work can also be separated by limiting roles of staff to specific tasks (i.e., a staff member who inactivates the sample also prepares the master mix but does not handle purified/amplified samples).
8. Samples that are positive should generate strong distinct products of the size indicated in Table 1. Any sample producing multiple bands or ones outside of the expected size range are unlikely to be true positives, but all results generating amplicons *must* be sequenced to confirm their identity.
9. *Ct* values can vary with each assay and platform used for amplification and detection, but the following guidelines can usually be used for evaluating the results. *Ct* values less than 36 are indicative of positive reactions, *Ct* between 36.1 and 39.9 are indeterminate results and require additional testing, and *Ct* values greater than 40 are indicative of negative reactions. Many laboratories/authorities require two independent targets to be positive to call a sample positive. Patients with negative results should be retested after 24–48 h to confirm the negative result and ensure that the original sample was not too early in the course of disease to demonstrate viremia. Patients with indeterminate results should also be retested after 24–48 h.

References

1. Sanchez A, Ksaizek TG, Rollin PE, Miranda MEG, Trappier SG, Khan AS et al (1999) Detection and molecular characterization of ebola viruses causing disease in human and nonhuman primates. J Infect Dis 179(Suppl 1):S164–S169
2. Towner J, Sealy T, Khristova M, Albariño C, Conlan S, Reeder S et al (2008) Newly discovered ebola virus associated with hemorrhagic fever outbreak in Uganda. PLoS Pathog 4(11):e1000212
3. Grolla A, Jones SM, Fernando L, Strong JE, Ströher U, Möller P et al (2011) The use of a mobile laboratory unit in support of patient management and epidemiological surveillance during the 2005 Marburg outbreak in Angola. PLoS Negl Trop Dis 5(5):e1183
4. Towner J, Rollin PE, Bausch DG, Sanchez A, Crary SM, Vincent M et al (2004) Rapid diagnosis of Ebola hemorrhagic fever by reverse transcription-PCR in an outbreak setting and assessment of patient viral load as a predictor of outcome. J Virol 78(8):4330–4341
5. Trombley A, Wachter L, Garrison J, Buckley-Beason V, Jarhling J, Hensley L et al (2010) Comprehensive panel of real-time TaqMan polymerase chain reaction assays for detection and absolute quatification of filoviruses, arenaviruses and New World hantaviruses. Am J Trop Med Hyg 82(5):954–960
6. Panning M, Laue T, Olschlager S, Eickmann M, Becker S, Raith S et al (2007) Diagnostic reverse-transcription polymerase chain reaction kit for filoviruses based on the strain collections of all European biosafety level 4 laboratories. J Infect Dis 196(Suppl 2):S199–S204
7. Pinsky BA, Sahoo MK, Sandlund J, Kleman M, Kulkarni M, Grufman P et al (2015) Analytical performance characteristics of the Cepheid GeneXpert Ebola assay for the detection of ebola virus. PLoS One 10(11):e0142216
8. Dreier J, Störmer M, Kleesiek K (2005) Use of bacteriophage MS2 as an internal control in viral reverse transcription-PCR assays. J Clin Microbiol 43(9):4551–4557
9. Boddicker JD, Rota PA, Kreman T, Wangeman A, Lowe L, Hummel KB et al (2007) Real-time reverse transcription PCR assay for detection of mumps virus RNA in clinical specimens. J Clin Microbiol 45(9):2902–2908
10. Smither SJ, Weller SA, Phelps A, Eastaugh L, Ngugi S, O'Brien LM et al (2015) Buffer AVL alone does not inactivate ebola virus in a representative clinical sample type. J Clin Microbiol 53(10):3148–3154

Chapter 28

Production of Antigens for ELISA

Robert W. Cross and Thomas G. Ksiazek

Abstract

Immunologic assays such as ELISA allow detection of either virus antigens or the host's immune response to antigens associated with prior infections and offer a powerful means to approach understanding the epidemiology and epizootiology of these agents. However, the success of these assays is highly dependent on the production of high-quality materials both to establish the assays (i.e., antigen target for antibody) and to serve as controls for establishing assay parameters and sensitivity. Here we describe methods for preparing ebolavirus antigens suitable for use as either reagents or controls in a variety of ELISA formats. Considering work with filoviruses is typically restricted to maximum containment laboratories; thus, we also provide methods for inactivating and safety testing of these antigens for safe use in properly established field laboratories.

Key words ELISA, Ebola, Surveillance, Detection, Diagnostics, Filovirus, Antigen production

1 Introduction

The use of enzyme-linked immunosorbent assay (ELISA) format for the detection of antigens or antibodies has proven to be instrumental to both management and surveillance of filovirus outbreaks in both humans and animals [1–12]. Both antigen and antibody detection methods have proven very important in the respect that assessments of current and/or previous exposure to these agents can be made based on the presence of antigen, immunoglobulin (Ig) M, or IgG. Samples submitted for testing may include sera from humans and possibly other species. However, the success of these methods relies upon the availability of appropriate high-quality antigen preparations both as reagents, for the development of assays, and as controls.

The production of antigen for use in both IgG and IgM serological assays is generally done using the virus released into the supernatant or using whole supernatant/cell preparations, which have been frozen and thawed or sonicated to release cell-associated antigen. This protocol describes for IgG ELISA the use of cellular material and the "extraction" of viral antigens through the use of

Thomas Hoenen and Allison Groseth (eds.), *Ebolaviruses: Methods and Protocols*, Methods in Molecular Biology, vol. 1628, DOI 10.1007/978-1-4939-7116-9_28, © Springer Science+Business Media LLC 2017

detergents to solubilize the viral components from the cells and cellular fragments. For IgM ELISA we recommend the use of infected cell slurry consisting of the entire supernatant and viral antigens contained in the cell. These protocols lead to a relatively dilute preparation, but the ease of performance and the benefits from utilizing the cellular-associated viral antigen usually result in an antigen that is useful at a dilution from 1:10 to 1:50 in the IgM capture format and at 1:1000–1:2000 in the indirect IgG assays described elsewhere in this book (*see* Chapter 29).

In order to allow antigens produced following this protocol to be used in our serology lab or in field laboratories, we use irradiated antigens. This reduces the danger to technical staff and avoids contaminating the environment with potentially hazardous agents; however, inactivation needs to be confirmed before using these samples outside of BSL4. Thus we have also provided a protocol describing the testing of the antigens for the presence of any residual infectious virus. Since inactivation of cell lysates for their protein content (antigen) or nucleic acid content (RNA or DNA) uses detergents or chaotropes, which themselves are deleterious to the cells that are used for safety tests, these contaminants must be removed prior to being able to perform the safety test. Alternatively, the material can be diluted beyond the harmful concentrations of the detergents or chaotropes. Indeed, the dilution of the material has been used as a means of performing these safety tests [13], but it reduces the sensitivity of the assays and therefore poses the issue that the virus load may have been reduced, but not altogether eliminated. To avoid this, we have devised a method to recover virus from material being inactivated by using the centrifugation of the material through a sucrose step gradient onto a sucrose cushion to separate it from the original detergent or chaotropic buffer in which it was inactivated [14]. This is the method that is presented here. Following removal of the inactivating chemical(s), the safety test protocol is basically a long-term culture of the antigen and observation of the material for the presence of cytopathic effect (CPE) or specific viral antigen. In addition, viral antigen can be stained for by direct or indirect fluorescent antibody testing to confirm results based on CPE or under conditions where CPE formation is not apparent or is unreliable.

2 Materials

2.1 Preparation of Infected Cell Slurry for IgM ELISA

1. Ebola virus stock.
2. Hank's Balanced Salt Solution (HBSS).
3. Permissive cell lines (e.g., Vero E6 cells).
4. 850 cm^2 roller bottles.

5. Roller bottle apparatus.
6. Incubator capable of accommodating roller bottle apparatus, if not using an integrated unit.
7. Maintenance medium: Eagle's Minimal Essential Medium (EMEM) with 2% fetal bovine serum (FBS) or Minimum Essential Medium modified with Hank's salts (HMEM) with 2% FBS, 20 mM tricine, 1.85 g/L bicarbonate, 100 U/mL penicillin, and 100 μg/mL streptomycin.
8. 3–4 mm glass beads or a long (ca. 50 cm) custom-made rubber policeman.
9. 1 L plastic bottles.
10. 1 M Tris, pH 8.5 (at room temperature).
11. Gamma irradiation unit.
12. Sonication apparatus with macro probe attachment.
13. Lyophilizer (optional).

2.2 Preparation of Viral Cell Lysates for IgG ELISA

1. Ebola virus stock.
2. HBSS.
3. Permissive cell lines (e.g., Vero E6 cells).
4. 850 cm^2 roller bottles.
5. Roller bottle apparatus.
6. Incubator capable of accommodating roller bottle apparatus, if not using an integrated unit.
7. Maintenance medium: EMEM with 2% FBS or HMEM with 2% FBS, 20 mM tricine, 1.85 g/L bicarbonate, 100 U/mL penicillin, and 100 μg/mL streptomycin.
8. 3–4 mm glass beads or a long, custom-made rubber policeman.
9. 200 mL Oak Ridge style centrifuge bottles.
10. Centrifuge and rotors capable of accommodating 200 mL Oak Ridge bottles.
11. Borate saline, pH 9.0: 80 mL 1.5 M NaCl, 100 mL 0.5 M H_3BO_3, 24 mL 1.0 M NaOH. Adjust the final volume to 1 L with distilled H_2O. Check the pH, dispense into 100 mL aliquots, and autoclave. Store at room temperature. Refrigerate when opened.
12. 50 mL polycarbonate Oak Ridge tubes.
13. Centrifuge and rotors capable of accommodating 50 mL Oak Ridge tubes.
14. 1% Triton X-100, in borate saline, pH 9.0.
15. Gamma irradiation unit.
16. Sonication apparatus with cup horn attachment and chilled water recirculator.

2.3 Safety Testing of Prepared Antigen

1. Product to be tested, representative sample (e.g., three vials). These should have been inactivated according to the standard operation procedures (SOPs) for preparation of virus cell lysate antigens (i.e., detergent extraction and/or irradiation as described in Subheadings 3.1 and 3.2).
2. Negative control cell extract (e.g., uninfected cell culture treated as described in Subheadings 3.1 and 3.2).
3. Positive virus control (vial of working stock of virus from which the material being tested was derived).
4. Sterile HBSS and/or Dulbecco's Phosphate-Buffered Saline (PBS).
5. 20% w/v sucrose in HBSS (sterilized by autoclaving).
6. 60% w/v sucrose in HBSS (sterilized by autoclaving).
7. Ultra-clear polyallomer centrifuge tubes for Beckman SW-28 swinging bucket rotor.
8. T-150 flasks with confluent or slightly sub-confluent Vero E6 cells monolayer.
9. Maintenance medium: EMEM with 2% FBS or HMEM with 2% FBS, 20 mM tricine, 1.85 g/L bicarbonate, 100 U/mL penicillin, and 100 μg/mL streptomycin.
10. Sterile 3–4 mm glass beads or rubber policeman.
11. 50 mL centrifuge tubes.
12. Teflon-coated 12-well spot slides.

3 Methods

Safety Precautions: The cells and supernatants from which this material is being prepared have been infected with viruses that have great potential for producing serious disease in man. The viruses will have grown to high concentration when this procedure is performed. Therefore, all work should be performed in a certified laminar flow safety cabinet and in a biosafety lab of an appropriate level (i.e., biosafety level 4, BSL4). Adherence to appropriate safety guidelines as outlined in institutional SOPs is mandatory.

Further, safety testing provides a high level of confidence that inactivation was successful, but good laboratory practices at the BSL2 level should still be used at a minimum for any material removed from the BSL4 laboratory.

3.1 Preparation of Infected Cell Slurry for IgM ELISA

1. A suitably permissive cell line should be chosen and either a high multiplicity of infection (MOI) and single step growth (e.g., MOI = 3) or low MOI and multistep growth (e.g., MOI = 0.001) of the virus seems to be appropriate. Cells are grown to near confluence after a split of approximately 1:8

(this will take approximately 6–7 days). We use a roller speed of 0.6 rpm, which is sufficient to allow attachment and consistent growth. In order to track the cytopathic effect in the inoculated cell culture vessel, one should also include a cell control bottle that is synchronized with a change of growth medium to maintenance medium and start it rolling at the same time than proceeding with the inoculation of the infected roller bottles.

2. Make up a sufficient volume of inoculum in HBSS. The volume for 850 cm^2 roller bottles should be approximately 10 mL for each bottle. This protocol will describe volumes for ten 850 cm^2 roller bottles of cells; for other amounts, adjust accordingly.
3. Pour off the growth medium and add 10 mL of the virus inoculum to each bottle.
4. Replace bottles on the roller apparatus and allow the virus to adsorb for approximately 1 h. Make certain the roller apparatus is leveled and that the entire monolayer is being bathed in the inoculum and that the roller apparatus doesn't have dead rollers (i.e., when placing the bottles in the device, the bottles must be adjacent to slots with roller bottles in them).
5. Add approximately 150 mL of maintenance medium to each bottle, including a mock-inoculated cell control. Make sure the caps are tightened to retain CO_2 produced by the cells if not using a CO_2 incubator.
6. Allow the bottles to incubate at ~35–37 °C for the predetermined incubation time or until the CPE is 3 to 4+ (i.e., severe CPE with most cells detached from the bottle). Feed the cells if the incubation period is prolonged, usually at ~7-day intervals, with the quantity of maintenance medium specified in **step 5**.
7. Remove the remaining attached cells from the walls of the roller bottles using 3–4 mm glass beads or a long, custom-made rubber policeman. Beads are sterilized by autoclaving (e.g., in a heat-resistant tube) and are then added in sufficient number (approximately 10–15 mL) to each roller bottle. The roller can be replaced in the roller apparatus or manually rolled back and forth inside the laminar flow biosafety cabinet until the cells are removed. This should be done with the roller bottle tightly sealed.
8. Decant the combined cells and maintenance medium into suitable 1 L bottles that will fit in the irradiator. Stabilize the pH by adding 100 mL of 1 M Tris, pH 8.5 (at room temperature), per liter of cell slurry, so that the pH is 9.0 in the resultant cell slurry at refrigerator temperatures. Make certain not to overfill the bottles and check to make sure the bottles are not cracked or showing other signs of physical damage, as this may indicate potential failure of the vessel (*see* **Note 1**).

9. Freeze the combined cells and supernatant in the bottle at −70 °C to facilitate cooling during inactivation using the gamma cell.
10. Inactivate the combined cell slurry in the gamma cell, keeping the sample cool with dry ice while administering an appropriate dose of gamma irradiation. We have been using 5 Mrad of gamma radiation, but independent validation of different doses for a specific virus and sample type is recommended.
11. Material must be safety tested if it is to be used outside of containment using institution-specific validated safety testing SOPs (*see* **Note 2**, Subheading 3.3 below for a sample protocol).
12. Once inactivation is confirmed, sonicate the irradiated cell slurry using a standard macro probe until no large visible cellular debris remains. Keep the bottles cooled by placing them in an ice/water bath and replenishing ice in the bath throughout the sonication process. Use settings on the sonicator that seem to yield maximum cavitation around the probe tip.
13. Store aliquots of the cell slurry at below −20 °C (*see* **Note 3**). Alternatively, if available, these antigens tolerate lyophilization and rehydration very well, and we consider this the preferred method of storage for these materials.

3.2 Preparation of Viral Cell Lysates for IgG ELISA

1. Follow the protocol described in Subheading 3.1 from **steps 1** to 7. Decant the combined scraped cells and maintenance medium into suitable 200 mL centrifuge bottles (e.g., Oak Ridge style bottles). Make certain not to overfill the bottles and check to make sure the bottles are not cracked or showing other signs of physical damage as this may indicate potential failure of the vessel.
2. Spin the bottles at approximately 8000 × *g* (e.g., at 8000 rpm in a GSA rotor in a RC-5B centrifuge or using a type 13 rotor in a Beckman medium speed centrifuge) at 4 °C for 10 min.
3. Decant the supernatant into a suitably sized container and discard if you have no other intended use for these supernatants (*see* **Note 4**).
4. Resuspend the cell pellets in borate saline, pH 9.0, and quantitatively transfer the cell pellets into two 50 mL Nalgene polycarbonate (PC) Oak Ridge tubes (*see* **Note 5**). To maximize the washing effect, bring the volumes in each of the two tubes to the maximum volume safely accommodated by the tubes (i.e., approximately 50 mL).
5. Pellet the samples at approximately 21,000 × *g* (i.e., at 12,000 rpm in a RC-5B centrifuge with a SA600 or SS34 rotor or using a type 21 rotor for a Beckman centrifuge) at 4 °C for 10 min. Discard the supernatant into a suitable discard container.

6. Resuspend the cell pellets in a total of 28 mL of cold 1% Triton X-100, in borate saline, pH 9.0. Use a vortex to dislodge pellets from tube walls and disperse into the detergent buffer.
7. Appropriately double bag, transfer samples out of BSL4 suite, freeze at −70 °C, and gamma irradiate on dry ice with 5 Mrad.
8. Sonicate the samples in the sealed tubes, using a cup horn, until no large visible cellular debris remains. Keep the tubes cooled by placing them in an ice/water bath and by using a chilled water recirculator with the cup horn (*see* **Note 6**).
9. Following sonication, pellet any remaining cell debris at approximately 21,000 × *g* (e.g., at 12,000 rpm in a RC-5B centrifuge using an SA600 (or SS34) rotor or a type 21 rotor for a Beckman centrifuge) at 4 °C for 10 min. Retain the supernatants by transferring them to disposable polypropylene centrifuge tubes. Distribute the prepared material in 0.2 or 0.5 mL aliquots with appropriate labels.
10. Store aliquots at below −20 °C (*see* **Note 3**). Alternatively, the antigens can be lyophilized and stored at −80 °C (or −20 °C for shorter periods of time) (*see* **Note 7**). We consider lyophilized material to be the preferred method of storage of these materials.
11. Safety test any materials if to be used outside of containment using institution-specific validated safety testing SOPs (*see* **Note 5**, Subheading 3.3 below for a sample protocol).

3.3 Safety Testing of Prepared Antigens

1. Take a representative amount of the lot of material to be tested (three vials), a vial of negative control (uninfected) cell extract prepared in the same way, and a vial of the stock virus from which the material being tested was derived, and subject them to the following procedure to separate virus from the chemicals present in the antigen preparation (*see* **Note 8**).
2. First pipette 4 mL 60% sucrose in each ultra-clear polyallomer tube, followed by gently overlaying with 10 mL of 20% sucrose. Mark the interface with a lab marking pen.
3. Add 20 mL HBSS by gently overlaying it onto the 20%/60% sucrose step gradient that was made as above. This can be done just prior to entering the BSL4 laboratory.
4. In the BSL4, add the material to be safety tested to the top of the HBSS and bring the final volume of each tube to 38 mL with additional HBSS.
5. Spin the samples at approximately 83,000 × *g* (e.g., in an SW28 (or SW32Ti) rotor at 25,000 rpm) for 2 h.
6. After the spin, remove the HBSS overlay and most of the 20% sucrose, by pipetting it from the top, and discard. Collect the 20–60% interface in a 4 mL volume. Make sure that by the

time you have collected 3 mL, the 20–60% interface is now being collected.

7. Gently mix each sample collected by vortexing.
8. Adsorb each of the collected fractions onto the monolayer in a T150 flask and allow adsorption to proceed for 45–60 min at 37 °C, with rocking every 20 min (*see* **Note 9**). Prepare the sample for the negative control antigen first, followed by the test lot, and finally the positive virus control, to prevent cross contamination.
9. Following adsorption, rinse each flask thoroughly once with 50 mL of Dulbecco's PBS or HBSS.
10. Add 25–30 mL of maintenance medium to the cells and incubate the cells at 37 °C.
11. Check the flasks for CPE every other day (*see* **Note 10**). Once CPE develops continue with preparing spot slides (**step 14**).
12. Change the maintenance medium at 7 days to fresh maintenance medium.
13. At 14 days, if no visible CPE has developed, prepare spot slides with the cells in the flasks. For the positive control, prepare slides once prominent CPE is visible.
14. To prepare spot slides, use sterile glass beads or a rubber policeman to detach the cell monolayer into the medium in the bottle.
15. Pipette 10 mL of the medium containing the scraped cells into a 50 mL centrifuge tube containing 40 mL of PBS, spin it at approximately 2000 × *g* (4000 rpm in a tabletop Eppendorf refrigerated centrifuge) for 5 min in a swinging bucket aerosol-tight rotor container, and open the sample buckets only after returning to the biosafety cabinet.
16. Remove the supernatant by decanting it into the waste container.
17. Resuspend the cell pellet in 5 mL of HBSS or Dulbecco's PBS. This will generate a cell density that should be sufficient to appear when the slide is viewed on an inverted microscope.
18. Dispense drops of cell suspension onto the spots of a Teflon-coated 12-well spot slide (*see* **Note 11**).
19. Allow the slides to air-dry in the biosafety cabinet or in an out-of-the-way location in the lab.
20. Remove the slides from the BSL4 laboratory according to the approved institutional SOPs and gamma irradiate the dried slides using 5 Mrad of radiation.
21. Fix the slides in Coplin jars containing cold acetone.

22. Label and store the slides in a slide box at −20 °C or lower until ready to proceed with antibody staining. The specific parameters for fluorescence assay have to be established based on the ebolavirus species in question and the available antibodies.

4 Notes

1. It is recommended at this point to make spot slides to assess the proportion of cells that are infected by immunocytochemistry (as described in Subheading 3.3, **steps 14–22**).
2. The protocols provided here recommend that the lysates are gamma irradiated with 5 Mrad of radiation. In our experience, this inactivates all the remaining infectivity (i.e., no lots of gamma-irradiated materials have failed safety tests). However, it is important to note that the material must have the detergent removed in order to allow the use of cell cultures to perform the safety test. In Subheading 3.3 of this protocol, we have used a sucrose cushion and step gradient (20% over 60%) method to remove the detergent from any potential virus and then test the material on top of the cushion for residual virus. This must obviously be performed in the containment suite, and the method should be controlled with non-inactivated virus.
3. We have kept some aliquots prepared using this method at ~−30 °C for in excess of 2 years with retention of antigenic activity.
4. If one wishes, the same centrifuge bottles may be used to collect additional cell/maintenance medium from additional roller bottles and spin again until all of the roller bottles have been harvested. Care should be taken not to contaminate the outside of the bottles and to disinfect them when moving to the centrifuge rotor.
5. The use of polycarbonate (PC) or polysulfonate tubes for the subsequent sonication step is important because glass and certain other plastics allow the sonic energy to pass through the wall of the tube, while softer plastics such as polypropylene adsorb much of the energy from the cup horn and polystyrene tubes are not stable enough to reliably endure the sonication process.
6. For sonication we are using either a 600 W Tekmar sonic disruptor with cup horn (max power settings, 10 min with 50% duty cycle) or a Heat Systems 500 W unit with similar settings.
7. The antigens which we have tested using this method include everything from filoviruses to Rift Valley fever virus, alphaviruses, flaviviruses, and Nipah virus, and all tolerate lyophilization and rehydration very well.

8. Antigen from the prepared slurry remains in the culture during safety testing despite inactivation. Therefore, a blind harvest, freeze cycle, and subsequent passage, following a simple centrifugation, must be employed to properly allow testing for antigen using an indirect fluorescent antibody test, where absence of visualization of the antigen contained in the inactivated antigen preparation serves as validation of the inactivation.
9. The cells do survive the 45–60 min of osmotic shock imposed by the concentrated sucrose, but do not adsorb virus longer than this.
10. Not all viruses cause readily apparent CPE. If this is the case, it will be necessary to check for virus antigen in spot slides (as described in Subheading 3.3, **steps 14–22**).
11. Make a sufficient number of slides (three or four) to allow several repeat IFA testing of the cells.

References

1. Ksiazek TG et al (1999) ELISA for the detection of antibodies to Ebola viruses. J Infect Dis 179(Suppl 1):S192–S198
2. Smith DW et al (2015) Virological diagnosis of Ebolavirus infection. Pathology 47(5):410–413
3. Li H et al (2015) Survey and visual detection of Zaire ebolavirus in clinical samples targeting the nucleoprotein gene in Sierra Leone. Front Microbiol 6:1332
4. Sobarzo A et al (2012) Profiling the native specific human humoral immune response to Sudan Ebola virus strain Gulu by chemiluminescence enzyme-linked immunosorbent assay. Clin Vaccine Immunol 19(11):1844–1852
5. Shoemaker T et al (2012) Reemerging Sudan Ebola virus disease in Uganda, 2011. Emerg Infect Dis 18(9):1480–1483
6. Hayman DT et al (2012) Ebola virus antibodies in fruit bats, Ghana, West Africa. Emerg Infect Dis 18(7):1207–1209
7. Amman BR et al (2012) Seasonal pulses of Marburg virus circulation in juvenile Rousettus aegyptiacus bats coincide with periods of increased risk of human infection. PLoS Pathog 8(10):e1002877
8. MacNeil A et al (2010) Proportion of deaths and clinical features in Bundibugyo Ebola virus infection, Uganda. Emerg Infect Dis 16(12): 1969–1972
9. Becquart P et al (2010) High prevalence of both humoral and cellular immunity to Zaire ebolavirus among rural populations in Gabon. PLoS One 5(2):e9126
10. Wauquier N et al (2009) Immunoglobulin G in Ebola outbreak survivors, Gabon. Emerg Infect Dis 15(7):1136–1137
11. Towner JS et al (2008) Newly discovered ebola virus associated with hemorrhagic fever outbreak in Uganda. PLoS Pathog 4(11): e1000212
12. Towner JS et al (2007) Marburg virus infection detected in a common African bat. PLoS One 2(8):e764
13. Blow JA et al (2004) Virus inactivation by nucleic acid extraction reagents. J Virol Methods 119(2):195–198
14. Towner JS et al (2007) High-throughput molecular detection of hemorrhagic fever virus threats with applications for outbreak settings. J Infect Dis 196(Suppl 2):S205–S212

Chapter 29

ELISA Methods for the Detection of Ebolavirus Infection

Robert W. Cross and Thomas G. Ksiazek

Abstract

Ebola viruses are high-priority pathogens first discovered in rural Africa associated with sporadic outbreaks of severe hemorrhagic disease in humans and nonhuman primates. Little is known about the disease ecology or the prevalence of past exposure of human populations to any of the five species of the genus *Ebolavirus*. The use of immunologic means of detection for either virus antigens or the host's immune response to antigen associated with prior infections offers a powerful approach at understanding the epidemiology and epizootiology of these agents. Here we describe methods for preparing antigen detection sandwich enzyme-linked immunosorbent assays (ELISAs) as well as IgG and IgM ELISAs for the detection of ebolavirus antigens or antibodies in biological samples.

Key words ELISA, Ebola, Surveillance, Detection, Diagnostics, Filovirus

1 Introduction

Filoviruses were first recognized in 1967 when several concurrent outbreaks of hemorrhagic fever erupted in laboratories that processed imported nonhuman primate tissues in Marburg and Frankfurt, Germany, and Belgrade, Yugoslavia [1]. All of these outbreaks were found to be caused by Marburg virus [1]. In 1976, two of the ebolaviruses were discovered in simultaneous outbreaks in Central Africa [2]. Since then a number of outbreaks of filovirus-related hemorrhagic fever have continued to sporadically surface and have ranged in size and duration from single cases to the recent outbreak in West Africa, due to the Makona variant of Ebola virus (EBOV), which is by far the largest and longest outbreak of filovirus-related hemorrhagic fever on record [3–5].

Epidemiological surveillance and clinical management of these outbreaks have involved the need for increased stringency in the triage, treatment, and tracing of both suspect and confirmed cases, as well as potential contacts, to improve infection control practices and reduce transmission from genuine Ebola virus disease cases to patients with non-Ebola virus disease. Confirmation of exposure has involved many methods including virus isolation, molecular

Thomas Hoenen and Allison Groseth (eds.), *Ebolaviruses: Methods and Protocols*, Methods in Molecular Biology, vol. 1628, DOI 10.1007/978-1-4939-7116-9_29, © Springer Science+Business Media LLC 2017

detection, and antigen/antibody detection. Antigen and antibody detection methods have proven very important in the respect that assessments of current and/or previous exposure to these agents can be made based on the presence of antigen, immunoglobulin (Ig)M, or IgG. Further, in cases where a new species of purportedly antigenically related ebolavirus is suspected, the use of antibodies known to be reactive with multiple species of ebolavirus may assist in at least providing confirmation at the genus level. The advent of more sensitive molecular detection methods (principally real-time RT-PCR assays) has greatly advanced clinical diagnosis in the field, but antigen detection utilizing point of care methods still has promise, particularly in filovirus infections in which the viremias and antigen loads are high, especially in the latter stages of the disease.

The use of the enzyme-linked immunosorbent assay (ELISA) format for the detection of antigen or antibody has proven to be instrumental in both management and surveillance of filovirus outbreaks in both humans and animals [6–17]. Samples submitted for serological testing may include sera from humans and possibly other species. The below protocols describe serological tests for the detection of antigen, IgM, and IgG in humans. Adjustments to account for different host species can be made by simply utilizing the desired host-specific reagents.

2 Materials

2.1 Antigen Detection and Titration

1. Phosphate-buffered saline (PBS, pH 7.4) with thiomersal (1:10,000) added as a preservative.
2. Wash buffer: PBS with 0.1% Tween-20.
3. Serum diluent (SerDil): Wash buffer with 5% dehydrated skim milk (*see* **Note 1**).
4. Polyvinyl chloride (PVC) microtiter plates (*see* **Note 2**).
5. Capture antibody: Mouse mAb mix against Ebola viral proteins, diluted 1:1000 in PBS (*see* **Note 3**).
6. Control capture antibody: Ascites produced with the myeloma parent line, diluted 1:1000 in PBS (*see* **Note 4**).
7. Unknown samples: Antigen-containing sample specimens (e.g., infected patient sera).
8. EBOV antigen cell slurry (prepared according to instructions in Chapter 28 in this book).
9. Negative control antigen cell slurry (prepared according to instructions in Chapter 28 in this book).
10. Detection antibody: Polyvalent hyperimmune rabbit anti-EBOV/Sudan virus/Reston virus, diluted 1:1500 in SerDil (*see* **Note 5**).

11. Conjugate: Goat anti-rabbit IgG conjugated to horseradish peroxidase (HRPO) (e.g., Biorad cat. no. STAR124P, diluted 1:8000 in SerDil).
12. Substrate: ABTS [2,2′-azino-bis(3-ethylbenzothiazoline-6-sulphonic acid)] substrate (prepared as directed by the manufacturer) (*see* **Note 6**).
13. Plate reader capable of reading at 414 or 410 nm (*see* **Note 7**).

2.2 IgM Detection and Titration

1. PBS.
2. Wash buffer.
3. SerDil (*see* **Note 1**).
4. PVC microtiter plates (*see* **Note 2**).
5. IgM capture antibody: Antihuman mu-chain (e.g., Thermo cat no. A18837, diluted 1:500 in PBS).
6. Test sera.
7. Positive control serum: Serum sample(s) from either a known human infection or an experimentally infected nonhuman primate.
8. Negative controls: Panel of at least 4–6 sera known to be negative for EBOV antibodies.
9. EBOV antigen cell slurry (prepared according to instructions in Chapter 28 in this book).
10. Negative control antigen cell slurry (prepared according to instructions in Chapter 28 in this book).
11. Normal human serum.
12. Detection antibody: Rabbit anti-EBOV (Special Pathogens Branch item identifier 703,355 diluted 1:4000 in SerDil) or other suitable anti-EBOV antibody (dilution must be empirically determined).
13. Conjugate: Goat anti-rabbit IgG (H&L)-HRPO (e.g., Biorad cat. no. 1721019, diluted 1:2000 in SerDil).
14. Substrate: ABTS substrate (prepared as directed by the manufacturer) (*see* **Note 6**).
15. Plate reader capable of reading at 414 or 410 nm (*see* **Note 7**).

2.3 IgG Detection and Titration

1. PBS.
2. Wash buffer.
3. SerDil (*see* **Note 1**).
4. PVC microtiter plates (*see* **Note 2**).
5. EBOV antigen cell slurry (prepared according to instructions in Chapter 28 in this book).

6. Negative control antigen cell slurry (prepared according to instructions in Chapter 28 in this book).
7. Test sera.
8. Positive control serum: Serum sample(s) from either a known human infection or an experimentally infected nonhuman primate.
9. Negative controls: Panel of at least 4–6 sera known to be negative for EBOV antibodies.
10. Conjugate: Mouse antihuman IgG (Fc)-HRPO (e.g., Gentaur product no. JMH035103, diluted 1:4000 in SerDil).
11. Substrate: ABTS substrate (prepared as directed by the manufacturer) (*see* **Note 6**).
12. Plate reader capable of reading at 414 or 410 nm (*see* **Note 7**).

3 Methods

Safety Precautions: The samples to be tested for the presence of anti-EBOV antibodies or antigens are potentially contaminated with viable EBOV or other agents for which a differential determination is being sought. Even where this is not the case, materials collected from human sources are also potentially contaminated with human viruses such as hepatitis B (HBV) or human immunodeficiency virus (HIV). With this in mind, if the perceived hazard is high, work should be conducted in an appropriate containment environment with stringent biosafety containment practices appropriate for manipulating live EBOV. Alternatively, irradiation of test samples, as performed for assay reagents (i.e., 5 Mrad of gamma radiation), will render these materials inert, and work can be undertaken at lower containment levels. Adherence to appropriate safety guidelines as outlined in institutional standard operating procedures is mandatory, and good laboratory practices at the biosafety level 2 should always be used at a minimum.

3.1 Antigen Detection and Titration

The basic antigen detection approach described here is that of a double-sandwich capture assay in which the antigen is captured by antibodies on a solid phase and then detected by a HRPO-coupled second antibody that binds captured antigen from the test samples. A detection system using ABTS is then applied to determine how much of the detection antibody has been retained on the solid phase of the system:

1. Coat the top half of the PVC plates (rows A–D) with 100 μL of capture antibody at an optimal dilution empirically determined for each capture antibody/antibody mixture. We use a mouse anti-Ebola mAb mix diluted 1:1000 in PBS. Coat the

bottom half of the plate (rows E–H) with normal myeloma ascites at the same dilution in PBS (negative coat wells). Incubate the plates overnight at 4 °C.

2. Wash the plates three times with 100 μL per well of wash buffer.
3. Prepare fourfold dilutions of unknown antigen-containing samples as well as positive and negative control antigens (*see* **Note 8**) in SerDil from 1:4 to 1:256 going down the plate (i.e., from row A to row D) (*see* **Note 9**).
4. Add 100 μL of each antigen dilution to the wells of the antibody-coated PVC plate.
5. Place plates in a humidified chamber (*see* **Note 10**).
6. Incubate the plates for 60 min at 37 °C.
7. Wash the plates three times with 100 μL per well of wash buffer.
8. Add 100 μL of detection antibody diluted to an optimal concentration (*see* **Note 11**). We use a rabbit anti-Ebola virus/Sudan virus/Reston virus diluted 1:1500 in SerDil.
9. Incubate the plates for 60 min at 37 °C.
10. Wash the plates three times with 100 μL per well of wash buffer.
11. Add 100 μL goat anti-rabbit IgG-HRPO, diluted 1:8000 in SerDil (*see* **Note 12**).
12. Place plates in a humidified chamber (*see* **Note 10**).
13. Incubate the plates for 60 min at 37 °C.
14. Wash the plates three times with 100 μL per well of wash buffer.
15. Prepare ABTS substrate as per the manufacturer's directions (e.g., dilute kit components 1:1 if necessary) and add 100 μL ABTS substrate per well.
16. Place plates in a humidified chamber (*see* **Note 10**).
17. Incubate the plates for 30 min at 37 °C.
18. Read plates at 414 (or 410) nm.

3.2 IgM Detection and Titration

The basic approach detailed here is that of an IgM capture assay in which the patient's IgM is captured onto the solid phase with anti-IgM. Specific antigen (and control antigen) is then applied and allowed to react with the patient's captured IgM. If the patient's serum contained specific anti-EBOV IgM, the captured antigen is then measured using a virus-specific antibody followed by a HRPO-coupled species-specific antibody with a detection system employing ABTS. This type of assay works because the proportion of specific IgM is high immediately following infection with the virus.

1. To coat 96 well PVC plates, dilute the IgM capture antibody (antihuman mu-chain antibody) 1:500 in PBS and add 100 μL per well. Incubate the plates overnight at 4 °C.

2. Wash the plates three times with 100 μL per well of wash buffer.
3. Prepare 1:100 dilutions in SerDil of test sera, as well as any control sera (positive and negative control sera). Prepare additional fourfold dilutions in SerDil going down the plate (*see* **Note 10**). Alternatively, the sera can be screened in duplicate at a 1:100 dilution only.
4. Place plates in a humidified chamber (*see* **Note 10**).
5. Incubate the plates for 60 min at 37 °C.
6. Wash the plates three times with 100 μL per well of wash buffer.
7. Prepare the EBOV and negative control (naïve VeroE6 cells) antigens for use by diluting the antigen slurries 1:10 in SerDil. Add normal human serum at a 1:50 ratio to the antigen slurries.
8. Add 100 μL of antigen slurry to each well, as desired.
9. Place plates in a humidified chamber (*see* **Note 10**).
10. Incubate the plates for 60 min at 37 °C.
11. Wash the plates three times with 100 μL per well of wash buffer.
12. Add 100 μL per well of anti-EBOV antibody diluted to an optimal concentration in SerDil. (e.g., rabbit anti-EBOV (Special Pathogens Branch item identifier 703,355), diluted 1:4000 in SerDil).
13. Place plates in a humidified chamber (*see* **Note 10**).
14. Incubate the plates for 60 min at 37 °C.
15. Wash the plates three times with 100 μL per well of wash buffer.
16. Add 100 μL per well of goat anti-rabbit IgM-HRPO (H&L), diluted 1:2000 in SerDil.
17. Place plates in a humidified chamber (*see* **Note 10**).
18. Incubate the plates for 60 min at 37 °C.
19. Wash the plates three times with 100 μL per well of wash buffer.
20. Prepare ABTS substrate as per the manufacturer's directions (e.g., dilute kit components 1:1 if necessary) and add 100 μL ABTS substrate per well.
21. Place plates in a humidified chamber (*see* **Note 10**).
22. Incubate the plates for 30 min at 37 °C.
23. Read plates at 414 (or 410) nm.

3.3 IgG Detection and Titration

The basic approach is that of an IgG assay in which EBOV antigen is applied onto the solid phase of a microtiter plate. EBOV-specific IgG is allowed the opportunity to bind to the antigen. After washing, an antihuman IgG conjugated to HRPO is applied and allowed to bind. This, in turn, is followed with ABTS substrate:

1. Coat 96 well PVC plates with 100 μL per well of either EBOV (top half; rows A–D) or negative control (naïve VeroE6 cells) antigens (bottom half, rows E–H) diluted 1:1000 in PBS. Incubate the plates overnight at 4 °C.
2. Wash the plates three times with 100 μL per well of wash buffer.
3. Prepare 1:100 dilutions in SerDil of test sera, as well as any control sera (positive and negative control sera). Prepare additional twofold dilutions in SerDil going down the plate. Alternatively, the sera can be screened in duplicate at the 1:100 dilution only.
4. Place plates in a humidified chamber (*see* **Note 10**).
5. Incubate the plates for 60 min at 37 °C.
6. Wash the plates three times with 100 μL per well of wash buffer.
7. Add 100 μL antihuman IgG-HRPO conjugate diluted 1:4000 in SerDil.
8. Place plates in a humidified chamber (*see* **Note 10**).
9. Incubate the plates for 60 min at 37 °C.
10. Wash the plates three times with 100 μL per well of wash buffer.
11. Prepare ABTS substrate as per the manufacturer's directions (e.g., dilute kit components 1:1 if necessary) and add 100 μL ABTS substrate per well.
12. Place plates in a humidified chamber (*see* **Note 10**).
13. Incubate the plates for 30 min at 37 °C.
14. Read plates at 414 (or 410) nm.

3.4 Evaluating ELISA Results

We control for sera that are "sticky" or have antibodies to the substrate (i.e., cell lines) used to make the antigen by using a mock-infected negative control antigen. The OD values for reactivity to the mock antigen are subtracted from those for positive antigen to give an adjusted OD value for each sample.

A standard positive control serum should be used and is run in a standard dilution series. This, in effect, provides a standard curve, which will determine the limits of detection of the assay.

A group of normal sera, negative for specific antibody, are also run to determine the background of the assay. This allows us to determine the limit at which the positive control was still positive and assess the reactivity of unknown samples. Determination of the background should be made using a panel of 5–6 geographically relevant (i.e., taken from community under investigation) normal sera, which are run each time the assay is used.

The mean and standard deviation of the background value (of the adjusted ODs) are calculated and used to calculate a value equal to the mean plus three standard deviations. This represents the cutoff value for identifying samples positive in the assay.

In performing IgM and IgG tests, sera that are specific for a particular agent rarely give values that are "marginal." Marginal values may represent either sera that are positive for some related, cross-reacting virus or noise in the system. In interpretation of IgG assays, one should be conservative in assigning positive meaning to sera with other than high OD values. Another confounding factor in performing assays on acute patients is the dynamics of the immune response, as very early responses will be low in specific antibody content, and they will rise with time since infection [18].

4 Notes

1. Serum diluent contains skim milk and Tween-20 to reduce nonspecific binding.
2. Other types of immunoassay plates will also work, but PVC plates are light and easily transported to field situations and have good adsorbing capacity.
3. Plates are coated with an antibody capable of capturing viral antigen from the test sample. This is a mixture of monoclonal Abs (mAb mix) selected for their ability to capture Ebola antigens derived from all five different species of *Ebolavirus* onto the solid phase plastic of a microtiter plate. The dilution given in the Materials and Methods has been optimized for the lot of mAb mix described. Empirical estimation of any capture antibody used will require independent optimization [19].
4. A duplicate set of wells on the lower half of the plates is coated with an ascites, produced with the myeloma parent line, which will not bind ebolavirus antigen. The difference between the wells coated with anti-Ebola mAbs and those coated with a nonspecific ascites provides what we call the adjusted OD value, which provides an indication of the specificity of the test results.
5. This polyclonal antibody was produced in rabbits sequentially immunized with live Ebola virus, Sudan virus, and Reston virus. Other suitable anti-EBOV antibody can be used, but the appropriate dilution must be empirically determined.
6. ABTS, in the presence of the enzyme HRPO and hydrogen peroxide, is converted from a colorless liquid to an intense green color with maximum light adsorption at 414 nm. The amount of color developed is proportional to the amount of IgG which has bound to the antigen on the solid phase.

7. Generally we used dual-wavelength recording with one wavelength being specific for the converted substrate and the other not reading the wavelength and used to correct for light path interference such as scratches or smudges on the plate.
8. Use 1:10 dilutions of cell slurries created from EBOV-infected cells and noninfected cells according to the instructions in Chapter 28 in this book as starting material, and further dilute them as described (i.e., so that the beginning dilution is 1:40 and the last dilution is 1:2560). Alternatively, recombinant EBOV proteins produced in vitro may be used, but this will require empirical titration to establish the optimal working dilution range.
9. This is best done by placing the undiluted specimens to be tested in a master plate (large microtiter format tube holder) and then transferring them from the master plate to the first row of the test plate 33 μL of the specimen. The same quantity of specimen is placed in the negative-coated wells of the plate. Fourfold dilutions are done by carrying 33 μL of antigen down the column series of both the positive and negative antigen. Make certain to include positive and negative controls in all tests at the same dilutions.
10. We recommend placing the plates in a humidified chamber consisting of a plastic tray in which moistened paper towels have been placed. Afterward, the tray is covered with a clear plastic cover, which is secured with rubber bands. The humidification minimizes issues with the milk diluent drying at the surface interface in the wells.
11. The detection antibody can be any antibody with a high titer for specific viral antigens, but must be of a species that, when an anti-detection species is placed on top of the sandwich, will not cross-react with the capture antibody on the solid phase. In our example, this is a rabbit anti-Ebola made against Ebola virus, Sudan virus, and Reston virus grown in RK13 (rabbit) cells with rabbit serum. The strategy here is to identify several antigenically distinct ebolaviruses in one assay. The dilution used has been optimized by cross-titration with the detector system.
12. This immunoglobulin has been conjugated to an enzyme (e.g., HRPO) that converts a substrate (e.g., ABTS) to a measurable (i.e., wavelength absorbing) color. The chief concept to remember in applying this general scheme for detection is that the bottom (capture) and top (detection) antibodies must be of dissimilar species so that the anti-species conjugate does not "see" the antigen-trapping antibody on the bottom of the plate. The anti-species conjugate will correspond to the species of detection antibody used.

References

1. Martini GA et al (1968) On the hitherto unknown, in monkeys originating infectious disease: Marburg virus disease. Dtsch Med Wochenschr 93(12):559–571
2. Report of an International, Commission (1978) Ebola haemorrhagic fever in Zaire, 1976. Bull World Health Organ 56(2): 271–293
3. Baize S et al (2014) Emergence of Zaire Ebola virus disease in Guinea. N Engl J Med 371(15):1418–1425
4. Bagcchi S (2014) Ebola haemorrhagic fever in west Africa. Lancet Infect Dis 14(5):375
5. Dixon MG et al (2014) Ebola viral disease outbreak—West Africa, 2014. MMWR Morb Mortal Wkly Rep 63(25):548–551
6. Ksiazek TG et al (1999) ELISA for the detection of antibodies to Ebola viruses. J Infect Dis 179(Supplement 1):S192–S198
7. Smith DW et al (2015) Virological diagnosis of ebolavirus infection. Pathology 47(5): 410–413
8. Li H et al (2015) Survey and visual detection of Zaire ebolavirus in clinical samples targeting the nucleoprotein gene in Sierra Leone. Front Microbiol 6:1332
9. Sobarzo A et al (2012) Profiling the native specific human humoral immune response to Sudan Ebola virus strain Gulu by chemiluminescence enzyme-linked immunosorbent assay. Clin Vaccine Immunol 19(11):1844–1852
10. Shoemaker T et al (2012) Reemerging Sudan Ebola virus disease in Uganda, 2011. Emerg Infect Dis 18(9):1480–1483
11. Hayman DT et al (2012) Ebola virus antibodies in fruit bats, Ghana, West Africa. Emerg Infect Dis 18(7):1207–1209
12. Amman BR et al (2012) Seasonal pulses of Marburg virus circulation in juvenile *Rousettus aegyptiacus* bats coincide with periods of increased risk of human infection. PLoS Pathog 8(10):e1002877
13. MacNeil A et al (2010) Proportion of deaths and clinical features in Bundibugyo Ebola virus infection, Uganda. Emerg Infect Dis 16(12): 1969–1972
14. Becquart P et al (2010) High prevalence of both humoral and cellular immunity to Zaire ebolavirus among rural populations in Gabon. PLoS One 5(2):e9126
15. Wauquier N et al (2009) Immunoglobulin G in Ebola outbreak survivors, Gabon. Emerg Infect Dis 15(7):1136–1137
16. Towner JS et al (2008) Newly discovered ebola virus associated with hemorrhagic fever outbreak in Uganda. PLoS Pathog 4(11):e1000212
17. Towner JS et al (2007) Marburg virus infection detected in a common African bat. PLoS One 2(8):e764
18. Ksiazek TG et al (1999) Clinical virology of Ebola hemorrhagic fever (EHF): virus, virus antigen, and IgG and IgM antibody findings among EHF patients in Kikwit, Democratic Republic of the Congo, 1995. J Infect Dis 179(Suppl 1):S177–S187
19. Ksiazek TG et al (1992) Enzyme immunosorbent assay for Ebola virus antigens in tissues of infected primates. J Clin Microbiol 30(4): 947–950

Chapter 30

Ebola Virus Field Sample Collection

Brian R. Amman, Amy J. Schuh, and Jonathan S. Towner

Abstract

Sampling wildlife for ebolaviruses presents the researcher with a multitude of challenges, foremost of which is safety. Throughout the methods described in this chapter, personal safety and personal protective equipment (PPE) will be reiterated for each methodology. The methods described here are those used to successfully detect and isolate marburgviruses from their natural reservoir, *Rousettus aegyptiacus*, and therefore should be applicable for diagnostic testing for ebolaviruses via RT-PCR, ELISA, and IHC techniques.

Although an ebolavirus natural reservoir has yet to be identified, the majority of disease ecologists believe the reservoir to belong to the order *Chiroptera* (bats). The methods presented in this chapter are presented with bats as an example, but all of these methods would be applicable to other species of wildlife with few or no modifications.

Key words Ebola, Ebolavirus sample collection, Bats, Personal protective equipment, Filovirus

1 Introduction

Sampling wildlife for ebolavirus diagnostics presents the researcher with a multitude of circumstances to be considered above and beyond normal wildlife sampling. When capturing and sampling wild animals for any reason, a modicum of personal protective equipment (PPE) is essential to prevent injury and potential transmission of pathogenic agents. When capturing and sampling animals for high-consequence human pathogens like ebolaviruses, wearing PPE, practicing safe animal handling, and preparing for all contingencies are of the utmost importance.

Collecting samples from captured wildlife for ebolavirus diagnostics is an important aspect of understanding the ecology of the virus and its, as yet unknown, reservoir(s). Based on limited ebolavirus PCR evidence [1] and several definitive field and laboratory studies with marburgvirus [2–6], bats are thought to be the most likely candidates for ebolavirus reservoirs. The remainder of this protocol on field sample collection for ebolaviruses is, therefore, written with bats as the example species. However, all of the methods described herein can be used for other wild animals with little or no modification.

Thomas Hoenen and Allison Groseth (eds.), *Ebolaviruses: Methods and Protocols*, Methods in Molecular Biology, vol. 1628, DOI 10.1007/978-1-4939-7116-9_30,

Ebolaviruses are transmissible via direct and indirect contact with bodily fluids of infected humans and animals [7–9]. Airborne droplets are also a concern, and, therefore, respiratory protection is an important consideration when handling bats during capture. Respirators provide a protective barrier preventing droplets from reaching the mucous membranes. To protect the hands from accidental bites while removing bats from nets or traps, bite-resistant gloves are worn (*see* **Note 1**). These protective gloves can be cumbersome when removing bats from nets, so it may become necessary to work in teams of two people, where one person wearing bite-resistant gloves holds and secures the bat while another person wearing only double latex or nitrile gloves disentangles the captured bat from the net (Fig. 1). Additional PPE such as rubber boots, snake chaps, and caving helmets may be required depending on the target species and its associated habitat.

Fig. 1 Two-person method of removing a bat from a mist net. One person wearing bite-resistant gloves holds and restrains the bat, while the other person, wearing only double latex gloves, extricates the bat from the net

The use of nonlethal methods for sample collection may be required for vulnerable or endangered species (*see* IUCN Red List, http://www.iucnredlist.org) or mandated by host country government agencies or by internal animal care and use committee (IACUC) regulations. Methods involving venipuncture, swabbing, and excrement collection are examples of nonlethal sampling.

Destructive sampling involves euthanasia under anesthesia of the captured bats. Terminal bleeding is performed under anesthesia and in accordance with institutionally approved IACUC protocols. Exsanguination under anesthesia is generally an acceptable method of euthanasia [10]. An overdose of inhalant anesthesia can also be used as the primary method of euthanasia, with subsequent exsanguination to ensure euthanasia. However, individual IACUC regulations may vary. Terminal bleeding requires the use of a needle and vacutainer system or syringes, so extreme care must be used when handling sharps.

Necropsies are performed in order to collect a variety of tissues for diagnostic testing. Tissues collected in past field and experimental studies that have yielded the highest number of filovirus and filovirus RNA positive samples include pooled liver-spleen, heart, lung, and kidney [2, 3, 6, 11]. Evidence of filovirus infection has never been found in sylvan bat neural tissue, and it is therefore no longer routinely sampled.

Samples collected in the field for RNA extraction and Ebola virus qRT-PCR analysis can be placed directly in an appropriate volume of guanidinium-containing virucidal lysis solution, homogenized (tissues only), and then frozen. Liquid (e.g., urine) or swab (e.g., oral) samples for virus isolation can be frozen for long-term storage in viral transport medium (e.g., BD Universal Viral Transport System). Carcasses or small pieces of tissue in histology cassettes can be fixed and inactivated in formalin for histological examination and permanent storage.

2 Materials

2.1 Personal Protective Equipment for Capture, Anesthesia, and Nonlethal Sampling

1. Surgical gown or Tyvek coverall.
2. Latex or nitrile gloves (double pair).
3. Bite-resistant or Kevlar gloves (*see* **Note 1**).
4. Half- or full-face respirator with P100 filters.
5. Disposable face shield (if not wearing a full-face respirator) (*see* **Note 2**).
6. Boots.

2.2 Personal Protective Equipment for Necropsy

1. Surgical gown.
2. Powered air-purifying respirator (PAPR) with HEPA filter and head cover assembly.
3. Latex or nitrile gloves (double).

2.3 Anesthesia

1. Isoflurane.
2. Gallon-sized Ziploc bag.
3. Tea strainer (ball style).
4. Gauze pads (2″ × 2″).

2.4 Measurement

1. Large ruler (measurement in mm).
2. Slide or dial caliper (measurement in mm).
3. Blunt forceps.
4. Digital scale or Pesola.
5. Clipboard with data recording sheet.
6. Identification (foot) tags.

2.5 Processing Equipment and General Supplies

1. Folding table.
2. Plastic sheeting.
3. Absorbent bench pads.
4. Spray bottles filled with 5% MicroChem.
5. 1000 mL plastic beaker filled with 5% MicroChem.
6. Squirt bottle filled with 70% ETOH.
7. Cryovials and racks.
8. Grinding vials.
9. Histology tissue cassettes.
10. Viral transport media.
11. Virucidal lysis buffer (e.g., MagMAX lysis/binding solution).
12. 10% buffered formalin.
13. Kimwipes.
14. Sharps containers.
15. 4 L Nalgene jars for formalin-fixed tissue storage.
16. 35 L (or larger) containers for formalin-fixed carcass storage.
17. Biohazard bag for waste.
18. Plastic beakers, 1000 mL, to hold clean necropsy tools.

2.6 Bleeding

1. Vacutainer blood collection tubes, holders, and sheathed needles.
2. Alcohol wipes (for nonlethal bleeding only).
3. Styptic powder or gel (for nonlethal bleeding only).
4. 2″ × 2″ gauze squares (for nonlethal bleeding only).
5. Lancet (for nonlethal bleeding only).
6. Disposable Pasteur transfer pipettes to transfer blood and serum from vacutainer to cryovial.
7. P200 pipette with tips for nonlethal bleeding.

2.7 Necropsy

1. Dissecting scissors.
2. Blunt dissecting forceps (not rat tooth).
3. Squirt bottle filled with 70% ethanol.
4. Absorbent bench pads.

2.8 Disinfection and Cleanup

1. Plastic beaker, 1000 mL.
2. Plastic tubs or buckets, 4–8 L.
3. Fingernail brush.
4. 5% MicroChem.
5. 10% bleach.
6. Spray bottles.

3 Methods

Collected bats can either be sampled according to nonlethal or destructive sampling workflows according to the provided protocols. Processing for nonlethal sampling proceeds in the following order: anesthesia (if required by institutional guidelines) (Subheading 3.2), oral swab collection (Subheading 3.3) (*see* **Note 3**), nonlethal venipuncture (Subheading 3.4), specific urine, and/or feces collection (Subheading 3.5). Nonlethal urine and feces sampling using a nonspecific approach (Subheading 3.6) does not require capture or handling of the bats and can be completed at any time.

The processing and tissue collection during destructive sampling described herein is designed to be performed in an assembly line fashion. Bats are processed in the following order: anesthesia (Subheading 3.2), terminal bleeding (euthanasia by exsanguination and overdose of inhalant anesthesia) (Subheading 3.7), oral swab collection (Subheading 3.8) (*see* **Note 3**), body measurement and data recording (Subheading 3.9), necropsy and tissue collection (Subheading 3.10), and disinfection of necropsy instruments (Subheading 3.11).

3.1 Personal Protective Equipment During Sample Collection

Personal protective equipment (PPE) should always be worn during sampling procedures. For nonlethal sampling, the PPE required is the same as that used for the capture, handling, and anesthesia (i.e., as listed in Subheading 2.1). Nonlethal sampling should be performed by two people, both wearing PPE, with the person holding the animal additionally wearing bite-resistant gloves (*see* **Note 1**).

The PPE used for necropsies and tissue collection can be the same as that used for capture and handling, with bite-resistant gloves worn for anesthesia procedures only. However, powered air-purifying respirators (PAPR) equipped with high-efficiency particulate air (HEPA) filters and head covers or hoods with built-in face shields tend to be more comfortable for the extended periods of time required by the necropsy process (Subheading 2.2). Conversely, PAPRs may also be used for nonlethal sampling and at the trapping sites under certain conditions, but can be cumbersome while moving around and do not comfortably allow for the use of a head lamp for night work.

3.2 Anesthesia of Animals

1. Put on PPE according to the pertinent segments of Subheading 3.1, including bite-resistant gloves.
2. Add cotton balls or 2″ × 2″ squares of gauze to tea strainer halves (Fig. 2).
3. Charge the tea strainer by adding isoflurane directly to the gauze or cotton in each half of the strainer until it is saturated, and put the two halves together.
4. Lock the halves of the tea strainer into place with a twist.
5. Put the tea strainer in a 4 L Ziploc bag, or similar, and seal the bag to prevent loss of isoflurane fumes. If a bag other than a Ziploc type is used, a tight seal can be created by folding the top few inches of the bag and securing it with a clip (Fig. 3).
6. Remove the bat from the containment vessel (bat bag), and place it in the Ziploc bag containing the tea strainer charged with isoflurane.
7. Quickly seal the bag and allow time for the isoflurane to sedate the bat. An adequate level of anesthesia has been achieved when the bat no longer responds to external stimuli, but respiration is still visible.

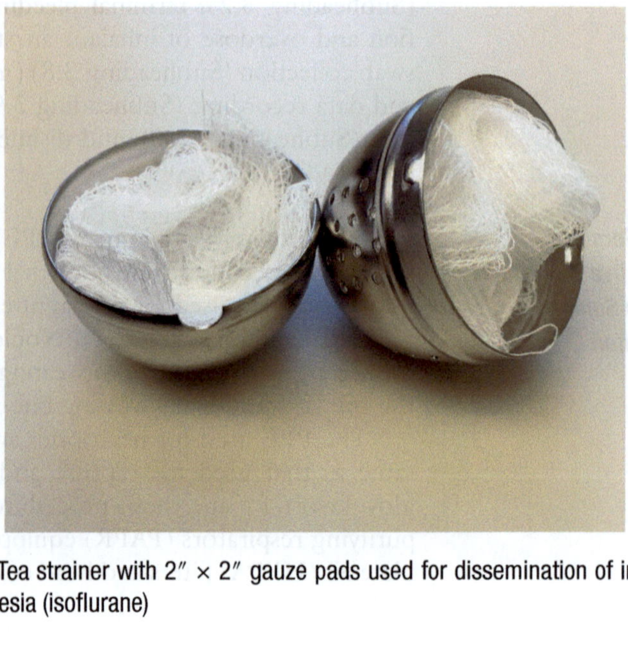

Fig. 2 Tea strainer with 2″ × 2″ gauze pads used for dissemination of inhalant anesthesia (isoflurane)

Fig. 3 A tea strainer fully charged with isolurane inside 4 L Ziploc bag sealed with a metal clip

8. Once anesthesia is achieved, remove the bat from the Ziploc anesthesia bag and place it on the absorbent bench pad at the bleeding station of the processing table.
9. Reseal the empty anesthesia bag to prevent loss of the isoflurane vapors.
10. If the bat regains consciousness, carefully place it back into the anesthesia bag, and repeat the steps until anesthetization is again achieved.
11. Recharge the tea strainer with isoflurane when anesthetization of bats begins to take longer than a few minutes before visible signs appear (cessation of struggling, relaxed demeanor) and before each new day of processing.

3.3 Nonlethal Oral Swab Sample Collection

The nonlethal collection of secretions from the oral cavity of a bat (if performed) is typically implemented just prior to venipuncture, while the bat is being restrained and can be done with or without anesthesia (Subheading 3.2, *see* **Notes 3** and **4**). The nonlethal swabbing procedure is safest with two people (a holder and sampler). One person wearing bite-resistant gloves (**the holder**) secures the bat so that it's laying with its back on a flat surface. The other person (**the Sampler**) gently swabs the cheeks and mouth.

1. Put on PPE according to the pertinent segments of Subheading 3.1, including bite-resistant gloves for **the holder**.
2. **Holder**: Place the bat on its back on the bench pad with its head toward **the sampler,** and secure it in place.
3. **Sampler**: Take two sterile polyester-tipped applicators, and insert them into the bat's mouth, swabbing the cheeks and mouth of the bat for 30–60 sec (Fig. 4). Care must be taken not to apply too much pressure, which could irritate the tissue of the oral cavity and cause bleeding.
4. **Sampler**: Once the swabbing is complete, place one swab in a cryovial containing virucidal lysis buffer (*see* **Note 5**). The swab should be raised just off the bottom of the cryovial and the plastic applicator stick broken off such that the entire remainder of the swab fits inside the cryovial. The entire polyester tip should be completely immersed in lysis buffer.

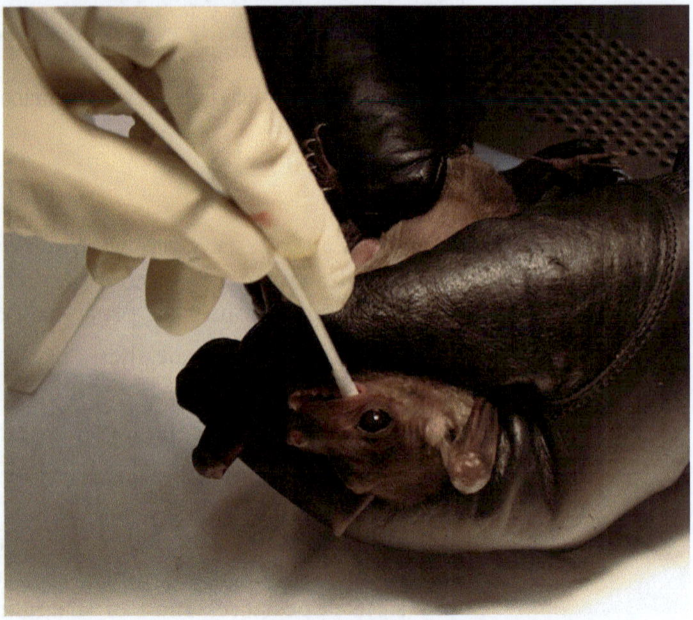

Fig. 4 Oral swab collection. The procedure shown here is the non-anesthetic method requiring two people. A holder, wearing bite-resistant gloves, secures and restrains the bat, and a bleeder, wearing double latex gloves, obtains the oral swab sample

5. **Sampler**: Insert the other swab into a vial containing 500 μL (or more) of viral transport medium (or a similar stabilizing medium) for future virus isolation attempts, breaking the plastic applicator stick off as mentioned above in Subheading 3.3, **step 4**.
6. Freeze all samples for downstream analysis.

3.4 Nonlethal Blood Sample Collection

Venipuncture is a nonlethal blood sample collection procedure that may or may not require anesthesia depending on institutional IACUC regulations and/or the level of difficulty involved in handling the animal during the procedure. For bats, anesthesia is generally not required when two people work together to obtain the sample. One person wearing bite-resistant gloves (**the holder**) secures the bat so that it's laying with its back on a flat surface with one wing accessible. The other person (**the bleeder**) gently spreads the wing to expose the propatagium, located anterior to the humerus, radius, and ulna (upper arm and forearm). The blood sample is obtained from the cephalic vein on the anterior edge of the propatagium (Fig. 5). **Never touch the unsheathed needles/ lancets or attempt to recap needles.**

1. Put on PPE according to the pertinent segments of Subheading 3.1.
2. **Holder:** Place the bat on its back on a bench pad laid out on a stable, flat surface such as a folding table.
3. **Holder:** Secure the bat in a manner in which the bleeder can access the wing without danger of being bitten. Specifically,

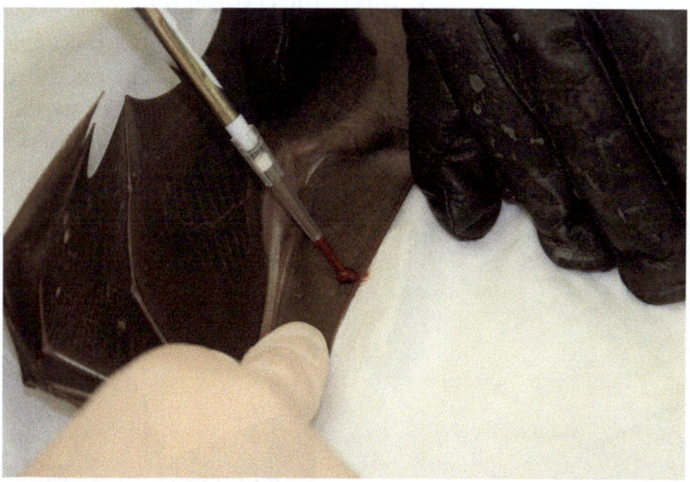

Fig. 5 Nonlethal venipuncture of the cephalic vein on the propatagium of a bat wing. The procedure shown here is the non-anesthetic method requiring two people. A holder, wearing bite-resistant gloves, secures and restrains the bat, and a bleeder, wearing double latex gloves, obtains the blood sample

the muzzle of the bat should protrude through the bite-resistant gloved index and ring finger of the holder, or remain under the fingers if the bat is small. The holder should tent their hand so that the bat is secured, but minimal pressure is applied to the body of the bat. The opposite hand is placed between the bat and the bleeder's hand as an extra measure of security.

4. **Bleeder:** Clean the area to be bled (the cephalic vein on the propatagium) with an alcohol wipe, and dry the area with a 2″ × 2″ gauze pad.
5. **Bleeder**: Place the gauze pad directly under the wing where the venipuncture will take place.
6. **Bleeder:** Puncture the cephalic vein with a sterile lancet (Fig. 6), allowing the vein to bleed and pool sufficiently to obtain the predetermined amount of blood necessary for testing (typically 20 μL for RNA extraction and a minimum of 40 μL for serology). This amount of blood is minimal, less than 1% of total blood volume (TBV) for a 100 g bat, and well under the allowable 10% of TBV required by many institutional IACUC programs. Additional blood can be taken for virus isolation attempts as long as it is within institutional IACUC blood draw parameters.
7. **Bleeder:** Use a pipette to transfer the pooled blood (Fig. 5) from the cephalic vein into a cryovial containing a virucidal lysis buffer at a ratio of sample-buffer known to be virucidal. The amount of sample can be scaled up or down as long as the minimum ratio is maintained. Blood samples for serology are transferred into a cryovial.
8. Invert the cryovial multiple times to mix sample and buffer.
9. **Bleeder**: Immediately upon completion of bleeding, apply pressure to the cephalic vein on the distal side of the perforation, administer a small amount of styptic gel or powder to the bleeding site, and apply a small amount of pressure to stop the bleeding.

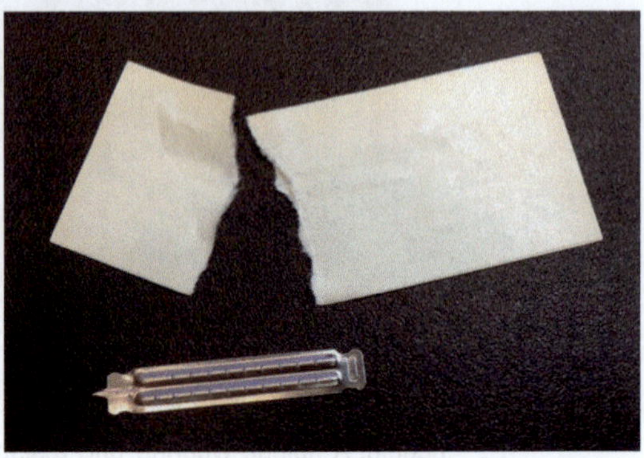

Fig. 6 A "feather"-type stainless steel lancet used for venipuncture

10. Check to make sure the bleeding has stopped. If bleeding continues, repeat **step 8**.
11. Once the bleeding has stopped, other nonlethal procedures can be performed or the bat can be released.
12. All samples are frozen for downstream analysis.

3.5 Nonlethal Fecal and Urine Sample Collection: Specific Collection

Urine and feces can also be collected in the field from individual bats using a specific collection method. The specific collection method takes a sample from one bat that can be identified as either Ebola virus positive or negative.

1. Put on PPE according to the pertinent segments of Subheading 3.1.
2. Place an individual bat in a Thoren Maximizer #9 Mouse Cage (Thoren Caging Systems) or similar, fitted with an inverted wire top (Fig. 7) (*see* **Note 6**).
3. Secure the inverted wire top to the plastic bottom with tape.
4. Remove the bat from the cage following urination or after 60 min (or whatever predetermined time frame is mandated by internal IACUC regulations) regardless of whether a sample has been produced.
5. Collect samples from the bottom of the cage using a sterile pipette, and transfer them to cryovials containing an appropriate amount of virucidal lysis buffer. This amount can be scaled up or down depending on sample size as long as the minimum lysis buffer to sample ratio is maintained.

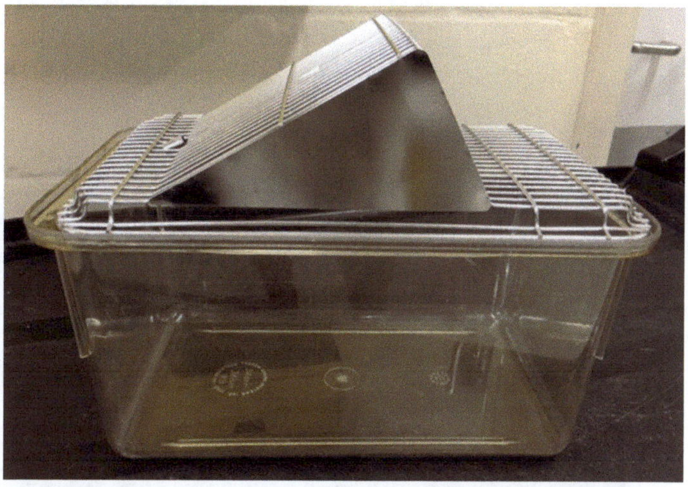

Fig. 7 Specific fecal and urine collection method. Shown here is a Thoren Maximizer #9 Mouse Cage. The bat is contained within a clean cage with the wire lid inverted until excreta is deposited. The sample is collected from the bottom of the cage

6. Invert the cryovial multiple times to mix sample and buffer.
7. Remaining sample can be placed in a cryovial containing 500 μL (or more) of viral transport medium.
8. Freeze all samples for downstream analysis.

3.6 Nonlethal Fecal and Urine Sample Collection: Nonspecific Collection

Urine and feces can be collected in the field from a population of bats using a nonspecific collection method. The nonspecific method, initially described by Chua [12] for Nipah virus sample collection, accumulates comingled samples, and therefore an individual bat cannot be identified as positive or negative, only a population.

1. Put on PPE according to the pertinent segments of Subheading 3.1.
2. For nonspecific collection, place a clean autoclave bag (or plastic sheeting) in or under the roosting site of bats, and leave it there for a predetermined amount of time (Fig. 8) (*see* **Note 7**).
3. Remove the bag or plastic from the roost after the desired time period.
4. Collect samples from the soiled plastic using individual sterile transfer pipettes.
5. Place samples in cryovials containing an appropriate amount of virucidal lysis buffer (minimum volume of 500 μL) (*see* **Note 5**). This amount can be scaled up or down depending on sample size as long as the minimum lysis buffer to sample ratio is maintained.

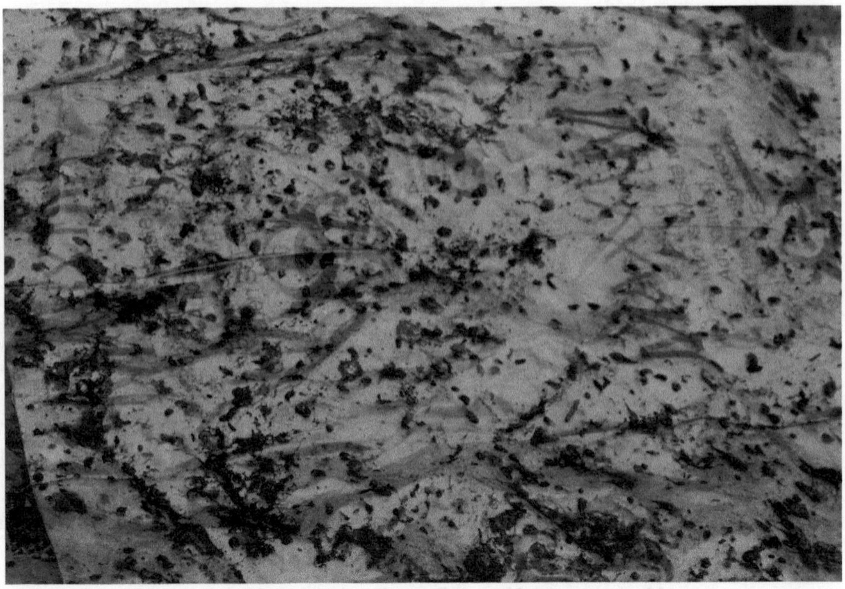

Fig. 8 Nonspecific fecal and urine sample collection. A clean plastic bag is placed within or around the roosting area to catch excreta. The samples are taken directly from the plastic

6. Invert or shake the cryovial multiple times to mix sample and buffer.
7. Freeze samples for downstream analysis.

3.7 Terminal Blood Sample Collection and Euthanasia

The protocol below describes cardiac puncture following entry into the thoracic cavity using a 1.5″ (3.8 cm) vacutainer needle, at a point just posterior to the sternum. On larger bats, a longer needle may be necessary to reach the heart. In the absence of a longer needle, the heart may be reached by entering the thoracic cavity through the intercostal space between the ribs on the left side of the bat. **Never touch the unsheathed needle, and use only forceps to remove the plastic sheath from the needle. It is also recommended that a one-use vacutainer collection device always be used during the bleeding process. Never attempt to recap needles.**

1. Put on PPE according to the pertinent segments of Subheading 3.1.
2. Assemble the vacutainer holder and sheathed needle (18–22 G) with a Luer adapter prior to the commencement of the procedure.
3. Loosen the sheath from the needle using forceps (not fingers!) so that the needle can be easily removed from the sheath by lifting the holder/needle assembly and allowing gravity to pull the sheath off.
4. Wet the sternum of the animal with 70% ethanol (*see* **Note 8**), and locate the xiphoid process on the sternum visually or by palpating.
5. Using the vacutainer holder/needle assembly, insert the needle (at approximately a 15° angle to the body plane) just posterior to the xiphoid process and into the thoracic cavity toward the heart (*see* **Note 9**).
6. Keep the vacutainer holder steady and push the vacutainer tube (typically 3–6 mL) onto the internal needle. If the needle insertion was done correctly, blood will flow into the vacutainer. If blood does not flow, rotating or slightly moving the needle in or out may locate the cardiac chamber to start the flow.
7. Collect the cardiac blood (typically 4–6 mL for a 150 g fruit bat) until exsanguination is achieved (Fig. 9).
8. After exsanguination, detach the vacutainer tube from the holder/needle assembly. Remove the needle/holder assembly from the bat, and dispose of the entire unit in an appropriate sharps container.
9. Place the exsanguinated bat inside a Ziploc bag containing isoflurane (as described in Subheading 3.2) to ensure euthanasia has been achieved (*see* **Note 10**).

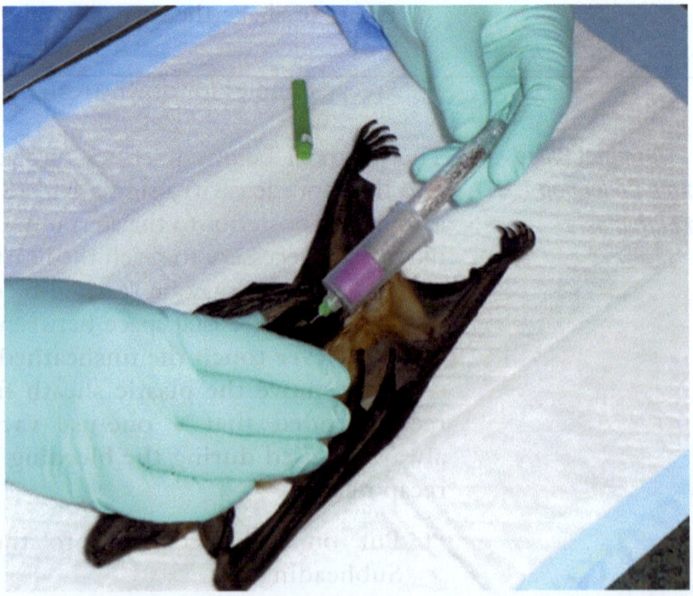

Fig. 9 Terminal bleeding procedure. Shown here are the vacutainer needle, holder, and container. The needle is inserted just posterior to the xiphoid process of the bat

10. Confirm euthanasia by absence of respiration (visual) and heartbeat (tactile).
11. For nucleic acid extraction, transfer blood to cryovials containing an appropriate amount of virucidal lysis buffer (*see* **Note 5**).
12. For serology, transfer blood or serum directly into cryovials using a sterile transfer pipette.
13. Freeze all samples for downstream analysis.

3.8 Terminal Oral Swab Sample Collection

Oral swabs taken during lethal sample collection should be taken after bleeding and euthanasia but before performing the necropsy. The swabbing procedure is performed in the same manner as for nonlethal oral swab collection (Subheading 3.3), but only one person is required to handle and sample the euthanized bat.

1. Put on PPE according to the pertinent segments of Subheading 3.1.
2. Take two sterile polyester-tipped applicators, and insert them into the bat's mouth, swabbing the cheeks and mouth of the bat for 30–60 s (Fig. 4). Care must be taken not to apply too much pressure, which could irritate the tissue of the oral cavity and cause bleeding.
3. Once the swabbing is complete, place one swab in a cryovial containing virucidal lysis buffer (*see* **Note 5**). The swab should be raised just off the bottom of the cryovial and the plastic

applicator stick broken off such that the entire remainder of the swab fits inside the cryovial. The entire polyester tip should be completely immersed in lysis buffer.

4. Insert the other swab into a vial containing 500 μL (or more) of viral transport medium (or a similar stabilizing medium) for future virus isolation attempts, breaking the plastic applicator stick off as mentioned above in Subheading 3.3, **step 4**.
5. Freeze all samples for downstream analysis.

3.9 Body Measurements

Weight, standard body measurements for mammals [13], sex, and reproductive observations should be made and recorded prior to necropsy. If relevant, some additional observations such as the early term of pregnancy can be determined during necropsy. Many of these measurements and observations are of diagnostic value and will aid in proper taxonomic identification and classification of the captured bats.

1. Put on PPE according to the pertinent segments of Subheading 3.1.
2. Affix a foot tag with a unique field identification number to the left leg of the bat, just above the ankle joint.
3. Measure the total length, tail length, hind foot length (typically the right hind foot), ear length (typically the right ear), and, for *Chiroptera*, also the forearm and tragus length, and record the measurements.
4. Weigh the bat. Weights on live bats can be obtained by wrapping the bat in a small bat bag and placing it on the scale. The weight of the bat bag can be subtracted from the total weight.
5. To measure total length, place the bat on the large ruler with the dorsal surface down so that the tip of the rostrum is at 0 mm. Measure the total length of the bat from rostrum to the tip of the tail. If the tail is very small or absent, measure to the last vertebra. Do not include excess skin or hair in this measurement.
6. The tail length (Fig. 10a) is the measurement from the base to the tip of the tail (last vertebra).
7. The hind foot measurement (Fig. 10b) is the length of the foot from the back of the ankle to the tip of the longest toe plus the claw. It is helpful to use forceps to bend the foot 90° at the ankle for this measurement (Fig. 10c).
8. The ear measurement (Fig. 10d) is taken from the notch at the base of the ear to the furthest most tip of the pinna.
9. For bats, the forearm measurement (Fig. 10e) is best taken using a caliper. Fold the phalanges in so that only the forearm is visible. Place the outside jaws of the caliper over both the distal and proximal ends of the forearm (radius and ulna) and lightly close until they stop. Read and record the measurement.

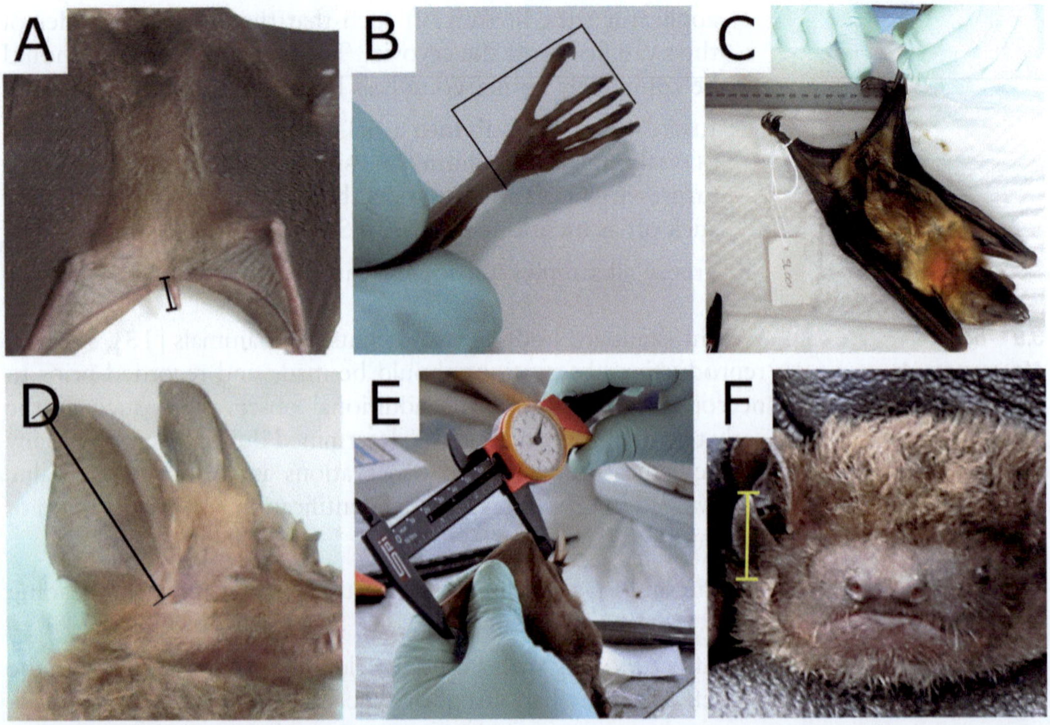

Fig. 10 Body measurements. (**a**) Tail length measurement. (**b**) Hind foot measurement. (**c**) Using forceps to bend the ankle 90° so that only the hind foot is measured. (**d**) Ear measurement. (**e**) Forearm measurement. (**f**) Tragus measurement

10. The tragus is a small, typically elongate flap-like structure projecting up from the base of the pinna on many, but not all, insectivorous bats. The tragus is measured from its base to the tip (Fig. 10f).

3.10 Necropsy and Tissue Collection

When sampling bats for filoviruses, the samples harvested should always be treated as though they are infectious. The following procedures should be performed with scissors and blunt forceps to reduce the use of sharps. At no time should fingers be used to hold, pull, or push any tissue or bone. The tissues collected during a field necropsy will either be placed in virucidal lysis buffer and subsequently pulverized to inactivate any virus that may be present [2, 3] or stored frozen in a cryovial or fixed in 10% buffered formalin for downstream analysis.

Preparation

1. Put on PPE according to the pertinent segments of Subheading 3.1.
2. Place an absorbent bench pad on the necropsy table, and tape edges of the pad with duct tape to prevent slippage. All necropsy activity should be performed on this bench pad.

3. Pre-label all tubes and vials with the date of the investigation, a unique identification number, and sample type (liver/spleen, heart, blood, etc.). Arrange tubes in a systematic order for each bat.
4. Prior to each necropsy, all dissecting instruments (scissors and forceps) should be sterile and/or thoroughly disinfected with 5% MicroChem and treated with 10% bleach (*see* **Note 11**).
5. Arrange the equipment and tubes around the working area within easy reach to maximize safety and efficiency (Fig. 11).
6. Place a euthanized bat on the clean bench pad on the table with the wings tucked behind and under the body.
7. Wet the ventral surface of the bat with 70% ethanol to reduce incorporation of hair into samples.
8. Make the first incision by pinching the skin and muscle with forceps at the most posterior aspect of the lower abdomen (Fig. 12). Care should be taken to use only blunt forceps and never fingers during this and all other cutting steps.
9. If subdermal tissues are to be collected, proceed to remove the skin on the on the abdomen and thorax prior to opening abdominal cavity.
10. Make a lateral cut, breaching the abdominal cavity.
11. From this point, open the body cavity with either two lateral cuts up both sides of the body or by one cut up through the midline of the body. In either case, the cuts should extend

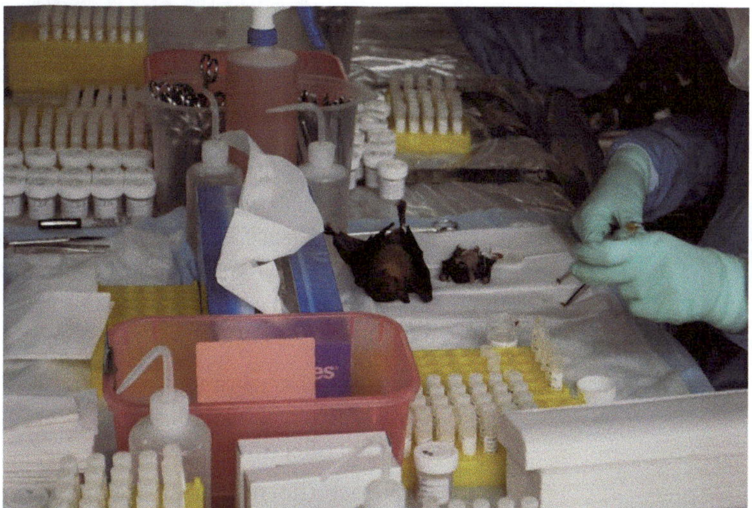

Fig. 11 The arrangement of cryovials and dissection tools during necropsies. The cryovials and tools should be arranged to maximize safety and efficiency during necropsies

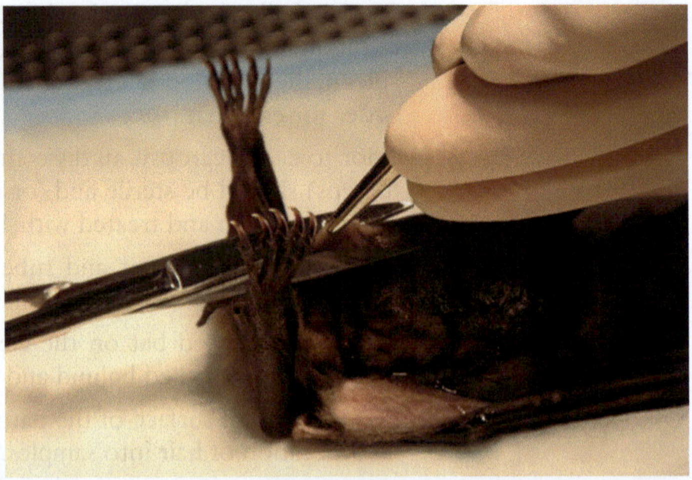

Fig. 12 Initial incision made to begin the necropsy. The cut is made through the dermis only if subdermal tissues are to be taken

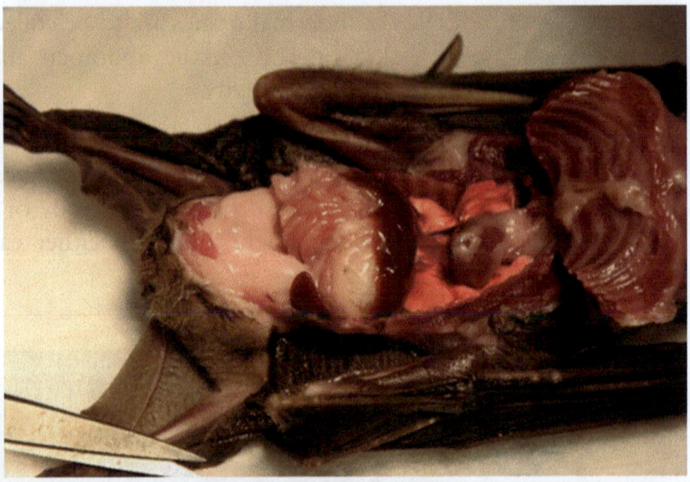

Fig. 13 Lateral cuts through the abdominal and thoracic cavities exposing tissues. This method allows for greater visibility and easy access to tissues and is the author's preference. A single incision up the midline through the *rectus abdominis* and sternum will also allow access to tissues

through the thoracic cavity and deep into the pectoral musculature. This will aid in keeping the cavity open during sample collection (Fig. 13).

12. Conduct the necropsy organ by organ, collecting tissues as needed. Keep in mind that when large blood vessels or well-vasculated tissues are cut, the body cavity can fill with pooled blood making it difficult to see smaller or more ventrally located organs.

13. The order of organ removal will depend on the desired tissues. It is typically best to start the removal of tissues with those that are closest to the recently made opening, then moving deeper into the visceral cavity. An example would be the liver, spleen, gonads, kidney, heart, and lung.
14. Place the sampled tissues in virucidal lysis buffer (for nucleic acid extraction), empty cryovials (for virus isolation), or 10% buffered formalin (for histology) (*see* **Note 12**). Small tissue samples collected for histology should be placed in histology tissue cassettes prior to submersion in 10% formalin. After collection of the tissues, the carcass can also be submerged in 10% buffered formalin (*see* **Note 13**).
15. Samples in virucidal lysis buffer (for nucleic acid extraction) or empty cryovials (for virus isolation) should be frozen on site as soon as practical for downstream analysis.
16. If grinding tissues on site, transfer tissue samples into a grinding vial containing 500 μL–1 mL solution for 50–100 mg tissue.
17. Pulverize the sample using a ball-mill tissue grinder for 2 min, at 1500 strokes/min (Geno/Grinder 2000, SPEX CentriPrep).
18. Transfer the resulting lysate to a cryovial, and freeze on site as soon as practical for downstream analysis.
19. Tissues and histology cassettes should remain in 10% formalin for a minimum of 4 days and then be switched to 70% ethanol for long-term storage. Carcasses should be fixed in 10% formalin for at least 7 days before being transferred to ethanol.

3.11 Disinfection of Necropsy Instruments

1. After the completion of the necropsy, any excess tissue or blood on the scissors and/or forceps should be removed using a Kimwipe (Kimberly-Clark) by placing a folded Kimwipe on the table and holding it in place with the soiled forceps. Do not hold the Kimwipe with hands while wiping instruments, or use fingers to secure the Kimwipe on the table.
2. While holding the soiled scissors (as one would during a necropsy), open the blades and wipe the excess tissue on the Kimwipe.
3. Repeat this step with the forceps while holding the Kimwipe with the scissors.
4. Place the scissors and forceps in a 1000 mL plastic beaker half-filled with 5% MicroChem located on the necropsy table.
5. When the beaker becomes full, or the necropsies for that day have been completed, the used instruments can be placed into a larger container (typically a plastic tub) of 5% MicroChem and fully submersed and allowed to soak for a minimum of 15 min.

6. After soaking, scrub off any remaining residue (tissue or blood) using a small fingernail brush (*see* **Note 14**).
7. Rinse the disinfected instruments in clean water (in another plastic tub) and placed them into a tub containing a 10% bleach solution.
8. Rinse the instruments again in clean water, and set out on an absorbent bench pad to dry, ideally in full sunlight. Instruments may also be rinsed with 70% ethanol to speed the drying process.

4 Notes

1. Examples of bite-resistant gloves include TurtleSkin NYCDoC leather gloves, the puncture proof Full Coverage series gloves (Warwick Mills Inc.), or Bitebuster gloves (BiteBuster LLC). It is extremely important to carefully weigh the puncture resistance of any individual product against the needs of the specific application and/or species to be handled.
2. To supplement half-face respirators, a disposable face shield should be worn over the respirator, protecting the eyes and uncovered portions of the face, as well as inhibiting inadvertent touching of the face with contaminated hands or other objects.
3. Oral swabs taken during nonlethal sample collection should be performed before bleeding. Oral swabs taken during lethal sample collection should be taken after bleeding and euthanasia but before performing the necropsy.
4. If the bat is anesthetized for nonlethal oral swab collection, the swabbing procedure is performed in the same manner as the two-person method, but only the bleeder handles the anesthetized bat. An anesthesia bag should be readily available should the bat begin to wake up and need more anesthesia.
5. MagMAX lysis-binding solution (Life Technologies) or TriPure (Roche Diagnostics) is recommended. If other lysis solutions are used, their virucidal capability should be verified.
6. The dimensions of the Thoren Cage (7.700″W × 12.170″L × 5.875″H) will allow most bats to comfortably assume hanging positions, larger bats will require larger cages.
7. Shorter sample collection periods result in fewer, but fresher samples, whereas longer periods of time result in less fresh, but more abundant samples. Environmental factors such as ground seepage (caves) or precipitation should be considered in determining the appropriate sample collection time.

8. This allows better visibility of the xiphoid process and minimizes microbial contamination of the bleeding needle.
9. The needle will meet some resistance as it penetrates the heart. Typically, the beating of the heart against the needle will be visible and indicates proper insertion/positioning.
10. Depending on the size of the bat and the amount of blood retrieved during bleeding, a second round of anesthesia may be necessary to achieve euthanasia.
11. During necropsy procedures, it is important to use a clean set of scissors and forceps for each necropsy and then disinfect the tools before they are used again. A bleaching step is also recommended for the removal of residual nucleic acids on disinfected tools.
12. Cryovials containing samples should not be filled completely to prevent rupture of the tube and cap upon freezing.
13. If carcasses are disposed of on site, this should be done discretely, typically via incineration.
14. This step can be performed without full PPE (double latex gloves only) once the instruments have been soaking and the processing area has been cleaned.

References

1. Leroy EM, Kumulungui B, Pourrut X et al (2005) Fruit bats as reservoirs of Ebola. Nature 438:575–576
2. Amman BR, Carroll SA, Reed ZD et al (2012) Seasonal pulses of Marburg virus circulation in juvenile *Rousettus aegyptiacus* bats coincide with periods of increased risk of human infection. PLoS Pathog 8(10):e1002877
3. Amman BR, Jones ME, Sealy TK et al (2015) Oral shedding of marburg virus in experimentally infected Egyptian fruit bats (*Rousettus aegyptiacus*). J Wildl Dis 51(1):113–124
4. Paweska JT, VJP v, Masumu J et al (2012) Virological and serological findings in Rousettus aegyptiacus experimentally inoculated with vero cells-adapted hogan strain of Marburg virus. PLoS One 7(9):e45479
5. Swanepoel R, Smit SB, Rollin PE et al (2007) Studies of reservoir hosts for Marburg virus. Emerg Infect Dis 13(12):1847–1851
6. Towner JS, Amman BR, Sealy TK et al (2009) Isolation of genetically diverse Marburg viruses from Egyptian fruit bats. PLoS Pathog 5(7):e1000536
7. Bausch DG, Towner JS, Dowell SF et al (2007) Assessment of the risk of Ebola virus transmission from bodily fluids and fomites. J Infect Dis 196(Suppl 2):S142–S147
8. Feldmann H, Wahl-Jensen V, Jones SM et al (2004) Ebola virus ecology: a continuing mystery. Trends Microbiol 12(10):433–437
9. Peters CJ, LeDuc JW (1999) An introduction to Ebola: The virus and the disease. J Infect Dis 179(Suppl 1):ix–xvi
10. Committee for the Update of the Guide for the Care and Use of Laboratory Animals, National Research Council (2011) Guide for the care and use of laboratory animals, 8th edn. National Academies Press, Washington, D. C.
11. Jones ME, Schuh AJ, Amman BR et al (2015) Experimental inoculation of Egyptian Rousette bats (*Rousettus aegyptiacus*) with viruses of the Ebolavirus and Marburgvirus Genera. Viruses 7(7):3420–3442
12. Chua KB (2003) A novel approach for collecting samples from fruit bats for isolation of infectious agents. Microbes Infect 5(6):487–490
13. Martin RE, Pine RH, DeBlase AF (2001) A manual of mammalogy: with keys to families of the world. McGraw-Hill, New York

Index

Thomas Hoenen and Allison Groseth (eds.), *Ebolaviruses: Methods and Protocols*, Methods in Molecular Biology, vol. 1628, DOI 10.1007/978-1-4939-7116-9, © Springer Science+Business Media LLC 2017

U

V

W

Y

MIX
Papier aus verantwortungsvollen Quellen
Paper from responsible sources
FSC® C105338

If you have any concerns about our products,
you can contact us on
ProductSafety@springernature.com

In case Publisher is established outside the EU,
the EU authorized representative is:
Springer Nature Customer Service Center GmbH
Europaplatz 3, 69115 Heidelberg, Germany

Printed by Libri Plureos GmbH
in Hamburg, Germany